OXFORD WORLD'S CLASSICS

DIALOGUE ON THE TWO GREATEST WORLD SYSTEMS

GALILEO GALILEI (1564–1642) is one of the most fascinating and controversial figures in the history of science. He was appointed professor of mathematics at the University of Pisa in 1589, and in 1592 he moved to the University of Padua where he taught for the next eighteen years. In 1610 he published a number of sensational telescopic discoveries, including that the lunar landscape is like that of a barren earth, and his book, the *Sidereal Message*, sold out in less than a week. The next year he was awarded the prestigious position of personal mathematician and philosopher to the Grand Duke of Tuscany in Florence. He next found that there are dark spots on the face of the sun and that gave rise to a lively international controversy that is recorded in his *Letters on the Sunspots*. He argued for a non-literal interpretation of the Bible, and he became involved in a dispute over the nature of comets with a Jesuit professor whom he lampooned in a witty essay, *The Assayer*. When the Roman Inquisition banned the Copernican theory in 1616, he refrained from writing about the motion of the Earth until a Florentine friend became Pope Urban VIII in 1623. His *Dialogue on the Two Chief World Systems*, which is not only a scientific masterpiece but an outstanding literary work, appeared in 1632. Summoned to Rome he was put on trial and condemned to house arrest in 1633. He nonetheless went on to write his *Discourse on Two New Sciences*, the work for which he is remembered as the forerunner of Newton. He died in Florence in 1642.

WILLIAM R. SHEA was Galileo Professor of History of Science at the University of Padua where Galileo taught for eighteen years. He has written extensively on the Scientific Revolution of the seventeenth century, and is currently working on a biography of Galileo.

MARK DAVIE has taught Italian at the Universities of Liverpool and Exeter, and has published studies on various aspects of Italian literature, mainly in the period from Dante to the Renaissance. He is particularly interested in the relations between learned and popular culture, and between Latin and the vernacular, in Italy in the Renaissance.

OXFORD WORLD'S CLASSICS

For over 100 years Oxford World's Classics have brought readers closer to the world's great literature. Now with over 700 titles—from the 4,000-year-old myths of Mesopotamia to the twentieth century's greatest novels—the series makes available lesser-known as well as celebrated writing.

The pocket-sized hardbacks of the early years contained introductions by Virginia Woolf, T. S. Eliot, Graham Greene, and other literary figures which enriched the experience of reading. Today the series is recognized for its fine scholarship and reliability in texts that span world literature, drama and poetry, religion, philosophy, and politics. Each edition includes perceptive commentary and essential background information to meet the changing needs of readers.

OXFORD WORLD'S CLASSICS

GALILEO GALILEI

Dialogue on the Two Greatest World Systems, the Ptolemaic and the Copernican

Translated by
MARK DAVIE

With an Introduction and Notes by
WILLIAM R. SHEA

OXFORD
UNIVERSITY PRESS

Great Clarendon Street, Oxford, OX2 6DP,
United Kingdom

Oxford University Press is a department of the University of Oxford.
It furthers the University's objective of excellence in research, scholarship,
and education by publishing worldwide. Oxford is a registered trade mark of
Oxford University Press in the UK and in certain other countries

First published as an Oxford World's Classics paperback 2022
Impression: 1

Published in the United States of America by Oxford University Press
198 Madison Avenue, New York, NY 10016, United States of America

British Library Cataloguing in Publication Data
Data available

Library of Congress Control Number: 2022935975

ISBN 978–0–19–884013–8

Printed and bound in the UK by
Clays Ltd, Elcograf S.p.A.

ACKNOWLEDGEMENTS

Our debts to colleagues and friends are too numerous to be recorded in full but we wish to express our special gratitude to Michele Camerota, who supplied us with illuminating information on science in Galileo's day, to Maurice A. Finocchiaro for helping us understand the clash between Galileo and the Church, and to Stefano Gattei for his scholarly study of Galileo's background. We owe a special thanks to John Heilbron for his outstanding biography of Galileo and for his helpful advice. We are grateful to Flavia Marcacci for a better understanding of the history of Copernicanism, to Jürgen Renn for pointing out to us some of Galileo's forerunners, to Gheorghe Stratan for his lucid critical advice, and to Gino Tarozzi who provided us with new insights.

We are deeply grateful to our wives, Evelyn and Grace, who have been admirably supportive and patient at all times, and to whom we owe more than one good idea.

CONTENTS

INTRODUCTION

GALILEO GALILEI was born in Pisa on 16 February 1564,[1] the eldest son of Vincenzio Galilei and Giulia Ammannati. His father was a musician and the author of an influential *Dialogue on Ancient and Modern Music* that contains a fierce attack on his former master, Gioseffo Zarlino, and shows a gift for polemics that his son was to display in his own writings. The Galilei were an old Florentine family, and Galileo's great-great-granduncle, also named Galileo Galilei, was a famous physician who had been twice elected officer of the Governing Body of Florence and, in 1445, filled the high office of Minister of Justice. Galileo was proud of his ancestry and he described himself as a 'noble Florentine' on the title page of his first printed book, a handbook that appeared in 1606 on how to use a geometrical and military compass that he had invented.

Vincenzio Galileo had settled in Pisa when he married in 1562 but the family moved to Florence in 1574. In 1580 his son returned to Pisa to attend the University. At the time philosophy and science were still deeply influenced by the writings of the fourth-century BC Greek philosopher Aristotle, whose works were rediscovered in the Middle Ages. Aristotle maintained that things on Earth were made of four basic elements (earth, fire, air, and water) that were mixed in ever-changing proportions. He thought, however, that celestial bodies were made of an entirely different kind of material, which was unchangeable, or to use the old expression 'incorruptible'. This physics was well adapted to the concept that the Earth is at rest, and it was seen as compatible with Christian theology. Young men studying for the priesthood were expected to grasp the main lines of Aristotelian cosmology. This did not hinder the development of observational astronomy or mathematics, to which the Jesuits, a religious teaching order founded in 1534, made significant contributions.

[1] Italians counted days from sunset before the calendar reform of 1582, and Galileo recorded his date of birth as follows: 15 February at 22.30 hours. The sun set at 5.30 p.m. on that day according to our modern way of reckoning, which means that he was born on 16 February at 4.00 p.m.

First Steps in an Academic Career

Galileo attended the University of Pisa for four and a half years, but he left before getting his degree. This practice was not uncommon, and it was not held against him when he later applied for a university post. For a short time Galileo attended the Drawing Academy in Florence where he studied with Ostilio Ricci, who taught not only painting and design but also perspective and geometry. Galileo displayed considerable proficiency and soon composed a short treatise, *The Little Balance*, in which he reconstructed the reasoning that he believed had led the Greek scientist Archimedes to devise a way of detecting whether a goldsmith had substituted a baser metal for gold in a crown that he had fashioned for King Hero, the ruler of Syracuse. Galileo went on to tackle problems related to the determination of the centre of gravity of solids, a hot topic in mathematics and physics at the time. His results were highly praised and he was appointed professor of mathematics at the University of Pisa in 1589. Three years later he left for the more prestigious Chair of Mathematics at the University of Padua. The number of faculty members in universities was small in the sixteenth century, and Galileo was acquainted with all his colleagues. There were only forty-seven of them, and Galileo was the only one to teach mathematics, astronomy, and physics. His duties were light: he had to give no more than sixty lectures per year. His classes were attended mainly by medical students who wanted to learn how to make horoscopes for their eventual patients. Without these students, Galileo would have had an empty classroom, as we know from the formal protest he lodged when Annibale Bimbiolo, a professor of medicine, decided to give his course at 3.00 p.m., the time that Galileo had chosen because no one else taught at that hour. Outside the university, he gave private lectures on fortifications and military engineering to young noblemen bent on a military career. As was not uncommon for professors, he also rented a large house and let out rooms to rich foreign students. Galileo supplemented his income by manufacturing and selling a mathematical instrument, a forerunner of the slide-rule, which he called the military and geometrical compass. Sales of Galileo's compass went on apace and in 1606 he wrote an Italian handbook for users.

The first indication of Galileo's commitment to heliocentrism dates from 1597. In August of that year, he received a book by the German astronomer Johann Kepler, who argued that Copernicus had been right.[2] After reading the preface, Galileo wrote back to voice his long-standing sympathies for the view that the Earth is in motion, but also to express his fear of making his position known to the public at large.[3] When Galileo read Kepler more carefully, he became aware of their profound methodological differences, and he later reproached him in his *Dialogue on the Two Greatest World Systems* because he 'still listened and assented to the notion of the Moon's dominion over the water, and occult properties, and similar childish ideas'.[4] This statement refers to Kepler's belief that the tides are caused by the Moon, a phenomenon that Galileo thought he could explain on purely mechanical grounds as the result of the combination of the Earth's diurnal and annual revolution.

Galileo never married but his common-law wife, Marina Gamba, bore him three children: Virginia (born 13 August 1600), Livia (born 18 August 1601), and Vincenzio (born 21 August 1606). Galileo carefully cast the horoscope of all of them, something he had done for himself. The two girls were placed in a convent in Arcetri near Florence in 1614; his son Vincenzio was legitimized in 1619.

Around 1602, Galileo began making experiments with falling bodies in conjunction with his study of the pendulum. He first expressed the correct law of freely falling bodies, which says that distance travelled is proportional to the time squared, in a letter to Paolo Sarpi in 1604,[5] but he claimed to have derived it from the erroneous assumption that speed is proportional to distance whereas, as he realized later, it is proportional to the square root of the distance. In the autumn of 1604, the appearance of a particularly bright star, a supernova, revived the debate on the incorruptibility of the heavens that had been so lively a generation earlier when the Dane Tycho Brahe had argued that a very bright star, which had appeared in 1572, proved that heavenly matter was subject to change. Galileo took his side against Aristotle in a book written in the Paduan dialect under the pseudonym of Cecco di Ronchitti. This was his first attempt to reach an audience outside university circles and poke fun at his opponents.

[2] Copernicus's *On the Revolutions of the Heavenly Spheres* had appeared in 1543 but it was generally considered a mere conjecture until Kepler argued that it was correct in his *Cosmographic Mystery* published in 1597.

[3] Letter to Kepler, 4 August 1597 (*Opere*, 10, pp. 67–8).

[4] *Opere*, 7, p. 486; p. 481 in this volume.

[5] Letter to Paolo Sarpi, 16 October 1604 (*Opere*, 10, p. 115).

A New World in the Heavens: A Sidereal Message

In July 1609, while visiting friends in Venice, Galileo heard that a Dutchman had invented a device to make distant objects appear nearer and he immediately attempted to construct such an instrument himself.[6] Others were at work on similar devices, but by the end of August 1609 Galileo had managed to produce a nine-power telescope that was better than those of his rivals. He returned to Venice, where he gave a demonstration of his spying glass from the highest bell towers in the city. The practical value for sighting ships at a distance greatly impressed the Venetian authorities, who confirmed Galileo's appointment for life and raised his salary from 520 to 1,000 florins, an unprecedented sum for a professor of mathematics.

Galileo had put a convex lens at one end of a tube roughly one metre long, and a concave lens at the other end where he applied his eye. He arrived at the kind of telescope we know as the opera glass. He had the great good fortune of having access to the excellent lenses that were made on the island of Murano, just off Venice. Without the unknown and unsung craftsmen who made the lenses that he needed, Galileo would not have achieved his lasting fame as the Columbus of a new heaven. The telescope changed the way of looking at the heavens when it was used by Galileo, whose eye was prepared to see new things and whose hand was able to depict what he saw. It was not only because Galileo was a gifted and persistent observer, but also because he was an exceptional draughtsman that he was able to describe what others had failed to see or lacked the ability to record. Early in January 1610, he turned an improved version of his telescope to the skies. What he saw and reported in the *Sidereal Message*, which appeared on 13 March 1610, was to revolutionize astronomy. First, the Moon was revealed as having mountains and valleys like those on Earth. This was exciting news, because if the Moon resembled the Earth, then it might be inhabited! Second, innumerable stars popped out of the sky and untold worlds were suddenly and unexpectedly disclosed. Third, the Milky Way, which looks like a whitish cloud when seen with the naked eye, showed itself to be a mass of starlets. Fourth, the faint

[6] Actually the telescope had been invented in Italy around 1590. On the technology and the problems involved, see *Galileo's Sidereus Nuncius or A Sidereal Message*, translated from the Latin by William R. Shea (Sagamore Beach, MA: Science History Publications, 2009), reprinted in Galileo's *Selected Writings*, 1–32.

luminosity, which is observed on the dark side of the Moon when its illuminated part is only a thin crescent four or five days old, was correctly interpreted by Galileo, who understood that it was caused by the reflection of sunlight bouncing off the surface of the Earth. So the Moon has 'earthshine', and the reflected light that reaches it is more powerful than moonlight on Earth because the Earth is four times as big as the Moon.[7] The fifth discovery was even more spectacular: four new celestial bodies were found to orbit Jupiter, something that had never been anticipated in the wildest dreams of philosophers or astronomers. Furthermore, it enabled Galileo to name them after the Medici, the ruling family of Tuscany, the state where he was born and where he soon hoped to be recalled.

The discovery of four satellites orbiting around Jupiter was particularly important in the Copernican debate, for if Jupiter could revolve around a central body (be it the Earth or the Sun) with four attendant satellites, then it was no longer absurd to suggest that the Earth could go around the Sun with just one satellite, the Moon. If Jupiter's satellites did not prove that Copernicus was right, they removed a major obstacle to the acceptance of his theory.

While the night sky provided startling news, mundane events continued to matter. One occurred in Florence in January 1609 when the Grand Duke Ferdinand died in Florence and was succeeded by his son, Cosimo II, Galileo's former pupil. Galileo, who had been wanting to return to Florence for some time, realized that his newly-won fame might assist him in effecting a change of residence. He christened the satellites of Jupiter 'Medicean stars' in honour of the Grand Duke and his family, and he began corresponding with Belisario Vinta, the Secretary of State. In May 1610, he formally applied and was awarded a special post with the title of 'Mathematician and Philosopher to the Grand Duke of Tuscany'.

Galileo's departure from the Venetian Republic has often caused surprise. 'Where can you find the freedom and the independence that you enjoyed in Venice?' wrote his friend Sagredo.[8] But Galileo did not see it in this light. To a Florentine correspondent he confided that

[7] See Mark Davie, 'Galileo and the Moon', in Stefano Jossa and Giuliana Pieri (eds), *Chivalry, Academy, and Cultural Dialogues: The Italian Contribution to European Culture* (Italian Perspectives, 37) (Cambridge: Legenda, 2016), 153–63.

[8] Letter to Galileo of 13 August 1611 (*Opere*, 11, p. 171).

a Republic made too many demands on his time, and that he could only hope to obtain the leisure needed to do his work from 'an absolute Prince'.[9]

The New Science Enters the Public Domain: The Letters on the Sunspots

The *Sidereal Message* created a sensation throughout Europe. In a characteristically generous and enthusiastic reaction, Kepler immediately hailed Galileo's achievement. Others wavered on theoretical grounds but mainly, as it seems, because their own telescopes were of poor quality. An influential Florentine nobleman, Francesco Sizzi, criticized Galileo's discoveries on astrological and hermetic grounds, but the most serious charge was made in a privately circulated manuscript written by the Florentine Ludovico delle Colombe and entitled *Against the Motion of the Earth*. It did not attack Galileo directly but ranted against Copernicans and accused them of going against Scripture. Galileo's extensive marginal notes to this tract show how incensed he was and how much he feared that he would be dragged into theological waters.

Meanwhile Galileo continued his telescopic observations in the hope of finding other satellites. The Queen of France, Marie de Médicis, who was a Florentine, wrote to say that she would be obliged if her husband, King Henry IV, could find a place in heaven. Galileo did his best, and in July 1610, he noticed that Saturn had an elongated shape. He conjectured that two satellites, one on each side of Saturn, cause this phenomenon. As was sometimes done to ensure priority without making an immediate disclosure, he sent out an anagram of thirty-seven words to announce that he had made a new discovery. No one worked out the meaning of the anagram, and Giuliano de' Medici, the Tuscan Ambassador in Prague, was charged by the Emperor Rudolph II to ask for the solution. Galileo then officially announced that Saturn had two satellites. Unfortunately, the two objects decreased in size and, at the close of 1612, vanished altogether. Were all of Galileo's celestial discoveries to suffer such a fate? Genuinely embarrassed but still capable of mocking his own discomfiture, Galileo wrote to a friend on 1 December 1612: 'Has Saturn

[9] Letter of Galileo to a correspondent in Florence in February 1619 (*Opere*, 10, p. 233).

devoured its children? Or was their original appearance an illusion produced by the lenses, which deceived me for so long, as well as the many others who observed it with me on many occasions?'[10] What Galileo had seen were the rings of Saturn that are sometimes seen edgewise, when they are hard to detect, and sometimes slanted, when they can be identified with a better telescope than the one Galileo had, something that was done many years later by the Dutch scientist Christiaan Huygens.

Galileo's telescope was powerful enough, however, to enable him to make another discovery that was both sensational and, this time, genuine, namely that Venus has phases, not unlike the Moon. This had to be so if Venus went around the Sun, but the phases cannot be seen with the naked eye. Galileo felt the need to double-check, and once again he protected his priority claim by sending an anagram to Giuliano de' Medici in Prague on 13 November 1610. Six weeks later he had convinced himself that he had been right, and on 1 January 1611 he gave the solution of the riddle and disclosed that Venus had phases and undoubtedly went around the Sun.[11]

The Jesuits in Rome confirmed Galileo's observations, and when he returned to the Eternal City in the spring of 1611, they gave him the equivalent of an honorary doctorate in a ceremony that included no less than three cardinals. The young Prince Federico Cesi, who was in attendance, immediately asked Galileo to join the prestigious Lincean Academy that he had founded. Everyone in Rome wanted to see the new sky for himself or herself, and Galileo was frequently invited to display the marvels of his telescope in the stately gardens of the most important Roman families. His success was so impressive that Cardinal Francesco Maria del Monte wrote to the Grand Duke Cosimo II in Florence: 'If we were still living under the ancient Roman republic, I am certain that a statue would have been erected to him on the Capitoline.'[12]

In the summer of 1611, friends met at the villa of Filippo Salviati at Le Selve near Florence. Galileo suffered from a number of physical complaints, which he blamed on the air of Florence, and he welcomed

[10] Letter to Mark Welser, the third *Letter on the Sunspots* (*Opere*, 5, pp. 237–8); extract in *Selected Writings*, 53.

[11] *Opere*, 10, p. 474 and 11, pp. 11–12.

[12] Letter of 31 May 1611 (*Opere*, 11, p. 119).

the opportunity to go to Salviati's country house. He was present when a discussion arose on the qualities of hot and cold and, specifically, why ice is lighter than water since the action of cold is to condense not rarefy. This led to a debate over the cause of floating bodies in general, with Galileo maintaining, along Archimedean lines, that the cause of floating was the relative density, whereas an Aristotelian opponent held that it was the shape. Shortly thereafter, Cardinal Maffeo Barberini (the future Pope Urban VIII) and Cardinal Ferdinando Gonzaga happened to be in Florence, and the Grand Duke invited Galileo and the Aristotelian philosopher to repeat their arguments before the distinguished visitors. Cardinal Barberini not only enjoyed the discussion but also sided with Galileo, and the two men became friends. Unfortunately, they were later to clash over the issue of Copernicanism, as we shall see.

In the autumn of 1611, Christoph Scheiner, a Jesuit professor at the University of Ingolstadt in southern Germany, made a number of observations of sunspots and was eager to make them known. He had a friend and patron in the person of Mark Welser, a wealthy merchant in Augsburg and an enthusiastic amateur of science, and he sent him three letters in which he described his findings. Welser had these letters printed and sent copies abroad, notably to Galileo and members of the Lincean Academy. As Scheiner was forbidden to use his name, lest he be mistaken and bring discredit on the Society of Jesus, he had concealed his identity under a pseudonym.[13] Galileo, however, identified him and, in two letters to Welser, took him to task for suggesting that the spots were small satellites orbiting around the Sun. Scheiner then wrote a reply that he entitled *A More Accurate Discussion of Sunspots and the Stars that Move around Jupiter*, and Galileo retorted with a third letter to Welser in December 1612 in which he declared that he had observed the sunspots before Scheiner. In an age when scientists were hypersensitive about priorities, this did not endear

[13] Scheiner signed himself *Apelles latens post tabulam* (Apelles hiding behind the painting), a reference to the famous Greek painter Apelles (fourth century BC) who hid behind his paintings to hear the comments of passers-by. When a cobbler found fault with sandals that he had drawn with one of the loops too few, Apelles made the corrections that very night. The next morning the cobbler was so proud that he began to criticize how Apelles portrayed the leg, whereupon the painter emerged from his hiding-place and said, '*Shoemaker, don't judge above the sandal!*' The source is Pliny's *Natural History*, book 35, paragraph 85.

him to his rival. It also opened a breach between Galileo and the Jesuits. Galileo's three letters were published in Rome the following year, under the auspices of the Lincean Academy.[14] Since planets had been mentioned in his *Letters on the Sunspots* Galileo had been able to bring up the matter of the Copernican system and, for the first time, to endorse it unequivocally in print.

Science and the Bible

Meanwhile the problem of Copernicanism did not lie dormant in Florence. In November 1612, Galileo heard that a well-known Dominican friar, Niccolò Lorini, had criticized the heliocentric theory. When asked for an explanation, he replied that he had meant no harm. When the motion of the Earth had been mentioned in a discussion he had said, 'a few words just to show I was alive. I said, as I still say, that this opinion of Ipernicus—or whatever his name is—would appear to be hostile to Divine Scripture.'[15] Lorini's ignorance of the very spelling of Copernicus's name makes it unlikely that he had much interest in astronomy, and it is a pity that after attacking the Jesuits, Galileo should have been led to antagonize the Dominicans. But things were to get worse. On 12 December 1613, objections were raised, on Scriptural grounds, against the notion that the Earth was a planet, at a dinner the Grand Duke gave in Pisa. Galileo was absent but his friend and former pupil, Benedetto Castelli, defended the Copernican theory when he was asked about it by Christina of Lorraine, the mother of the Grand Duke. A couple of days later Castelli wrote to Galileo to inform him that he had behaved as gallantly as a knight, but Galileo felt that the battle could not be entrusted to such an inexperienced defender. He immediately sent Castelli a letter in which he argued that the motion of the Earth was not at variance with the Bible.[16] Peace seemed restored, but a year later, in 1614, another Dominican, Tommaso Caccini, preached an inflammatory sermon against the Copernican system from the pulpit of Santa Maria Novella in Florence on 21 December, the Fourth Sunday of

[14] Galileo Galilei and Christoph Scheiner, *On Sunspots*, translated by Eileen Reeves and Albert Van Helden (Chicago: Chicago University Press, 2010). Relevant excerpts can be found in *Selected Writings*, 33–54.

[15] Letter of Niccolò Lorini, 5 November 1612 (*Opere*, 11, p. 427).

[16] See the *Letter to Don Benedetto Castelli* in *Selected Writings*, 55–61.

Advent. News of his outburst soon reached Rome, and Galileo's friend, Monsignor Giovanni Ciampoli, raised the matter with Cardinal Barberini, the future Urban VIII, who urged 'greater caution in not going beyond the arguments used by Ptolemy and Copernicus', as Ciampoli told Galileo in a letter. Ciampoli went on to explain that great care should be exercised because someone might claim that the telescope provided evidence for life on the Moon, and this could lead to someone else asking if the lunar inhabitants 'were descended from Adam or how they came out of Noah's ark'. It was wise, therefore, 'to declare frequently that one placed oneself under the authority of those who have jurisdiction over the minds of men in the interpretation of Scripture'.[17]

Not to be outdone by Caccini, the Dominican Niccolò Lorini forwarded a copy of Galileo's *Letter to Castelli* to the Inquisition in Rome, and in order to cope with this new challenge Galileo expanded his *Letter to Castelli* into the *Letter to the Grand Duchess Christina*, which contains his most detailed pronouncement on the relations between science and Scripture.[18] Borrowing the *bon mot* of Cardinal Cesare Baronio, 'the intention of the Holy Spirit is to teach us how to go to heaven, not how the heavens go',[19] Galileo stressed that God speaks through the Book of Nature as well as the Book of Scripture, and that care must be exercised lest metaphorical expressions in the Bible be interpreted as scientific facts.

Further complications arose from the publication early in 1615 of a book by Paolo Antonio Foscarini, a Carmelite monk, who also wanted to defend Copernicanism from the charge that it was at variance with what the Bible taught. Foscarini sent a copy of his book to Cardinal Robert Bellarmine, requesting his opinion. The Cardinal's reply was friendly but firm: 'It seems to me that both you and signor Galileo are acting prudently in confining yourselves to speaking hypothetically and not in absolute terms', and he added that 'to demonstrate that the appearances are saved by the hypothesis that the Sun is at the centre and the Earth is in the heavens, is not the same as demonstrating that the Sun really is at the centre and the Earth in the heavens. I believe that the first can be demonstrated, but I have very

[17] Letter of Giovanni Ciampoli to Galileo, 28 February 1615 (*Opere*, 12, p. 146).
[18] See the *Letter to the Grand Duchess Christina* in *Selected Writings*, 61–94.
[19] Ibid. 84.

great doubts about the second.'[20] Bellarmine admitted that should such a proof become available then Scripture would have to be reinterpreted with the utmost care.

A copy of the Cardinal's letter was sent to Galileo, who was not unduly troubled because he believed that he had just such a proof. This was his clever (but unfortunately wrong) theory that the tides would not occur if the Earth were not in motion. This theory eventually became the subject of the Fourth Day of the *Dialogue on the Two Greatest World Systems*, which he had originally entitled *A Discourse on the Tides*. Armed with this argument, which he believed to be decisive, Galileo marched off to Rome at the end of 1615. He talked to all and sundry about his proof, but the only result was that he got the Holy Office of the Inquisition to take the heliocentric theory seriously, something they had not done before. The eleven theological experts who acted as consultants to the Holy Office were asked on 19 February 1616 to examine the two following propositions: (1) The Sun is at the centre of the world and hence completely motionless, and (2) The Earth is not the centre of the world and motionless but moves as a whole and also with diurnal motion. Five days later they unanimously condemned the first proposition as 'foolish and absurd in philosophy, and formally heretical since it explicitly contradicts sentences found in many places in Sacred Scripture according to the proper meaning of the words and according to the common interpretation and understanding of the Holy Fathers and of learned theologians'. The second proposition received the 'same censure in philosophy' and, with respect to theology, was considered 'at least erroneous'.[21] Their report was approved by the Pope and the members of the Holy Office on 25 February and the next day, acting on the orders of Pope Paul V, Cardinal Bellarmine informed Galileo that he had to abandon the Copernican theory, and stop teaching it orally or in writing.

As was current practice, the ruling of the Holy Office was passed to the Congregation of the Index, which was in charge of censoring books, and on 5 March, Copernicus's *On the Revolution of the Heavenly Spheres* was proscribed 'until revised' and Foscarini's book banned outright. Galileo himself was not mentioned but he was worried

[20] Letter from Roberto Bellarmine to Paolo Antonio Foscarini, 12 April 1615, in *Selected Writings*, 94–5.

[21] Minutes of the meeting of 24 February 1616 (*Opere*, 19, p. 321).

and, before he left Rome, he called on Cardinal Bellarmine on 26 May to inform him that it was rumoured in Florence that he had been condemned by the Inquisition. The Cardinal gave him a certificate in which it is stated that he had not been condemned but merely informed of the decision of the Congregation of the Index. Galileo kept this precious document, which he showed to no one.

The Problem of Longitude at Sea and the Quarrel over Comets in The Assayer

Upon his return to Florence Galileo applied himself to a much discussed but non-controversial topic: the determination of longitudes at sea. He hoped that accurate tables of the periods of revolutions of the satellites of Jupiter would enable him to predict how their relative positions would appear from any point on earth, and hence allow seamen to know their location merely by looking through the telescope. The method was sound and it kept Galileo busy over a long period of time but it could not be used because of the practical difficulties of applying it on a ship at sea.

In the autumn of 1618 great excitement was generated over the appearance, in rapid succession, of three comets. Galileo was bedridden at the time and unable to make observations. He was nonetheless asked to comment on accounts given by others, and he chose to attack a lecture delivered by Orazio Grassi, the Jesuit professor of mathematics at the Roman College, who had located the comet beyond the Moon, as we do today. This irked Galileo, for he approached the problem from a very different standpoint since he thought that comets were merely optical phenomena caused by refraction in atmospheric vapour or in the clouds. He set out this view, which caused surprise since it had been that of Aristotle in antiquity, in a *Discourse on Comets* that he published under the name of Mario Guiducci, a young lawyer who enjoyed no scientific reputation. It was clear to Grassi that Galileo was the author, and he prepared a rejoinder that appeared in print in 1619. Entitled *The Astronomical and Philosophical Balance*, it purported to weigh the arguments of the *Discourse on Comets*. Galileo prepared a rebuttal in which he offered to 'ponder' Grassi's arguments with a more delicate weighing instrument, known as an 'assayer', which became the title of his book. *The Assayer* appeared in 1623 and was widely acclaimed as a devastatingly witty masterpiece.

It is here that we find the celebrated passage about the book of nature: 'Philosophy is written in this great book which is continually open before our eyes—I mean the universe—but before we can understand it we need to learn the language and recognise the characters in which it is written. It is written in the language of mathematics, and its characters are triangles, circles, and other geometrical figures, without which it is humanly impossible to understand a word of what it says.'[22] Just before it emerged from the press, Cardinal Maffeo Barberini became Pope and took the name Urban VIII. Galileo was delighted and proudly dedicated the book to him.

The Dialogue on the Two Greatest World Systems

The election of Urban VIII looked to Galileo like a turning-point in his career, and he went to Rome in 1624 to pay his respects to the new pontiff. During the six weeks of his stay, he was admitted to no fewer than six audiences by the Pope, who gave him a painting, two medals, several Agni Dei,[23] and the promise of a pension for his son. It does not seem that Galileo was able to ask Urban VIII whether he could now write about the motion of the Earth, but the friendliness of the Pope led him to believe that he could as long as he presented it not as an ascertained fact but as a mere scientific hypothesis. His impression was strengthened by a meeting with Cardinal Frederic Eutel Zollern, who offered to broach the Copernican question with the Pope. When he saw Urban VIII the Cardinal pointed out that German Protestants were all in favour of the new system, and that it was necessary to proceed with the utmost caution before attempting to resolve the matter. The Pope replied that the Church had never declared the view of Copernicus to be heretical and would not do so, but that he saw no reason to suppose that a proof of the Copernican system would ever be forthcoming. Unfortunately, Cardinal Zollern died in 1625, and

[22] See the extracts from *The Assayer* in *Selected Writings*, 115–21 (p. 115).

[23] An Agnus Dei is the name given to discs of wax impressed with the figure of a lamb and blessed at stated seasons by the Pope. The lamb usually bears a cross or flag, and a figure or the name and arms of the Pope are commonly impressed on the reverse. They are made of the remnants of the preceding year's paschal candle, and in the Middle Ages the Popes sent them as presents to sovereigns and distinguished personages. They were considered a protection against blights and tempests. In the penal laws of Queen Elizabeth Agni Dei are mentioned among 'popish trumperies', the importation of which into England was strictly forbidden.

Galileo lost someone who could have been a key witness at his trial eight years later.

When Galileo returned to Florence, he immediately set to work on his *Dialogue on the Two Greatest World Systems*, but poor health meant that between 1626 and 1629 he was unable to work with any regularity, and it was only in January 1630 that he managed to finish his long-awaited book. The three interlocutors of the *Dialogue* are the Florentine Filippo Salviati (1583–1614), the Venetian patrician Giovanfrancesco Sagredo (1571–1620), and the Aristotelian Simplicio, an imaginary character. They are presented as having gathered in Sagredo's Palace at Venice for four days to discuss the arguments for and against the heliocentric system. Salviati is a militant Copernican, Simplicio an avowed defender of geocentrism, and Sagredo an intelligent amateur already half-converted to the new astronomy.

The First Day belongs to the long history of anti-Aristotelianism, and Galileo borrowed extensively from his predecessors' criticism of Peripatetic philosophy. What must be considered significant about his attack, however, is the skill with which it is conducted. Never before had any critic of Aristotle been so gifted as a writer, so apt at convincing an opponent by the sheer brilliance of his presentation, and so masterful at laughing him off the stage when he refused to be persuaded. Galileo draws from the literary resources of his native Italian to convey insights and to stimulate reflection, but his style does not possess the bare factualness of the laboratory report or the unflinching rigour of a mathematical deduction. Words are more than vehicles of pure thought. They are sensible entities, they possess associations with images, memories, and feelings, and Galileo knew how to use these associations to attract, hold, and absorb attention. He does not present his ideas in the nakedness of abstract thought, but clothes them in the colours of feeling, intending not only to inform and to teach but also to move and entice to action. He wished to bring about nothing less than a reversal of the 1616 decision against Copernicanism, and the dialogue form seemed to him the most conducive to this end. It is true that the written dialogue is deprived of the eloquence of facial expression and the emphasis of gestures, of the support of modulated tone and changing volume, but it retains the effectiveness of pauses, the suggestiveness of questions, and the significance of omissions. Galileo makes the most of these techniques, and it is important to keep this in mind when assessing his arguments, for too often passages of the *Dialogue* have been paraded without sufficient regard for their highly rhetorical content.[24]

The First Day put an end to Aristotelian cosmology by showing that terrestrial and celestial bodies should be explained in the same way. The remaining three days of the *Dialogue* apply this conviction to the problem of whether the Earth moves: the daily rotation of the Earth is discussed in the Second Day, its annual revolution around the Sun in the Third, and the Fourth attempts to show that the tides could not be explained if the diurnal and annual motions of the Earth were denied.

The Second Day opens with the affirmation that the motion of the earth, if it exists, must be altogether imperceptible to its inhabitants. Consequently, opinion turns on what is more plausible. Salviati, speaking for Galileo, argues that it is easier to believe that the Earth rotates on its axis than that all the planets and the stars revolve around the Earth. The diurnal motion of the Earth does away with a host of complexities in the geocentric system. First, it would remove the anomaly of the stars moving westward when all the planets move eastward. Secondly, it would explain the apparent variation in the orbits and periods of the planets and, thirdly, it would dispense with the solid crystalline sphere that was said to carry the stars around in the Ptolemaic system.

The Third Day addresses itself to the central issue of the Earth's annual motion. It begins with a vigorous denunciation of the crude errors of Chiaramonti, a philosopher who had attempted to show that recent astronomical data did not favour locating the Sun at the centre of the universe. Galileo's objective is to persuade his readers that the major objections against Copernicus have been incontrovertibly removed by the telescope. It is now for all to see that Venus has phases like the Moon, that the apparent diameters of Mars and Venus vary as much as 40 and 60 times, and that a planet, Jupiter, orbits with not only one but four moons. Furthermore, since the telescope does not magnify the distant stars but reduces them to tiny dots, there should be no fear, as some had declared, that the stars would have to be gigantic in their size. Among other advantages, the Copernican hypothesis accounts for apparent irregularities in the motions of the planets without cluttering the heavens with deferents and epicycles as in the Ptolemaic system.

[24] On the subtle correspondence between saying and meaning, see Bernard J. F. Lonergan, *Insight* (London: Longmans, Green & Co., 1958), 177–8.

In the Fourth Day Galileo argues that the tides could not occur if the Earth were at rest, an argument that he had made known as early as 1616 and that he considered as his decisive, physical proof that the Earth was in motion. He was so convinced of his argument that he wanted to call his book *A Dialogue on the Tides*. It was only with considerable reluctance that he bowed to objections made by the Roman censors in 1632 and replaced it by *Dialogue on the Two Greatest World Systems.*

The observational evidence about the tides was well known: the *daily tide* with high and low tides recurring at intervals of 12 hours; the *monthly cycle* whereby the tides lag behind 50 minutes each day until they have gone round the clock and are back to that original position; the *half-monthly cycle* with high tide at new and full Moon and low tide at quadratures, and finally the *half-yearly cycle* with greater tides at the equinoxes than at the solstices.

Of the many ways that water can be made to flow, Galileo considers particularly suggestive the to-and-fro motion of water at the bottom of a boat that is alternately speeded up and slowed down. He likens the piling of the water now at one end and now at the other to the action of the tide. The analogy is not entirely satisfactory, however, since the acceleration or retardation is shared uniformly by the whole boat whereas the flux and reflux of the tide is not uniform throughout the sea basins in which they occur. Galileo parries this criticism by introducing a more sophisticated model familiar to contemporary mathematicians and astronomers. He asks his readers to imagine that the ecliptic and the equator coincide. A point on the surface of the Earth can be considered to move on an epicycle attached to a deferent representing the Earth's orbit, as in the figure below.

The cycle revolves once daily. For half the day, the speed at a point is greater than that of the epicycle's centre (the centre of the Earth); for the other half, the speed is less. Maximum and minimum velocities occur when a given point is collinear with the centres of both the epicycle and deferent. The idea is ingenious, but Galileo makes the mistake of mixing two different frames of reference. Whereas the motion of the Earth is considered relative to the Sun, the motion of the water is considered relative to the Earth. But relative to the Earth, the water can receive no acceleration due to the Earth's annual motion, and the water must therefore be at rest relative to the Earth. This non-technical criticism of Galileo was expressed as early as 1633

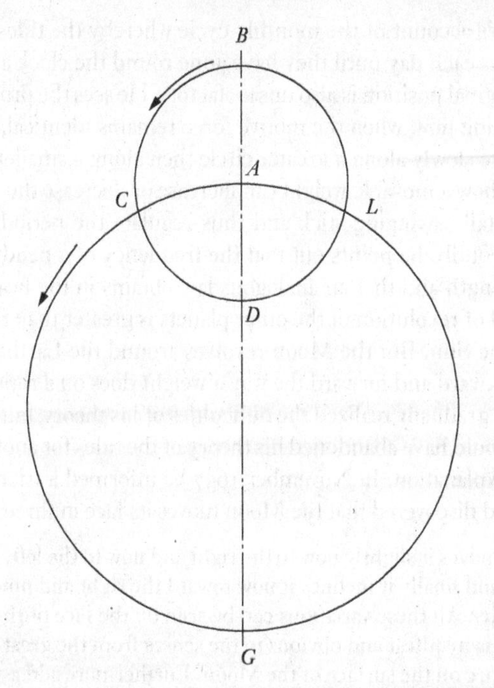

by a group of French physicists.[25] In other words the centripetal acceleration arising from the earth's rotation is everywhere constant and can have no effect in generating the tides. Galileo's failure to distinguish centripetal acceleration from linear acceleration explains this oversight. But there is a direct observational consequence entailed by his model which it is surprising that he overlooked. The axial and orbital speeds of the Earth are so combined that a particle on the surface of the Earth moves very fast once a day when both revolutions are in the same direction (at point *B* in the diagram) and once very slowly when they are going in opposite directions (at point *D*). It follows immediately that high water should occur at noon, the time of the highest retardation, and low water at midnight, the time of greatest acceleration. Galileo does not mention this consequence of his theory.[26]

[25] See the letter of Jean-Jacques Bouchard to Galileo, 5 September 1633 (*Opere* 14, p. 378).

[26] For further details see William R. Shea, *Galileo's Intellectual Revolution*, 2nd edition (New York: Science History Publications, 1977), 172–89.

Galileo's account of the monthly cycle whereby the tides lag behind 50 minutes each day until they have gone round the clock and are back to that original position is also unsatisfactory. He sees the problem as one of explaining how, when the motive force remains identical, a body will move more slowly along a greater circle then along a smaller one. First, he shows how a movable weight can increase or decrease the vibration of a horizontally swinging stick and thus regulate the period of a wheel clock. Secondly, he points out that the frequency of a pendulum varies with its length and that an analogous law obtains in the heavens where the period of revolution of the outer planets is greater than that of those close to the Sun. But the Moon revolves around the Earth; it does not move backward and forward the way a weight does on a rigid bar.

Galileo gradually realized the difficulties of his theory, but the irony is that he should have abandoned his theory of the tides for another equally ill-fated explanation. In November 1637 he informed a friend in Venice that he had discovered that the Moon moves its face in three ways:

> namely it moves it slightly now to the right and now to the left, it raises and lowers it, and finally it inclines it now toward the right and now toward the left shoulder. All these variations can be seen on the face of the Moon and what I say is manifest and obvious to the senses from the great and ancient spots that are on the surface of the Moon. Furthermore add a second marvel: these three different variations have three different periods, for the first changes from day to day and so has its diurnal period, the second changes from month to month and has a monthly period, and the third has an annual period whereby it completes its cycle. Now what will you say when you compare these three lunar periods with the three diurnal, monthly and annual periods of the motions of the sea, of which, by unanimous consent, the Moon is arbiter and superintendent?[27]

Galileo's startling willingness to discard his original model of epicycle and deferent was not taken seriously by all his friends in Venice.

Getting Published

Galileo hoped that the manuscript would be steered through the shoals of Roman censorship by his friends Giovanni Ciampoli and the Dominican Niccolò Riccardi, who had become Master of the Apostolic Palace, and whose duty it was to authorize the publication

[27] Letter to Fulgenzio Micanzio, 7 November 1637 (*Opere*, 17, pp. 214–15).

of books. In the spring of 1630, Galileo set off once more for Rome, where he personally handed over his manuscript to Riccardi, who passed it on to a fellow Dominican, Raffaello Visconti, who knew some astronomy, but was mainly interested in astrology. He was a personal friend of Orazio Morandi, the Abbot of Santa Prassede in Rome, who was considered an authority on horoscopes. Shortly before Galileo's arrival in Rome on 3 May 1630, Morandi had published a number of prophecies based on astrological computations, among them one that predicted the early death of the Pope. Galileo was almost certainly unaware of this incident when he received, on 24 May, an invitation to dine with Morandi in the company of Visconti. Roman gossip lost no time in linking his name with the astrologer, and this was to prove a more delicate matter than Galileo could have foreseen. Galileo left Rome on 26 June, and less than three weeks later Morandi was imprisoned by the Inquisition. When Galileo requested information from a mutual friend, he was told that the trial was so secret that there was no way of knowing what was happening.[28] An *Astrological Discourse on the Life of Urban VIII* bearing Visconti's name was brought forward at the trial, but Visconti was partly successful in his plea of innocence since he was only banished from Rome, while several others received heavy sentences. Morandi himself died in prison on 7 November 1630, before the completion of his trial. The next year, Urban VIII issued a papal edict against astrologers who claimed that they could know the future and set in motion secret forces for the good or harm of the living. Urban ordered his staff to be on the lookout for prognostications that were directed against his own life or that of his relatives down to the third degree. Guilty parties were to be punished with death and confiscation of property.

That Galileo's name should have been associated with those of Morandi and Visconti was unfortunate, to say the least. Little did he suspect that his intimacy with another Vatican official, Giovanni Ciampoli, was to prove even more damaging. Urban VIII was a poet in his leisure hours and he enjoyed the company of intellectuals, one of whom was Ciampoli, who handled his international correspondence. Ciampoli was impatient to secure the Cardinal's hat that Urban

[28] Letter of Vincenzo Langieri to Galileo, 17 August 1630 (*Opere*, 14, pp. 134–5). On the Morandi affair, see Brendan Dooley, *Morandi's Last Prophecy and the End of Renaissance Politics* (Princeton, NJ: Princeton University Press, 2002).

VIII distributed to men whom Ciampoli considered his inferiors. In his frustration he became reckless and allowed himself to be drawn into the circle of the Spanish Cardinal, Gaspare Borgia, the spokesman of Philip IV, the King of Spain and a thorn in Urban's flesh. After Cardinal Borgia publicly protested against the Pope's position in the struggle between France and the House of Hapsburg in a stormy consistory on 8 March 1632, Urban decided to purge his entourage of pro-Spanish elements. He was particularly incensed upon hearing of Ciampoli's behaviour. He stripped him of his considerable powers and ordered him to leave Rome and take up residence in the small town of Montalto in the countryside.

Ciampoli's downfall was to have important consequences for Galileo. Although Visconti had informed Riccardi that Galileo's *Dialogue* only needed a few minor corrections, it was Ciampoli who told Riccardi that the Pope raised no objection. Ciampoli had no warrant for saying this, and the Pope was incensed when he heard about it later. The outbreak of the plague in 1630 had rendered communications between Tuscany and Rome difficult, and Riccardi allowed the *Dialogue* to be printed in Florence. He insisted, however, that the preface and the conclusion be sent to him prior to publication, and he expected Galileo to return to Rome to discuss the final draft. When Galileo objected that the plague made travel between Florence and Rome dangerous, Riccardi agreed that it would be enough if a copy of the manuscript were sent to Rome 'to be revised', he wrote, 'by Monsignor Ciampoli and myself'.[29] Even this requirement was eventually waived, and thereafter Riccardi heard no more of the book until a printed copy reached him in Rome. Above the Florentine *imprimatur* he discovered, to his horror, his own approbation. As Urban VIII expostulated, 'The name of the Master of the Holy Palace has nothing to do with books printed elsewhere.'[30] Summoned to account for his behaviour, Riccardi excused himself by saying that he had received orders to license the book from Ciampoli himself.[31]

The *Dialogue* had gone to press in June 1631, but Galileo had received no financial assistance and had to agree to pay the publisher for the very

[29] As reported by Benedetto Castelli in his letter to Galileo, 21 September 1630 (*Opere*, 14, p. 150).

[30] This was said to the Tuscan Ambassador, Francesco Niccolini, who mentions it in his letter to the Secretary of State in Florence, 5 September 1632 (*Opere*, 14, p. 384).

[31] Account given by Giovanfrancesco Buonamici in 1633 (*Opere*, 19, p. 410).

large run of a thousand copies. The work was completed on 21 February 1632 but copies did not reach Rome until the end of May, thus bursting on to the Roman scene only a few weeks after the consistory in which Cardinal Borgia had attacked Urban VIII. The Roman *imprimatur* on a Florentine publication created a stir, and Riccardi was instructed to have a ban placed on the sale of Galileo's book pending further notice.

An Astronomer on Trial

In the summer of 1632, Urban VIII ordered a Preliminary Commission to investigate the licensing of the *Dialogue*. In the file on Galileo in the Holy Office they found a memorandum of 1616 that enjoined him not to hold, teach, or defend that the Earth moves. This was enough to conclude that Galileo had contravened a formal order of the Holy Office, but he was also indicted for what may seem trivial to the modern reader but was considered important at the time, namely having inserted the Roman *imprimatur* without formal approval and having printed the Introduction in a different typeface. He was summoned to Rome in September 1632, but he pleaded ill health and managed to postpone his trip for another five months. He arrived in Rome on 13 February 1633 and he was allowed to stay with Francesco Niccolini, the Tuscan Ambassador and an old friend, while three theologians read his *Dialogue* to determine whether he had presented the Copernican system as demonstrated. These censors were Niccolò Riccardi, the Master of the Sacred Palace, whose office included the responsibility of licensing books to be printed, Agostino Oreggi, who was the papal theologian, and a Jesuit by the name of Melchior Inchofer, who had had one of his own books placed on the Index. They agreed that there could be no doubt that Galileo argued in favour of the motion of the Earth, although he made an apparent disclaimer in the closing paragraph of the *Dialogue* where the highest authority, namely Pope Urban VIII, is quoted as having said that 'it would be the height of presumption to try to limit or restrict the divine power and wisdom to any one particular fantasy'.[32] Unfortunately, the one who says this is Simplicio, the Aristotelian pedant who had cut such a poor figure throughout the whole *Dialogue*. The theologians were quick to spot

[32] See Fourth Day below, p. 483 (*Opere*, 7, p. 488).

this, and the Pope, when it was called to his attention, was deeply affronted.

On 12 April Galileo was driven to the headquarters of the Holy Office, where he was not placed in a cell but provided with a three-room suite. He could stroll in the garden, and his meals were brought in from the Tuscan embassy, which had one of the best chefs in Rome. It is sometimes said that Galileo's trial was conducted before the Cardinals of the Holy Office, but this is not the case. The 'court' consisted of two officials: Vincenzo Maculano, a Dominican scholar and engineer who had recently been appointed Commissioner General of the Holy Office, and Carlo Sinceri, who had been working there since 1606. Galileo met them four times: on 12 April, 30 April, 10 May, and 21 June 1633. No one else was present. After the first meeting and formal interrogation that took place on 12 April, Commissioner Maculano decided on a private meeting with Galileo. They met on 27 April and worked out an informal compromise: Galileo would admit that he had presented the evidence for the motion of the Earth in too strong a light and, in return, he would be treated with leniency. In practice, this meant that he would recant his 'errors', and the Inquisition would condemn him to imprisonment, which would be immediately commuted to house arrest. After the second official meeting on 30 April, at which Galileo confirmed that he now realized that he had overstepped the limits imposed upon him, he was allowed to leave the Vatican and return to the Tuscan embassy. He was summoned back to the Holy Office on 10 May to be informed, as ecclesiastical law required, that he had eight days to reconsider. Galileo said that this was not necessary, and he returned to the Tuscan embassy until 21 June when he met Maculano and Sinceri for the fourth and last time. He vigorously denied that he had willingly infringed the ban on advocating the heliocentric system, but he was informed that he would nonetheless be condemned on the very next day for having contravened the orders of the Church. At no time was he tortured or molested. He spent the night at the Holy Office, and on the morning of 22 June 1633 he was taken to a hall in the convent of Santa Maria Sopra Minerva in Rome where he was made to kneel while the sentence was read, condemning him to imprisonment. Still kneeling, Galileo formally abjured his errors. As had been agreed, he was not formally imprisoned but allowed to leave for Siena where the Archbishop, Ascanio Piccolomini, had invited him to be his guest. In

1634, he obtained permission to return to Florence, where he was confined to his country house on the outskirts of the city.

Galileo sought comfort in work, and within two years he completed the *Two New Sciences*, the book to which his lasting fame as a physicist is attached. Here Galileo worked out the mathematical and physical implications of ideas he had adumbrated in his *Dialogue*. The most important is his discovery that all bodies fall at the same speed regardless of their weight. This was historically important because it led Newton to realize that new laws of motion were required to explain why this should be the case. The story that Galileo dropped balls from the Leaning Tower of Pisa is probably apocryphal but he showed great ingenuity in devising experiments with rolling balls along an inclined plane. He carefully measured the distance they travelled and the time it took. The outcome was the law that relates distance to the square of the time (expressed as: $s = \frac{1}{2} gt^2$, where s stands for distance, g for the acceleration caused by gravity, and t for time). The insight behind Galileo's reasoning is the surprising fact that the vertical and horizontal components of projectile motion are independents. He illustrated this by showing that when balls are projected horizontally from the same height they go further if impelled with a greater force but that, regardless of the force, they strike the ground at exactly the same time. A ball dropped vertically from the same height when the balls were projected will also strike the ground at the same time.

When Galileo cast about for a publisher, he came up against a new problem: the Church had issued a general prohibition against printing or reprinting any of his books. His manuscript was smuggled out of the country and reached the publisher Louis Elsevier in Holland, a Protestant country over which the Roman Church had no power. Galileo feigned surprise and pretended not to know how the manuscript had left Italy. Although it is unlikely that anyone believed his story, the Church let the publication of the *Two New Sciences* in 1638 go unchallenged. Galileo, however, never succeeded in obtaining the pardon he longed for. Urban VIII was adamant and Galileo remained under house arrest even after he became blind in 1638.

Galileo lived for four more years with the young Vincenzio Viviani as his assistant and, for the last few months, with Evangelista Torricelli. He died on 8 January 1642, five weeks before his seventy-eighth birthday. The sternness of Urban VIII was unabated. Although the

Grand Duke wished to erect a suitable tomb for Galileo in the Church of Santa Croce he was warned not to do so. Nearly a century elapsed before Galileo's body was placed in the main body of the church, and it was not until 1822 that the *Dialogue on the Two Greatest World Systems* was removed from the Index of Proscribed Books.

NOTE ON THE TEXT AND TRANSLATION

THE text of the *Dialogue* used for this translation is taken from volume 7 of the standard edition of Galileo's works: Galileo Galilei, *Opere*, edited by Antonio Favaro, 20 vols (Florence: Barbera, 1890–1909, with subsequent reprints; available online at http://portalegalileo.museogalileo.it). For ease of reference in the body of the translation, page numbers in the Favaro edition are shown in square brackets in the margin of the text. References in the Introduction and Explanatory Notes to this and to Galileo's other works are shown as *Opere* followed by volume and page number.

In preparing this translation and notes we have been greatly assisted by the critical edition and commentary of Ottavio Besomi and Mario Helbing: G. Galilei, *Dialogo sopra i Due Massimi Sistemi del Mondo, Tolemaico e Copernicano*, 2 vols (Padua: Antenore, 1998), which we cite in the notes as Besomi-Helbing followed by volume and page number.

A translation of the first and fourth days of the *Dialogue*, with extracts from the second and third days, was included in our volume of *Selected Writings*: Galileo Galilei, *Selected Writings*, translated by William R. Shea and Mark Davie, with an Introduction and Notes by William R. Shea (Oxford World's Classics, 2012). The translation has been revised for this new edition. Where the earlier volume contains relevant passages from Galileo's other works we have cited it as *Selected Writings* followed by page number.

A key text in the background to the *Dialogue* is Copernicus's *On the Revolutions of the Heavenly Spheres*, first published in 1543. We have used the translation by Charles Glenn Wallis (Great Books of the Western World, 16) (Chicago: Encyclopaedia Britannica, 1952), cited in the notes as Copernicus, *On the Revolutions*, followed by book and chapter number.

Translations have played an important part in the reception of the *Dialogue*, and we have found useful the following annotated translations: in German, by Emil Strauss, *Dialog über die beiden hauptsächlichsten Weltsysteme, das Ptolemäische und das Kopernikanische* (Leipzig: Teubner, 1891); in Spanish, by Antonio Beltrán Marí, *Diálogo Sobre los dos Máximos Sistemas del Mundo Ptolemaico y Copernicano* (Madrid: Alianza

Editorial, 1994); and in French, by René Fréreux and François De Gandt, *Dialogue sur les Deux Grands Systèmes du Monde* (Paris: Éditions du Seuil, 1992).

There have been two substantial translations into English. The first, by Thomas Salusbury, first published in 1661, was reissued in a revised version by Giorgio de Santillana as *Dialogue on the Great World Systems* by the University of Chicago Press in 1953.[1] The second, by Stillman Drake, was published by the University of California Press as *Dialogue Concerning the Two Chief World Systems—Ptolemaic and Copernican* in 1962. Drake's translation is reliably accurate but occasionally somewhat elliptical. We have also benefited from Maurice A. Finocchiaro's translation of important parts of the *Dialogue* in his *The Essential Galileo* (Indianapolis: Hackett, 2008).

In making this new translation I have been conscious of the implications of Galileo's decision, after the success of his *Sidereal Message* (1610), to write in his native Tuscan vernacular rather than in the scientific lingua franca of Latin. In making this choice he was placing himself in a tradition going back ultimately to Dante, of writing for an educated lay public rather than for a narrowly academic readership. This tradition provided a model of Italian prose which combined analytical rigour with the forceful, sometimes colloquial directness of the vernacular. Galileo took this to unprecedented lengths with passages of closely argued scientific exposition alongside others which are informal and humorous. The result is a work of sometimes startling originality, and I have tried to capture some of this quality in my translation. This has meant, for example, simplifying Galileo's syntax and breaking up many of his long and complex sentences, or finding English equivalents for some of his idioms rather than attempting to translate them literally. On the other hand, I have kept his references to topical controversies such as the rival claims of Ariosto and Tasso to pre-eminence in Italian narrative poetry, with a brief account of the background in the explanatory notes. I hope that the effect will be to make the translation accessible, as Galileo's text was, to a non-specialist modern reader, while at the same time maintaining its place in its

[1] A facsimile reprint of the original edition in two tomes is available with a preface by Stillman Drake: Thomas Salusbury, *Mathematical Collections and Translations* (London: Dawsons of Pall Mall, 1967).

historical context. Translating a 400-year-old text inevitably involves negotiating such tensions between accessibility and authenticity. In this I have benefited greatly from the expert advice of my co-author William R. Shea, especially in keeping me on sound scientific ground, for which I am very grateful. Our collaboration on every aspect of this edition has, indeed, been mutually enlightening and rewarding.

M.D.

SELECT BIBLIOGRAPHY

Editions of the Dialogue

(A) IN ITALIAN

Dialogo sopra i due massimi sistemi del mondo, Tolemaico e Copernicano, in Galileo Galilei, *Opere*, edited by Antonio Favaro, 20 vols (Florence: Barbera, 1890–1909, with subsequent reprints; available online at http://portalegalileo.museogalileo.it), vol. 7.

Galileo Galilei, *Dialogo sopra i due massimi sistemi del mondo, Tolemaico e Copernicano*, edited by Ottavio Besomi and Mario Helbing, 2 vols (Padua: Antenore, 1998).

(B) IN ENGLISH

Translation by Thomas Salusbury (first published in 1661): Galileo Galilei, *Dialogue on the Great World Systems in the Salusbury Translation*, revised, annotated and with an introduction by Giorgio de Santillana (Chicago: University of Chicago Press, 1953). A facsimile reprint of the original edition in two tomes is available with a preface by Stillman Drake: Thomas Salusbury, *Mathematical Collections and Translations* (London: Dawsons of Pall Mall, 1967).

Galileo Galilei, *Dialogue Concerning the Two Chief World Systems—Ptolemaic and Copernican*, translated with introduction and notes by Stillman Drake (Berkeley and Los Angeles: University of California Press, 2nd edition, 1967).

English translations of other works by Galileo

Bodies that Stay Atop Water or Move in it, translated with introduction and notes in Stillman Drake, *Cause, Experiment and Science* (Chicago: University of Chicago Press, 1981).

Controversies on the Comets of 1618, translated by Stillman Drake (Philadelphia: University of Pennsylvania Press, 1960).

On Motion and On Mechanics, translated by I. E. Drabkin and Stillman Drake (Madison, WI: University of Wisconsin Press, 1960).

On Sunspots, letters of Galileo Galilei and Christoph Scheiner translated and introduced by Eileen Reeves and Albert Van Helden (Chicago: University of Chicago Press, 2010).

Sidereus Nuncius or A Sidereal Message, translation with introduction and notes by William R. Shea and Tiziana Bascelli (Sagamore Beach, MA: Science History Publications, 2009). Also translated as *The Sidereal*

Messenger, with introduction and notes by Albert Van Helden (Chicago: University of Chicago Press, 1989), and as *The Starry Messenger* in Stillman Drake, *Telescopes, Tides and Tactics* (Chicago: University of Chicago Press, 1983).
Discoveries and Opinions of Galileo, selections from Galileo's works edited and translated by Stillman Drake (New York: Anchor Books, 1957).
The Essential Galileo, selections from Galileo's works edited and translated by Maurice A. Finocchiaro (Indianapolis, IN: Hackett, 2008).
The Little Balance, translated by Laura Fermi and Gilberto Bernardini, in *Galileo and the Scientific Revolution* (Greenwich, CT: Dover, 1965).
Two New Sciences, translated with introduction and notes by Stillman Drake (Madison, WI: University of Wisconsin Press, 1974).

Works on Galileo

Bagioli, Mario, *Galileo Courtier* (Chicago: University of Chicago Press, 1993).
Bagioli, Mario, *Galileo's Instruments of Credit* (Chicago: University of Chicago Press, 2006).
Bucciantini, Massimo, Camerota, Michele, and Giudice, Franco, *Galileo's Telescope* (Cambridge, MA: Harvard University Press, 2015).
Büttner, Jochen, *Swinging and Rolling. Unveiling Galileo's Unorthodox Path from a Challenging Problem to a New Science* (Dordrecht: Springer, 2019).
Camerota, Michele, *Galileo Galilei e la cultura scientifica nell'età della controriforma* (Rome: Salerno Editrice, 2004). In Italian.
Cohen, H. Floris, *How Modern Science Came into the World* (Amsterdam: Amsterdam University Press, 2010).
Drake, Stillman, *Galileo at Work: His Scientific Biography* (Chicago: University of Chicago Press, 1978).
Drake, Stillman, *Essays on Galileo*, 3 vols (Toronto: University of Toronto Press, 1999).
Fantoli, Annibale, *Galileo, for Copernicanism and for the Church* (Notre Dame, IN: Notre Dame University Press, 1996).
Finocchiaro, Maurice A., *The Galileo Affair: A Documentary History* (Berkeley and Los Angeles: University of California Press, 1989).
Finocchiaro, Maurice A., *Retrying Galileo 1633–1992* (Berkeley and Los Angeles: University of California Press, 2005).
Finocchiaro, Maurice A., *The Routledge Guidebook to Galileo's Dialogue* (London: Routledge, 2014).
Hall, Crystal, *Galileo's Reading* (Cambridge: Cambridge University Press, 2013).
Heilbron, J. L., *Galileo* (Oxford: Oxford University Press, 2010).
Machamer, Peter (ed.), *The Cambridge Companion to Galileo* (Cambridge: Cambridge University Press, 1998).

McMullin, Ernan (ed.), *The Church and Galileo* (Notre Dame, IN: Notre Dame University Press, 2005).

Numbers, Ronald L., *Galileo Goes to Jail and Other Myths About Science and Religion* (Cambridge, MA: Harvard University Press, 2009).

Palmieri, Paolo, *Re-enacting Galileo's Experiments* (Lewiston, NY: Edwin Mellen Press, 2008).

Redondi, Pietro, *Galileo: Heretic* (Princeton, NJ: Princeton University Press, 1987).

Reeves, Eileen, *Galileo's Glassworks* (Cambridge, MA: Harvard University Press, 2008).

Renn, Jürgen, ed., *Galileo in Context* (Cambridge: Cambridge University Press, 2001).

Roland, Wade, *Galileo's Mistake: The Archaeology of a Myth* (Toronto: University of Toronto Press, 2001).

Sharratt, Michael, *Galileo Decisive Innovator* (Cambridge: Cambridge University Press, 1996).

Shea, William R., *Galileo's Intellectual Revolution*, 2nd edition (New York: Neale Watson Academic Publications, 1977).

Shea, William R., *Conversations with Galileo* (London: Watkins, 2019).

Shea, Willliam R., and Artigas, Mariano, *Galileo in Rome: The Rise and Fall of a Troublesome Genius* (Oxford: Oxford University Press, 2003).

Shea, Willliam R., and Artigas, Mariano, *Galileo Observed: Science and the Politics of Belief* (Sagamore Beach, MA: Science History Publications, 2006).

Sobel, Dava, *Galileo's Daughter* (London: Penguin, 1999).

Vergara Caffarelli, Roberto, *Galileo Galilei and Motion* (Berlin: Springer, 2009).

Wallace, William A., *Galileo and His Sources: The Heritage of the Collegio Romano in Galileo's Science* (Princeton, NJ: Princeton University Press, 1984).

Wootton, David, *Galileo: Watcher of the Skies* (New Haven, CT: Yale University Press, 2010).

Further Reading in Oxford World's Classics

Ariosto, Ludovico, *Orlando Furioso*, ed. and trans. Guido Waldman.

Aristotle, *Physics*, trans. Robin Waterfield, ed. David Bostock.

Bacon, Francis, *The Major Works*, ed. Brian Vickers.

Galilei, Galileo, *Selected Writings*, ed. and trans. William R. Shea and Mark Davie.

Tasso, Torquato, *The Liberation of Jerusalem*, trans. Max Wickert, ed. Mark Davie.

Vasari, Giorgio, *The Lives of the Artists*, ed. and trans. Julia Conaway Bondanella and Peter Bondanella.

A CHRONOLOGY OF GALILEO

1543 Nicolaus Copernicus (1473–1543) publishes *On the Revolutions of the Heavenly Spheres* and Andreas Vesalius (1514–64), *On the Fabric of the Human Body.*

1545 Council of Trent convenes and will be in session off and on for eighteen years until 1563.

1551 The Roman College (now the Pontifical Gregorian University) founded by the Jesuits in Rome.

1559 First worldwide Index of Prohibited Books promulgated by the Roman Catholic Church.

1564 Galileo is born in Pisa, on 15 or 16 February; Michelangelo Buonarroti dies in Florence on 18 February; William Shakespeare is born in England on 23 April.

1581 Galileo enrols at University of Pisa.

1585 He abandons studies at Pisa without taking a university degree. He considers becoming a painter but decides to study mathematics.

1587 Galileo goes to Rome to discuss his essay on the centre of gravity of solids with Christopher Clavius, a Jesuit professor at the Roman College and the most famous astronomer of his day. During this period he also writes on the balance and on Archimedes.

1589 Appointed professor of mathematics at the University of Pisa; develops a rudimentary thermometer; begins to study falling bodies. His lecture notes indicate that he read the works of Aristotle and those of Christopher Clavius.

1591 Around this time, Galileo drafts a work, *On Motion*, in which he is critical of Aristotelian philosophy. Galileo's father, Vincenzio Galilei, dies.

1592 Galileo becomes professor of mathematics at the University of Padua, where he will spend eighteen years.

1593 He writes a *Treatise on Fortifications and Architecture* for young noblemen to whom he gives private lesssons.

1594 Drafts a *Treatise on Mechanics* that he will revise and enlarge over the next few years.

1597 Writes a *Treatise on the Sphere* for his students.

1600 Giordano Bruno is burnt at the stake in Rome. Galileo's first daughter, Virginia, is born in Padua.

1601 Birth of his second daughter, Livia.

1603 Prince Federico Cesi founds the Lincean Academy in Rome.

1604 Galileo gives three public lectures on a new star that appeared in the heavens.

1605 The teenage Prince Cosimo de' Medici takes instruction from Galileo in the summer in Tuscany.

1606 Galileo publishes sixty copies of the *Operations of the Geometric and Military Compass*, an instrument that he manufactured and sold. His son, Vincenzio, is born in Padua.

1607 Baldessar Capra publishes a pirated Latin edition of Galileo's *Operations of the Geometric and Military Compass*. Galileo sues him and publishes an account of the incident.

1609 The Grand Duke Ferdinando I dies in Florence; Cosimo II succeeds him. Galileo devises a new telescope, observes and measures mountains on the Moon.

1610 He discovers four satellites around Jupiter and publishes *A Sidereal Message*. He is appointed chief mathematician and philosopher to the Grand Duke of Tuscany and leaves Padua for Florence.

1611 Galileo visits Rome and is made a member of the Lincean Academy.

1612 He publishes a book on floating bodies.

1613 Prince Cesi publishes Galileo's *Letters on the Sunspots* in Rome. Galileo's daughters, Virginia and Livia, enter the Convent of San Matteo in Arcetri outside Florence.

1614 Galileo writes a *Letter to Benedetto Castelli* in which he argues that the Copernican theory is not at variance with Catholic doctrine.

1615 Galileo expands his *Letter to Benedetto Castelli* into the *Letter to the Grand Duchess Christina* that is widely circulated in manuscript but will only be published in 1636 in Holland. He writes his *Observations on the Copernican Theory* and leaves for Rome where he arrives in December and will stay until June 1616.

1616 Galileo writes his *Discourse on the Tides*. On 26 February Galileo is given a formal warning forbidding him from holding, teaching or defending Copernicanism. On 5 March Copernicus's *On the Revolutions of the Heavenly Spheres* is placed on the Index of Prohibited Books.

1618 Three comets appear in rapid succession, generating interest and debate. Galileo goes on a pilgrimage to Loreto.

1619 Galileo publishes anonymously a *Discourse on the Comets* criticizing the interpretation given by the Jesuit professor Orazio Grassi who replies in a work entitled *The Philosophical and Astronomical Balance*.

1623 Maffeo Barberini becomes Pope Urban VIII. Galileo dedicates *The Assayer* to him.

1624 Galileo travels to Rome and sees the Pope six times in so many weeks.

1629 Bubonic plague enters northern Italy from Germany.

1630 Galileo returns to Rome to obtain a printing licence for his *Dialogue Concerning the Two Greatest World Systems—Ptolemaic and Copernican*. Prince Cesi dies. Bubonic plague strikes Florence where the deaths will amount to ten thousand.

1631 Galileo's brother, Michelangelo, dies of plague in Germany.

1632 The *Dialogue Concerning the Two Greatest World Systems—Ptolemaic and Copernican* is published in Florence.

1633 Galileo stands trial in Rome. His *Dialogue* is prohibited and he is sentenced to house arrest.

1634 His daughter Virginia (in religion Suor Maria Celeste) Galilei dies in Arcetri.

1637 Galileo loses his eyesight.

1638 Louis Elsevier publishes Galileo's *Two New Sciences* in Leiden, Holland.

1641 Vincenzio Galilei draws his father's design for a pendulum clock.

1642 Galileo dies in Arcetri on 8 January.

1644 Pope Urban VIII dies.

The frontispiece of the first edition of the Dialogue (Florence, Landini, 1632) shows Aristotle (left) and Ptolemy (centre) debating with Copernicus (right, separated from the other two figures by a ship in the background).

DIALOGUE

BY

GALILEO GALILEI, LINCEAN ACADEMICIAN

SUPERORDINARY MATHEMATICIAN

OF THE UNIVERSITY OF PISA

and Philosopher and first Mathematician of

THE MOST SERENE

GRAND DUKE OF TUSCANY

Where in meetings during four days are discussed the two

GREATEST WORLD SYSTEMS

THE PTOLEMAIC AND THE COPERNICAN,

putting forward, without deciding, the philosophical and physical arguments

for both one side and the other.

WITH PRIVILEGES

IN FLORENCE, by Giovanni Batista Landini, 1632

Licensed by the Authorities.

[26] Permitted to be printed, if seen by the Reverend Father the Master of the Holy Apostolic Palace.

Antonio, Bishop of Belcastro, Vicegerent.

Permitted to be printed.

Fr. Niccolò Riccardi, Master of the Holy Apostolic Palace.*

Permitted to be printed in Florence, the customary orders having been observed.

Pietro Niccolini, Vicar General in Florence.

Permitted to be printed. 11 September 1630.

Fr. Clemente Egidi, Inquisitor General in Florence.

Permitted to be printed. 12 September 1630.

Niccolò dell'Antella.

DEDICATION

[27]

Most Serene Grand Duke,*

Great though the difference is between humankind and other animals, it may not be entirely misguided to say that differences between persons themselves are hardly less. What comparison is there between one and a thousand? And yet there is a common saying that one man is worth a thousand, where a thousand are not worth as much as a single one. The difference derives from their different intellectual capacities, which I equate with whether or not they are philosophers; for philosophy being their food, it separates those who can receive nourishment from it from the common crowd, in varying degrees depending on the degree of nourishment they take from it. Those who aspire to the highest degree will distinguish themselves the most; and their highest aspiration is to turn to the great book of nature,* which is the proper subject of philosophy. And although everything in that book is perfectly proportioned, being the handiwork of the omnipotent Maker, nonetheless the part which is most resplendent and worthy is that in which the greatness of his artistry is most clearly visible to our sight. The structure of the universe, among all those things in nature which we can apprehend, is that which I believe should take pride of place; for as it exceeds everything else in size, since it contains the whole universe, it should also exceed everything else in nobility, since it regulates and maintains all things. So if anyone has ever stood out intellectually above other men, Ptolemy and Copernicus are those who have scaled the heights in studying, scrutinizing, and reflecting on the structure of the world. Since these Dialogues of mine deal principally with their works, it seemed to me that they should be dedicated to none other than your Highness; for as their content rests on these two, whom I consider the greatest thinkers [28] to have left their works to us on these matters, it was only appropriate that I should commend them to the favour of the one who has given me the greatest support, so that they might receive from him both praise and patronage. And if Ptolemy and Copernicus have so illuminated my understanding that this work can be said to be as much theirs as mine, it can also be said to belong to your Highness, for your generosity has given me leisure and tranquillity in which to write, and your effective help, which has never tired of honouring me, has enabled it finally to see the light of day. So I ask your Highness to accept it with your customary kindness; and if you find in it anything which can bring greater understanding and benefit to lovers of the truth, recognize it as your own, for your rule is so beneficial that no one in your happy dominions is troubled by the universal miseries which prevail in the world. Praying that you may prosper so that your pious and magnanimous custom may ever increase, I offer you my humble reverence.

Your Serene Highness's humble and devoted servant and subject,

Galileo Galilei.

TO THE DISCERNING READER* [29]

A salutary edict was published some years ago in Rome which, as a defence against the dangerous scandals of our present age, imposed a timely silence on the opinion of Pythagoras concerning the mobility of the Earth.* There were some who were so bold as to assert that this decree was the product not of judicious consideration but of ill-informed prejudice, and there were protests that adjudicators who were totally inexperienced in astronomical observations* had no business clipping the wings of speculative thinkers by means of an over-hasty ban. I could not remain silent in the face of such presumptuous complaints, and as I was fully informed about this entirely prudent decision I resolved to appear on the stage of world opinion as a witness to the unvarnished truth. I was in Rome at the time,* where I was received and indeed applauded by the most eminent prelates of the Curia, and I had prior information about the publication of this decree. So it is my intention in this present work to show to all foreign nations that as much is known about these matters in Italy, and especially in Rome, as has been imagined by anyone working north of the Alps.* I shall also bring together all the speculations concerning the Copernican system, to make clear that they were all already known to the Roman censorship, and that this climate can produce ingenious insights to delight the mind as well as dogmas for the salvation of the soul.*

To this end I have taken the side of Copernicus in the argument, purely as a mathematical hypothesis, and I have adopted every artifice to show that it is superior to the hypothesis that the Earth is* [30] *fixed—not as an absolute position, but as it is defended by some who call themselves Peripatetic philosophers, although they are Peripatetics only in name.* For they do not walk about but are content to worship shadows, basing their philosophy not on their own insight but on their memory of a few principles which they have only half understood.*

Three main topics will be discussed. I shall aim to show, first, that none of the experiments that can be carried out on Earth are sufficient to prove that the Earth moves, but that they are equally compatible with its moving or being fixed; and here I hope to make *known many observations which were unknown in antiquity. Second, I shall examine celestial phenomena so as to strengthen the Copernican hypothesis as if it were absolutely proved to be superior, and I shall add new considerations, but only to facilitate the work of astronomers, not as necessities in nature. In the third place, I shall put forward an ingenious speculation. I had occasion to observe many years ago that some light might be shed on the unsolved problem of the cause of the tides* if we were to accept the movement of the Earth. This observation of mine came to be generally known, and some people were so charitable as to adopt it as a brainchild of their own. So now, to forestall any foreigner who might take over our arguments and then criticize our lack of insight in such a fundamental question as the tides, I have decided to set out the suppositions which might make this plausible if one assumed that the Earth moves. I hope that these considerations will show the world that, if other nations have sailed further, we have speculated no less than them, and that if we submit to asserting that the Earth is fixed and that the contrary hypothesis is purely a mathematical invention, this is not because we are not aware of what others have thought on the matter, but, if nothing else, for reasons which piety, religion, and a recognition of divine omnipotence and the limitations of human understanding dictate.*

It seemed to me, next, that it would be appropriate to set these ideas out in the form of a dialogue, which is not bound always to follow the strict rules of mathematics and so gives scope for occasional digressions, which are sometimes no less interesting than the main argument itself.

Many years ago, in the marvellous city of Venice, I had several occasions to converse with signor Giovan Francesco Sagredo, a man of distinguished birth and great intelligence.*

[31] *He was visited from Florence by signor Filippo Salviati,* whose ancient lineage and magnificent wealth were the least of his marks of distinction; a man of sublime intellect, there was nothing he enjoyed more avidly than subtle philosophical speculations. I often discussed these matters with them both, along with a Peripatetic philosopher* for whom the greatest obstacle to perceiving the truth was the fame he had acquired as an interpreter of Aristotle.*

Now that cruel death has deprived Venice and Florence of these two great luminaries, both in the prime of their life, I have resolved *to perpetuate their fame in these pages, as far as my modest talents allow, by making them the participants in this discussion. The good Aristotelian, too, will have his place here; but given his great attachment to the commentaries of Simplicius,* I have thought it tactful to give him the name of his revered author rather than calling him by his own name. So may these two great spirits, who still have an honoured place in my heart, accept this public monument to my undying affection, and may the memory of their eloquence help me to expound these speculations to posterity.*

These gentlemen had already had some fragmentary discussions in the course of casual meetings, which had only whetted their appetite to learn more; so they wisely decided to meet on several consecutive days when they could set aside all their other concerns and devote themselves more systematically to admiring and speculating on the wonders of God's creation in heaven and on Earth. So, when they were met in signor Sagredo's palace, after the customary brief formalities signor Salviati began as follows.

FIRST DAY

[33]

SPEAKERS: SALVIATI, SAGREDO, AND SIMPLICIO

SALVIATI. We concluded and agreed yesterday that we should meet today to discuss, as systematically and in as much detail as we could, the physical reasons which have so far been brought forward by the advocates of the Aristotelian and Ptolemaic system on the one hand, and the Copernican system* on the other, and the effectiveness of each. And since Copernicus, by placing the Earth among the mobile bodies in the heavens, makes it a sphere similar to a planet, we should begin our investigation by examining the force of the arguments the Peripatetics use to show that such an assumption is impossible, on the grounds that nature necessarily has two different kinds of substance, one heavenly, which is impassible and immortal, the other made up of the elements, which is changeable and transient. Aristotle discusses this in his book on the Heavens, drawing first on arguments derived from certain general assumptions, and then backing these up by reference to experience and to detailed demonstrations. I will follow the same procedure, putting forward my view and then frankly declaring where I stand, and in doing so I will lay myself open to your criticism and especially to that of Simplicio, who is such a vigorous champion and upholder of Aristotelian teaching.

Copernicus considers the Earth to be a sphere similar to a planet.

According to Aristotle, nature requires heavenly substances to be unchangeable and substances made up of the elements to be changeable.

The first stage in the Peripatetics' argument is where Aristotle proves the integrity and perfection of the world by showing that it is neither a simple line, nor a pure surface, but a body endowed with length, width, and depth; and since there are only these three dimensions, and the world has all three, it is complete and therefore perfect. Starting with the simple dimension of length we obtain the unit of magnitude which is called a line, adding width we obtain a surface, and finally adding height or depth we produce a body. We cannot pass from these three dimensions to any other, since they constitute completeness and, so to speak, totality. I wish Aristotle had shown this by necessary demonstration, especially since this could be done quite clearly and easily.

[34]

Aristotle considers the world to be perfect because it has three dimensions.

Aristotle's demonstrations that there are three dimensions, and no more.

The number three is celebrated by the Pythagoreans.

SIMPLICIO. Surely there are splendid proofs in the second, third, and fourth parts of the text, after the definition of what is meant by 'continuity'?* Is there not, first of all, the proof that there is no other dimension beyond these three because the number three is everything, and is to be found everywhere? Is this not confirmed by the authority and teaching of the Pythagoreans, who state that everything has a threefold definition—a beginning, a middle, and an end—and that three is the number of totality? And what of the other argument, that the number three is used in sacrifices to the gods, as if by a natural law? And that, taking our cue from nature, we use the word 'all' to refer to three things but not to less—for we refer to two things as 'both', not 'all' as we do for three? All these arguments can be found in the second part of Aristotle's text. In the third part, pursuing the matter in more depth, we read that 'everything', 'all', and 'perfection' are formally the same, and that therefore the body is the only perfect figure, since it alone is defined by three, which is everything, and is divisible in three ways, that is, in all directions. The others, in contrast, are divisible in either one or two ways, because their division and continuity depend on the number which applies to them—so one is continuous in one direction, another in two, but only the body is continuous in all three. Moreover, in the fourth part, after various other arguments, does he not give another proof of the same point, namely that since passing from one dimension to another implies some lack or deficiency (so, for instance, we pass from a line to a surface area because a line lacks width), and since perfection, being in all directions, cannot lack anything, we cannot pass from a body to another unit of magnitude? Now, from all these texts, do you not think it has been sufficiently established that there can be no passing from the three dimensions of length, width, and height to any other dimension, and that therefore the body, which has all three, is perfect?

[35] SALVIATI. To tell the truth, the only thing I feel obliged to concede in all these arguments is that something which has a beginning, a middle, and an end can and should be called perfect; but I see no reason why it should follow that there is therefore anything perfect about the number three, or that the number three can confer perfection on anything which possesses

it. Nor do I believe that, for example, there is any more perfection in having three legs rather than four or two; or that because there are four elements they are in any way imperfect, or that the perfection would be greater if there were three. So it would have been better if he had left these fanciful ideas to the rhetoricians and established his case with necessary demonstrations,* which is the proper way to proceed in the demonstrative sciences.

SIMPLICIO. I think you are using these arguments facetiously; but it was the Pythagoreans who attributed such qualities to numbers, and yet you, as a mathematician and, as I believe, a Pythagorean in many of your opinions, now seem to despise their mysteries.

SALVIATI. I'm well aware that the Pythagoreans had the highest regard for the science of numbers, and that Plato himself admired the human intellect and regarded it as sharing in divinity simply because it could comprehend the nature of numbers. My own view is not very different, but I don't accept for a moment that the mysteries which caused Pythagoras and his followers to have such veneration for the science of numbers are the same as the foolish beliefs which are spoken and written about among the common people. Indeed, I know that the Pythagoreans were so concerned that these wonders should not be exposed to the scorn and contempt of the crowd that they condemned it as sacrilege to publish the hidden properties of numbers and the incommensurable and irrational quantities which they studied, and they declared that anyone who revealed these properties would be punished in the next world. I think it was for this reason that one of them said, to satisfy the crowd and free himself from their demands, that their numerical mysteries were the same as the trifles which were then widely repeated among the common people. He showed the same astuteness as the wise young man who, to rid himself of the importunate questions of his mother or his inquisitive wife who wanted him to divulge the secret deliberations of the Senate, made up the story which led to her and many other women being made a laughing-stock, to the great amusement of the Senate itself.*

Plato's opinion that the human intellect shares in the divine nature because of its comprehension of numbers.

The number mysteries of the Pythagoreans are fables.

SIMPLICIO. I have no wish to be counted among those who would pry into the mysteries of the Pythagoreans; but to come back to the point at issue, I reply that the reasons which Aristotle

[36] gives to prove that there are and can be only three dimensions seem to me to be conclusive; and I think that, if any further proof had been necessary, Aristotle would not have failed to provide it.

SAGREDO. You might add, if he had known it, or if it had come to his mind. But Salviati, I would very much like to hear some evident proof of this, if you have any which is clear enough for me to understand.

SALVIATI. I do indeed, for you and for Simplicio as well; and it is one which you can not only understand, but which you know already,* although you may not be aware of it. To make it easier to understand, we shall take pen and paper, which I see are ready here for just this purpose, and do a little drawing. First let us mark these two points, A and B; and then join them up with these curved lines, ACB and ADB, and this straight line, AB. Now tell me which line in your view determines the distance between A and B, and why.

Geometrical demonstration that there are three dimensions.

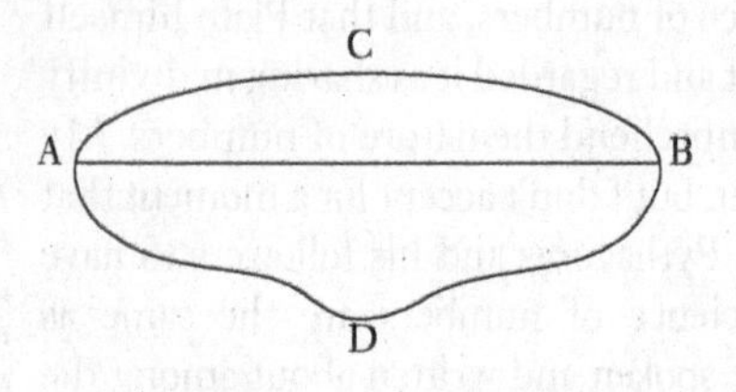

SAGREDO. I would say it is the straight line, not the curved ones, because the straight line is the shortest, and also because there is only one, and its position is determined; whereas there is an infinite number of other lines, all longer and of different lengths, and it seems to me that the distance should be determined by the line of which there is only one and whose position is fixed.

SALVIATI. So the straight line determines the distance between two points. Now let us add another straight line, CD, parallel to the line AB, so that there is a surface area between them, of which I would like you to tell me the width. So, starting from A, tell me where and how you would draw a line to meet the line CD so as to establish the width of the area between

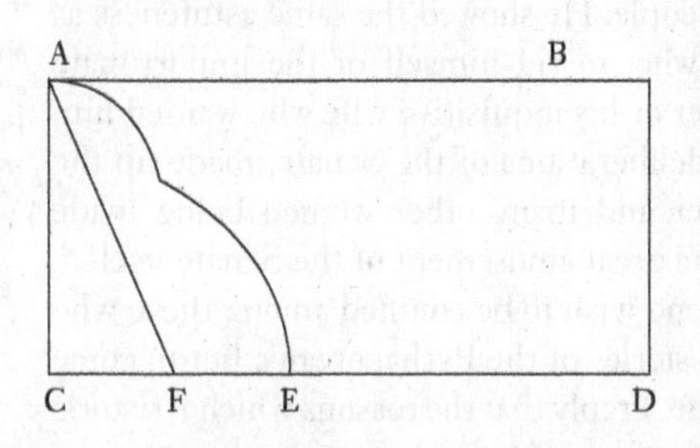

these two lines. Would you use the curved line AE, or the straight line AF, or . . . ?

SIMPLICIO. I would use the straight line AF, not the curved one, as we have already eliminated curves for this purpose.

SAGREDO. I wouldn't use either of them, because the straight line AF runs obliquely. I would draw a line at a right angle above CD, because this seems to me to be the shortest, and it would be unique among the infinite number of longer lines, all of unequal length, which could be drawn from A to any number of points on the opposite line CD.

SALVIATI. Your choice and the reasoning you give for it are [37] perfect. So we have established that the first dimension is determined with a straight line; the second, namely width, with another line, also straight but, as well as being straight, at right angles to the line which determined length. Thus we have defined the two dimensions of a surface area, namely length and width. What if, now, you had to determine a height—for example, the height from the ceiling to the floor beneath our feet: given that from any point on the ceiling you could draw an infinite number of lines, both curved and straight, all of different lengths, to an infinite number of points on the floor underneath, which of these lines would you use?

SAGREDO. I would attach a thread to a point on the ceiling with a lead weight on the end, and would let it hang freely until it touched the floor. I would say that the length of this thread, being the shortest straight line that could be drawn from this point to the floor, would be the correct height of the ceiling.

SALVIATI. Excellent; and if, from the point on the floor which was reached by this hanging thread (assuming that the floor is level and not sloping) you were to draw two other straight lines, one to determine the length and the other the width of the surface of the floor, at what angle would they be to the thread?

SAGREDO. They would certainly be at right angles, if the thread was perpendicular and the floor was smooth and properly levelled.

SALVIATI. So, if you establish a starting and finishing point for your measurements, and draw a straight line to determine the first measurement, namely length, then the line to determine width must necessarily be at right angles to the first; and the line to determine the third dimension, namely height, if it starts from the same point, must form right angles and not oblique angles with the other two. Thus from these three perpendiculars, being three lines all of which are unique, fixed, and the shortest possible, you will have determined the three [38] dimensions: AB for length, AC for width, and AD for height.

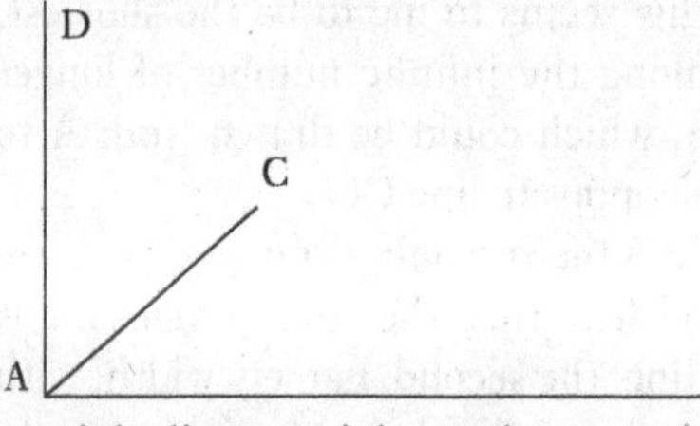

And since it is clear that no other line can run to this point at right angles with these three, and the dimensions must be determined solely by straight lines at right angles to each other, it follows that there cannot be more than three dimensions, and any body which has three dimensions has them all; if it has all three dimensions it is divisible on all sides, in which case it is perfect; and so on.

SIMPLICIO. But who is to say that it isn't possible to draw other lines? Why can I not draw another line coming up to point A from below, which is at a right angle to the others?

SALVIATI. It is certain that you can't make more than three lines converge on a single point which are at right angles to each other.

SAGREDO. Yes, because what Simplicio proposes is simply the line DA extended downwards; and indeed, in this way one could draw two more lines, but they would still be the same three original lines, the only difference being that whereas now they simply meet, they would then intersect each other, but they wouldn't introduce any new dimensions.

Geometrical precision is not to be sought in proofs regarding the natural world.

SIMPLICIO. I won't deny that this proof of yours may be conclusive, but I will follow Aristotle* in saying that the truths of the natural world do not necessarily have to be demonstrated with the exactitude of mathematical proofs.

SAGREDO. Maybe not, if no such proof is available; but if it exists, why should we not use it? But let us not expend more words on this point, as I think Salviati will concede to you and to Aristotle without further proof that the world is a body, and is perfect, indeed is the height of perfection, as the greatest of God's works.

SALVIATI. Yes indeed. So let us move from the general consideration of the whole to a discussion of the parts, which according to Aristotle's first division are twofold, different and in some respects opposed to each other, namely the heavenly and the elemental. The former is ingenerable, incorruptible, inalterable, impassible, etc.; the latter exposed to continual alteration, mutation, etc. This difference he derives, as its first cause, from the difference in their local motion, and he proceeds to argue as follows.

Aristotle divides the world into two parts, the celestial and the elemental, which are contraries to each other.

Leaving, so to speak, the world of the senses and withdrawing to the world of ideas, Aristotle begins to reason like an architect. Given that nature is the principle of motion, it follows that natural bodies have local motion. He goes on to say that local motions are of three kinds, namely circular, rectilinear, [39] and a mixture of rectilinear and circular. The first two he calls simple motions, because of all the different kinds of line only a circle and a straight line can be called simple. He then narrows down the argument and specifies that, of these simple motions, circular motion is motion around a centre and rectilinear motion can be either towards or away from a centre—upward if it is moving away from the centre and downward if it is moving towards the centre. From this he deduces that all simple motions are limited to these three, namely towards, away from and around a centre; and he finds a certain pleasing correspondence between what was said above about a body, that it too is perfected in three ways, and its motion. Having established these motions he goes on to say that, since natural bodies are either simple or composite (simple bodies being those which have their own natural principle of motion,* such as fire and earth), simple motions belong to simple bodies and mixed motions to composite bodies, with the proviso that composite bodies follow the motion which belongs to the predominant element in their composition.

Three kinds of local motion: rectilinear, circular, and mixed.

Rectilinear and circular motions are simple, because they follow simple lines.

SAGREDO. Please stop there for a moment, Salviati. Listening to this argument, I feel so many questions pressing in on me from all sides that I must either voice them, if I am to listen attentively to what you say next, or else cease paying attention to what you say in order not to forget my questions.

SALVIATI. I'll gladly stop here, because I too am running the same risk: I am constantly on the point of getting lost, and I feel as if I'm steering a course between rocks and stormy waves which threaten to make me lose my bearings. So do explain your difficulties, rather than letting them accumulate any further.

SAGREDO. You began, with Aristotle, by drawing me away from the world of the senses to show me the architecture underlying its construction; and I appreciated your explanation of how natural bodies are by their nature mobile, since it was established elsewhere that nature is the principle of motion. This prompted a doubt in my mind, and it was this: Aristotle's definition states that nature is the principle of motion and rest, so why did he not say that some natural bodies are by their nature mobile and others immobile? If natural bodies all have a principle of motion, then either it was superfluous to include 'rest' in his definition of nature, or it was not appropriate to cite this definition in this context.

Aristotle's definition of nature is either faulty, or inappropriate in this context.

Then, I have no problem with your explanation of what [40] Aristotle means by simple motions, and how he derives them from their movement in space, defining simple motions as those which follow simple lines, namely a straight line and a circle. I won't quibble over the example of a cylindrical screw, since this is the same in all its parts, and so could be included among examples of simple lines. But I do object to the way he narrows the discussion, while implying that he is simply repeating the same definition with different words when he equates circular motion with movement around a centre and rectilinear motion as movement *sursum et deorsum*, that is, up and down. These terms make sense only in the physical world, and presuppose not only that it is a physical world but that it is inhabited by us humans. For if rectilinear motion is simple because a straight line is simple, then it must be applicable to a simple natural body regardless of its direction, whether up or down, forward or back, to the left or to the right, or in any other conceivable direction, as long as it is in a straight line; if not, then Aristotle's supposition is false. He then goes on to state that there is only one circular motion in the world, and therefore only one centre, to which all references to rectilinear motion as being 'upward' or 'downward' relate.

The helix around a cylindrical screw can be defined as a simple line.

All these points suggest that Aristotle is resorting to sleight of hand, and is adapting his architecture to the building rather than constructing his building according to architectural principles. For, if I were to state that there could be a thousand circular movements, and therefore a thousand centres, in the natural world, then it would follow that there would also be a thousand upward and downward movements. Or again, we have seen that he postulates simple and mixed motions, defining simple motions as those which are either circular or rectilinear, and mixed motions as those which are a compound of these two. He defines natural bodies as either simple (those which have a natural principle of simple motion) or compound, and he attributes simple motions to simple bodies and mixed motions to compound bodies. But in saying this, he no longer takes mixed motion to mean a mixture of straight and circular motions—something which can exist in the real world—but instead introduces a mixed motion which is as impossible as it would be to mix motions in opposite directions in the same straight line, to produce a motion which is partly upward and partly downward. To mitigate such an absurdity and impossibility, he is reduced to saying that composite bodies move according to the simple element which predominates in their composition; and this makes it necessary to say that even motion in the same straight line is sometimes simple and sometimes mixed, so that the simplicity of the motion no longer depends solely on the simplicity of the line.

Aristotle adapts the principles of architecture to the construction of the world, not the construction to his principles.

According to Aristotle, rectilinear motion is sometimes simple and sometimes mixed.

SIMPLICIO. Isn't it a sufficient difference to say that the simple and absolute motion is much faster than the motion which is derived from the predominant element? After all, how much faster does a piece of solid earth fall than a piece of wood? [41]

SAGREDO. That's a valid point, Simplicio. But if simplicity of motion is affected by its velocity, not only will there be a hundred thousand kinds of mixed motion, but how will you be able to identify which motions are simple? What's more, if simplicity of motion depends on greater or lesser velocity, then no simple body will ever move with simple motion, because in all natural rectilinear motions the velocity constantly increases, and hence its 'simplicity' is constantly changing—yet, as simplicity, it ought to be immutable. More seriously, you will be burdening Aristotle with another error, because in his definition of mixed motion he makes no mention of greater or lesser velocity, and now you are saying that this is a necessary and essential factor. What's more, such a rule would serve no useful purpose, because there will be mixed motions—plenty of them—some of which will be slower and others faster than simple motion: the movement of lead or wood in comparison with earth, for example. So how would you decide which of these motions to call simple, and which mixed?

SIMPLICIO. I would define simple motion as that of a simple body, and mixed motion that of a compound body.

SAGREDO. Exactly. But think what this means, Simplicio. You argued earlier that simple and mixed motions serve to distinguish between simple and compound bodies, and now you are saying that simple and compound bodies will enable me to differentiate between simple and mixed motions. That sounds like a recipe for never being able to identify either motions or bodies. What's more, you are now saying that greater velocity is not enough, and you're adding a third condition for your definition of simple motion, which Aristotle was content to define with just one, namely simplicity of space. But now, according to your definition, simple motion is motion in a simple line, at a specified velocity, by a simple mobile body. Well, that's up to you, but I think we should go back to Aristotle, who defines mixed motion as being compounded of rectilinear and circular motion, although he was unable to give any examples of a body which naturally moves in such a way.

[42] SALVIATI. So I shall return to Aristotle. Having begun his exposition clearly and methodically, he now goes off at a tangent, being more concerned to arrive at a conclusion he had already reached in his own mind than to follow through the logic of his argument. He states as something well known and self-evident that upward and downward rectilinear motions belong naturally to fire and earth respectively, and that therefore there must be, in addition to these bodies which are close at hand, another body to which circular movement belongs naturally; and that this body is as much more excellent as circular motion is more perfect than rectilinear motion. He defines the greater perfection of circular movement by reference to the perfection

of a circular line compared to a straight line, calling the former perfect and the latter imperfect. He considers a straight line imperfect because, if it is infinite, it has no termination or end, and if it is finite, there is something beyond it to which it can be extended. This is the cornerstone and foundation of the whole Aristotelian structure of the world, on which all its other qualities depend—of having neither weight nor lightness, of not being susceptible to generation or corruption, of being exempt from all change except local motion, etc. He attributes all these properties to the simple body which moves naturally in circular motion, while the opposite qualities—of weight and lightness, corruptibility, etc.—belong to bodies which move naturally in a straight line.

Aristotle's view that a circular line is perfect and a straight line imperfect, and why.

It follows that if there is found to be any flaw in what has been established up to this point, there is reason to have doubts about everything else which is built on it. I don't deny that what Aristotle has introduced here in general terms, based on universal first principles, is confirmed later in his exposition with more detailed arguments and experiments, which must all be considered and evaluated in their own right. But given that there are many difficulties—and not insignificant ones—in what has been expounded thus far, and that the first principles and foundations need to be absolutely firmly established, so that we can build on them with confidence, I suggest that it would be a good thing, before we accumulate any more doubts, to see whether we can (as I think) follow another path which might be safer and more direct, and establish our basic foundations on more carefully considered architectural principles. So, leaving aside Aristotle's argument for now, which we shall [43] return to examine in detail later, I can say that of the points he has stated so far, I agree with him that the world is a body endowed with all the dimensions, and therefore absolutely perfect; and I would add that it is therefore necessarily perfectly ordered, that is, having its parts disposed with supreme and perfect order among themselves. I don't imagine that anyone will disagree with this assumption.

The author's supposition that the world is perfectly ordered.

SIMPLICIO. How could anyone deny it—first because it comes from Aristotle, and second because the order which the world perfectly contains is the basis of the name 'ordered world' itself?*

SALVIATI. So having established this principle, we can immediately conclude that, if the integral bodies* of the world must in their nature be mobile, their motion cannot possibly be rectilinear, or indeed anything but circular, for a very obvious reason. Anything which moves in a straight line changes its location, and the longer it goes on moving the further away it gets from its starting point and from all the other points it has passed through; and if this is its natural motion, then it was not in its natural place when it started, and therefore the parts of the world were not disposed in perfect order. But we presuppose that they are in perfect order, in which case it cannot possibly be in their nature to change their location, and therefore to move in a straight line. Moreover, rectilinear motion is by its nature infinite, because a straight line is infinite and has no predetermined end; and so no mobile body can possibly have a natural principle of rectilinear motion, as this would be motion towards a point which is impossible to reach, since it has no predetermined end. And Aristotle himself rightly says that nature does not undertake to do anything which is impossible, or to move towards a point which cannot be reached.

Rectilinear motion is impossible in a perfectly ordered world.

Rectilinear motion is by nature infinite; it is impossible in nature.

Nature does not undertake to do anything which is impossible.

It may be objected that even if a straight line, and hence rectilinear motion, can be extended to infinity, i.e. has no predetermined end, nonetheless nature has, so to speak, arbitrarily set limits on them and has endowed natural bodies with a natural instinct to move towards those limits. To this I would reply that we might fantasize that such a situation existed in the primeval chaos, where indistinct matter moved erratically in a confused and random way, and that nature quite properly used rectilinear motion to put it in order. For if rectilinear motion produces disorder when it moves [44] bodies which are rightly disposed, it produces order when it moves those which are wrongly disposed. But once everything had been optimally distributed and put in place, no body could possibly still have a natural inclination to move in a straight line, which could only mean moving out of its natural and proper place, and therefore producing disorder. So we can say that rectilinear motion can serve to move materials in the work of construction, but that once the construction is complete, it either remains motionless or, if it is mobile, moves only in a circular motion.

Rectilinear motion perhaps existed in the primeval chaos. It is suitable for setting in order bodies which are wrongly ordered.

We might, however, follow Plato in saying that even the heavenly bodies, once they had been created and fully established, were temporarily endowed by their Maker with rectilinear motion, and that when they reached their assigned places they began one by one to revolve in a circle, exchanging rectilinear for circular motion, which they have maintained ever since. This is an inspired idea, well worthy of Plato,* about which I recall our mutual friend the Lincean Academician* speaking. His argument, if I remember rightly, was as follows: any body which for whatever reason is constituted in a state of rest, but which is by nature mobile, will move when it is free to do so, provided it has a natural inclination to some particular place. If it had no such inclination it would remain in a state of rest, as it would have no more reason to move to one place than to another. The fact that it has such an inclination necessarily means that its motion will continually accelerate; and, starting with the slowest degree of motion, it will only reach a given degree of velocity by passing through all the lower degrees of velocity or, rather, the higher degrees of slowness. For, starting from a state of rest (which is the infinite degree of slowness), there is no reason for it to acquire a given degree of velocity without first passing through a lower degree, and an even lower degree before that. Indeed, it seems entirely reasonable that it should pass first through the degrees which are closest to the one from which it started, and from these to the degrees which are furthest from from it—and the degree from which a mobile body begins to move is the highest degree of slowness, namely a state of rest. Such acceleration will not happen unless the mobile body gains something by moving; what it gains is to come closer to its desired place, that is, the place to which its natural attraction draws it; and it will take the shortest route to that point, i.e. a straight line. So we can reasonably conclude that nature, to [45] confer a given velocity on a mobile body which has been constituted in a state of rest, does so by making it move, in a given interval of time and space, in a straight line. If this reasoning is correct, we may imagine that God created a heavenly body, e.g. Jupiter, and determined to confer on it a given velocity, which it should then maintain constantly in perpetuity. Following Plato's argument, we can say that God initially endowed it with a straight and accelerated motion, and that once it had reached that intended degree of velocity he converted its straight motion to a circular one, the velocity of which is naturally constant.

According to Plato, the heavenly bodies were moved initially with rectilinear, and then with circular motion.

A mobile body in a state of rest will not move unless it has an inclination towards a particular place. Its motion accelerates as it moves towards the place to which it is inclined.

In changing from a state of rest, a mobile body passes through every degree of slowness. A state of rest is the infinite degree of slowness.

A mobile body accelerates only when it draws closer to its goal.

To confer on a mobile body a given degree of velocity, nature moves it in rectilinear motion.

Constant velocity is a property of circular motion.

SAGREDO. I like this reasoning very much, and I will like it even better when you have resolved a difficulty for me. I don't quite understand how it must necessarily be the case that when a mobile body changes from a state of rest to its natural motion it has to pass through all the intermediate degrees of slowness, of which there must be an infinite number. Why could nature not endow Jupiter with its circular motion, at a given velocity, the moment it was created?

Between a state of rest and any intermediate velocity there are infinite lesser degrees of velocity.

SALVIATI. I didn't say, and I wouldn't presume to say, that it was impossible for God or nature to endow a body with that velocity immediately; I say only that in practice nature does not do so, so that this would be something outside the course of nature, and therefore miraculous. [If a body of whatever weight, moving at whatever speed, encounters a body at rest, however weak and yielding, it will not immediately impart its velocity to that body. Clear evidence of this is the sound of the impact, which would not be heard (or rather, would not exist) if the body at rest were to acquire the same velocity as the moving body at the moment of impact.]*

Nature does not immediately endow bodies with a specific degree of velocity, although it could do so.

SAGREDO. Is it your view, then, that when a stone changes from a state of rest to its natural motion towards the centre of the Earth, it passes through every degree of slowness up to a given velocity?

[46]

SALVIATI. I'm quite sure of it, so much so that I can make you sure of it as well.

SAGREDO. If understanding this was the only thing I learnt from today's discussion I would be well satisfied.

SALVIATI. As far as I understand your thinking, your difficulty consists largely in the fact that a mobile body has to pass in a very short time through all the infinite degrees of slowness leading up to whatever speed it reaches in that time. So before going on to anything else, let me try to resolve this doubt. This should not be difficult, if I say that the mobile body passes through all these degrees without pausing in any of them, taking only an instant to do so. And since any interval of time, however small, contains an infinite number of instants, there will always be enough instants for it to pass through every degree, however short the time may be.

When from a state of rest a mobile body passes through every degree of velocity, it does not pause in any of them.

SAGREDO. I follow you so far; but I still find it extraordinary that a cannon ball—for that is how I imagine this falling body—which we see falling so fast that it travels more than two hundred *braccia** in less than ten pulse beats, can still have passed through such a small degree of velocity that, if it had continued to move at that speed without accelerating further, it would not have travelled this distance if it had continued all day.

SALVIATI. Or all year for that matter, or ten years, or a thousand years, as I will try to show you, I hope without your contradicting a few simple points which I shall put to you. So tell me first, do you have any problem in agreeing that the cannon ball continually gains impetus* and speed as it falls?

SAGREDO. Not at all; I'm quite sure that it does.

SALVIATI. And would you agree that the impetus it had gained at any point in its motion would be sufficient to take it back to the height from which it started?

SAGREDO. I would certainly agree, provided the whole of the impetus it had gained was directed without any obstacle solely to taking the body, or another equal to it, back to its [47] original height. So, if there were a hole through the centre of the earth, and the cannon ball was dropped from a hundred or a thousand *braccia* above its surface, I don't doubt that it would pass the centre of the earth and rise as far as it had fallen. I deduce this from what happens when a pendulum is moved from the perpendicular, which is the state in which it is at rest, and then allowed to fall freely: it falls back to the perpendicular and then goes past it by the same amount, apart from such impetus as it loses through the resistance of the air, the cord and any other accidental impediments. The same can be seen when water flowing into a siphon rises by the same amount as it falls.

A falling body gains sufficient impetus to take it back to the height from which it has fallen.

SALVIATI. Your reasoning is impeccable. Now, I know you have no doubt that a body gains impetus by moving away from its starting-point towards the centre to which it tends. So can you agree that two equal bodies will, if they are not impeded, acquire equal impetus if their distance to the centre is equal, even if they move along different lines?

SAGREDO. I'm not sure that I understand the question.

SALVIATI. I can explain more clearly with a drawing, like this: Let the line AB be parallel to the horizon, and BC perpendicular to it. Add the inclined line CA. Now take the line CA to be a hard, perfectly polished inclined plane, on which a perfectly round, hard ball is rolled down, while another similar ball is dropped down the perpendicular CB. My question is whether you agree that the impetus gained by the ball which has rolled down the plane CA is the same, when it reaches A, as that gained by the other ball when it reaches B having been dropped down the perpendicular CB.

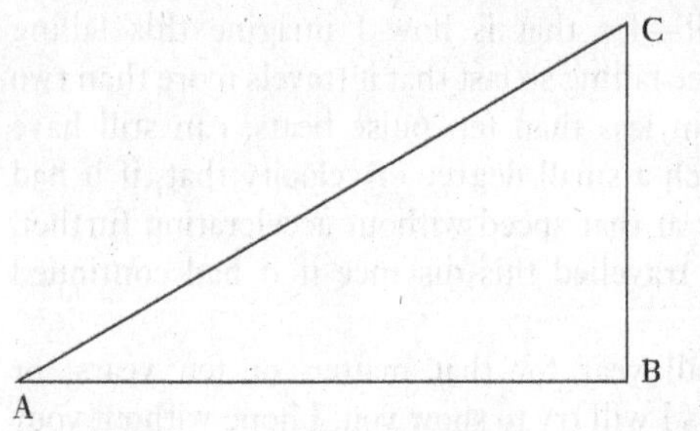

Impetus is equal for falling bodies that are equally distant from the centre.

SAGREDO. It seems clear that the answer is yes, because both balls have moved towards the centre by the same amount, and so, on the basis of what I have already granted, both will have gained impetus sufficient to take them back to the same height.

SALVIATI. Now tell me what you think this same ball would do if it were placed on the horizontal plane AB.

Mobile bodies remain motionless on a horizontal plane.

SAGREDO. It would not move, provided the plane was not inclined in any way.

SALVIATI. But on the inclined plane CA it would descend, [48] but with a slower motion than if it fell along the perpendicular CB.

SAGREDO. I was about to reply definitely that it would, because movement along the perpendicular CB seems necessarily to be faster than along the incline CA. But in that case how can the body which has reached A along the inclined plane have the same impetus, i.e. the same degree of velocity, as the one which has moved along the perpendicular to B? These two propositions seem to be contradictory.

SALVIATI. In that case, you will certainly not agree with me when I say categorically that the velocity of a falling body along an inclined plane and along a perpendicular are equal. But this is an absolutely true proposition; just as it is also a true proposition that a falling body moves more rapidly along a perpendicular than along an incline.

Velocity along an inclined plane is the same as velocity along a perpendicular, but motion along a perpendicular is faster than along an incline.

SAGREDO. These sound to me like contradictory propositions; what do you think, Simplicio?

SIMPLICIO. They seem contradictory to me as well.

SALVIATI. I think you are teasing me by pretending not to understand something which you understand better than I do. Tell me, Simplicio: when you imagine one moving body to be faster than another, how do you picture that in your mind?

SIMPLICIO. I picture it as one body covering a greater distance than the other in the same time, or else covering the same distance in a shorter time.

SALVIATI. Fine; and how do you picture two bodies moving at the same speed?

SIMPLICIO. As covering the same distance in the same time.

SALVIATI. Is that all?

SIMPLICIO. This seems to me to be the correct definition of equal motion.

SAGREDO. I think we could add a further statement: if we also say that velocities are equal when the distances covered are in the same proportion to the time taken to cover them, this will be a more universally valid definition.

Velocities can be said to be equal when the distances covered are in proportion to the time taken.

SALVIATI. So it will, because it includes equal distances covered in an equal time, and also unequal distances covered in times which are unequal but proportional to the distances. Now look again at the figure, and using your concept of faster motion, tell me why you think the velocity of the body falling along CB is greater than that of the body descending along CA.

SIMPLICIO. Because I think that in the same time it takes for the falling body to pass the whole of CB, the descending body will have passed less of CA than the total distance CB. [49]

SALVIATI. Indeed it will, which confirms that a body moves more rapidly along a perpendicular than along an incline. Now consider whether this same figure can be used to confirm the proposition that the bodies moved with equal velocity along both CA and CB.

SIMPLICIO. I can't see how this can be so; on the contrary, it seems to contradict what you said earlier.

SALVIATI. What about you, Sagredo? I've no wish to teach you what you already know, since you've just given me a definition of it.

SAGREDO. The definition I gave you was that moving bodies can be said to have equal velocity when the distances passed are proportional to the times which they take to pass them. For this definition to apply in the present case, the time taken for a body to descend along CA would have to be in the same proportion to the time taken for one to fall along CB as the lines CA and CB are to each other. But I don't see how that can be the case given that the motion along CB is more rapid than that along CA.

SALVIATI. You must see, surely: both movements continually accelerate, don't they?

SAGREDO. Yes, they do, but the acceleration is greater on the perpendicular than on the incline.

SALVIATI. But is the acceleration on the perpendicular such that if you took two sections of equal length anywhere on these two lines, the perpendicular and the incline, the motion along the perpendicular would always be faster than along the corresponding section of the incline?

SAGREDO. No; on the contrary, if I took a section of the perpendicular at the end nearest to C and a section of the incline at the opposite end from it, the velocity would be much greater on the incline than on the corresponding section of the perpendicular.

SALVIATI. So you see that the proposition 'motion along a perpendicular is more rapid than on an incline' is only universally true if both motions start from the same state, namely a state of rest. If this condition is not met, then the proposition is so flawed that its opposite could also be true, [50] namely that motion on an incline is more rapid than along a perpendicular, since it's perfectly possible to take a space on an incline which a moving body would cover more rapidly than an equivalent space on a perpendicular. So, if motion on an incline is more rapid than on a perpendicular in some places and less rapid in others, it follows that in some places the proportion between the time taken by a mobile body on an incline and the time taken on a perpendicular will be greater than the proportion between the distance covered in each case.

In other places the reverse will be true, i.e. the ratios of the times will be less than that of the distances travelled. So, for example, if two bodies move from a state of rest at C, one along the perpendicular CB and the other along the incline CA, in the time that the body on the perpendicular covers the whole distance CB the other will have covered only the shorter distance CT. Therefore the ratio of the time taken to travel CT to the time taken to travel CB (which are equal) will be greater than that of the line TC to the line CB, since the same measurement is greater in proportion to a smaller measurement than to a larger one. Conversely, if we took a section of the line CA, extended if necessary, which was equal to CB but could be passed in a shorter time, then the ratio of the time along the perpendicular to the time along the incline would be less than the ratio of the distances. So, if it is possible to conceive of an incline and a perpendicular with distances and velocities such that the ratio between distances and times can be both greater and less, then we can reasonably allow that there can also be inclines and perpendiculars with distances where the times taken to cover them have the same ratio as the distances themselves.

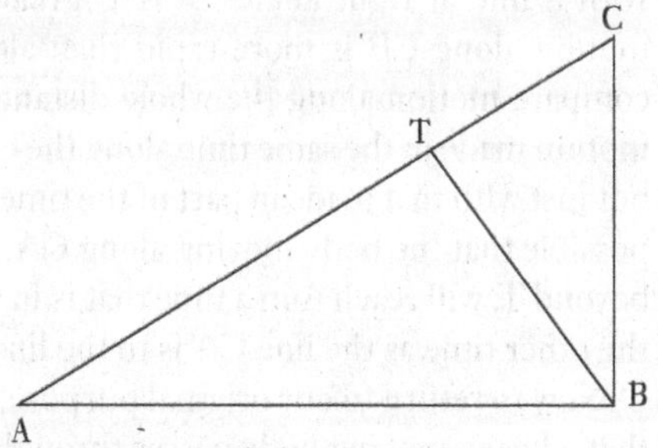

C
T
B
A

SAGREDO. You've resolved my main doubt, and I can see now that what I originally thought was a contradiction is in fact possible and indeed necessary. But I still don't see how one of these possible or necessary cases applies to what we are trying to establish here, namely that the times taken to descend along CA and to fall along CB have the same ratio as the lines CA and CB themselves, so that it can be said without contradiction that the velocities along the incline CA and the perpendicular CB are equal.

SALVIATI. Let it suffice for now that I've dispelled your disbelief. The scientific explanation will become clear when you read our Academician's proofs concerning local motion.* As you

[51] will see, he shows that in the time which it takes for a body to fall the whole of the distance CB, a body descending along CA will reach T, which is the point from which a line at right angles to CA reaches B. To find out the point which the body falling along CB has reached when the other body reaches A, draw a line from A at right angles to CA and extend CB until it meets this line; this will be the point you are seeking. This shows how motion along CB is more rapid than along the incline CA (assuming that in both cases the moving body starts from a state of rest at C), because the distance CB is greater than CT, and the distance from C to the point at which CB, extended, intersects with a line at right angles to A is greater than CA; therefore motion along CB is more rapid than along CA. But when we compare motion along the whole distance CA, not with all the motion made in the same time along the extended perpendicular but just with that made in part of the time from C to B, it is quite possible that the body moving along CA, continuing to descend beyond T, will reach A in a time that is in the same proportion to the other time as the line CA is to the line CB.

Now to return to our original purpose, which was to establish that a heavy moving body passes through every degree of slowness in falling from a state of rest to whatever velocity it acquires, let us look again at the same figure. We have agreed that a body falling along the perpendicular CB and another descending along the incline CA will acquire the same degree of velocity by the time they reach points B and A. So I imagine you will have no difficulty in conceding that on another plane less steep than AC, as for example DA, a moving body would descend even more slowly than on CA. Clearly we could add other planes with such a small degree of elevation above the horizontal AB that the time taken by the moving body, namely the ball we imagined, to reach point A was greatly prolonged; for on the plane AB itself that time would

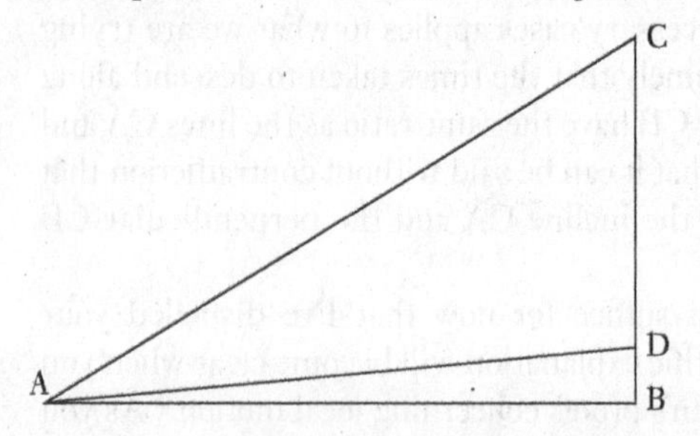

be infinite, and the smaller the degree of inclination the slower motion becomes. So it must be possible to fix a point such a small distance above B that, if you made a plane from that point to A, the ball would not traverse it even in a year. Next you [52] should know that the impetus, or degree of velocity, which the ball has gained when it reaches point A is such that, if it continued to move uniformly with the same velocity, without accelerating or slowing down, it would cover twice the distance in the same time as it had covered so far; so, for instance, if the ball had taken an hour to cover the distance DA, then if it continued to move uniformly at the velocity it had gained when it reached A, it would cover twice the distance DA in the next hour. We have seen that the degree of velocity gained at points B and A by bodies moving from any point on the perpendicular CB, one along the inclined plane and the other along the perpendicular, is always equal. Hence it would be possible for a body to fall along the perpendicular from a point so close to B that the velocity it had gained when it reached B would not be sufficient for it to cover a distance double the length of the inclined plane in a year, or for that matter in ten years or a hundred years. So we can conclude as follows. First, that in the course of nature, and in the absence of any external or accidental impediments, a body moves along an inclined plane at a degree of slowness determined by the degree of inclination, and that the degree of slowness becomes infinite if the plane is no longer inclined but horizontal. Second, the degree of velocity gained at any point on an inclined plane is equal to that gained by a body falling along the perpendicular to the point intersected by a horizontal line which passes through that point on the inclined plane. Hence it must be granted that a falling body, starting from a state of rest, passes through all the infinite degrees of slowness; and consequently, to gain a given degree of velocity it must first move in a straight line, descending by a greater or lesser distance depending on the velocity it is to acquire and on the degree of inclination of the plane along which it is moving. So there could be a plane with so little inclination that, to gain a given degree of velocity, a body would have to move a very great distance over a very long period of time; and on a horizontal plane it would never naturally gain

[53] any velocity at all, as the body would not move. But motion along a horizontal line which is not inclined either upwards or downwards is circular motion around a centre. Therefore circular motion can never be acquired naturally unless it is preceded by rectilinear motion; but once acquired, it will continue perpetually at a uniform speed.

Circular motion can never be acquired naturally unless it is preceded by rectilinear motion.

Circular motion is perpetually uniform.

I could explain and even demonstrate these truths with other arguments, but it would be a very long digression, and I would rather leave it for another occasion and return to the main theme of our discussion—particularly as it is relevant here not for any necessary demonstration, but as an illustration of a concept of Plato's. And here let me add another observation of our Academician's which is indeed marvellous. Let us imagine that one of the decrees of the divine Architect was that he would create the universe with these spheres which we see continually revolving, and that he placed the Sun as the fixed centre of their revolutions. Let us suppose, further, that he created all these spheres in the same place, and that he conferred on them an inclination to move by descending towards the centre until they acquired that degree of velocity which it pleased the divine Mind to give them; and that once they had reached it their motion should become circular, each in its own orbit and maintaining its allotted speed.* And let us ask what would be the height and distance from the Sun at which these spheres were created, and whether it is possible that they were all created in the same place. For this we need to take the calculations of the most expert astronomers for the size of the planets' orbits and their orbital periods, from which we can work out, for example, how much the velocity of Jupiter is greater than that of Saturn. Once we have established that Jupiter moves more rapidly than Saturn (as indeed it does), it follows that, if they both started from the same height, Jupiter must have fallen further than Saturn, as indeed we know to be the case, since Jupiter's orbit is below Saturn's. But we can go further: from the ratio of the velocities of Jupiter and Saturn, the distance between their orbits and the rate of acceleration of natural motion, we can discover at what height and at what distance from the centre of their revolutions is the place from which they started. Having found and established this we can see whether, if Mars [54] descends from the same place, the size of its orbit and its velocity correspond to what this calculation suggests; we find that they do, and that the same is true of the Earth, Venus and Mercury, all of whose orbits and velocities accord so closely with the result of these computations that it is a marvel to behold.

The size of the planets' orbits and the velocity of their motion are proportional to the distances they have descended from the same place.

SAGREDO. I am delighted with this idea, and if I didn't think that making such calculations accurately would be a long and laborious undertaking and beyond my powers of comprehension, I would ask you to illustrate it for me.

SALVIATI. It would indeed be a long and difficult task, and I'm not sure I could reproduce it here and now; so I suggest we leave it for another time.

[SIMPLICIO. Please make allowance for my lack of expertise in the mathematical sciences if I say that your explanations, based on greater or lesser proportions and other terms which I don't sufficiently understand, haven't resolved my doubt, or rather my incredulity. How can a heavy lead ball weighing a hundred pounds, when it is dropped from a height, pass from a state of rest through every extreme degree of slowness, when it falls more than a hundred *braccia* in four pulse beats?* I find it completely incredible that there could be any moment when it was moving with such a degree of slowness that, if it continued to move at that rate, it would take more than a thousand years to travel half an inch. But if this is really the case, I'd be glad to have it explained to me.

SAGREDO. Salviati is such a learned man that he often thinks that terms which are well known and familiar to him are equally familiar to everyone else, so he sometimes forgets that when he's talking to us he needs to help our lack of understanding by using less abstruse terms. Since I don't rise to such heights, with his leave I will try at least to mitigate Simplicio's incredulity by referring to the evidence of our senses. So, still staying with our example of the cannon ball, please tell me, Simplicio: do you not agree that in passing from one state to another it is naturally quicker and easier to pass to one which is closer at hand than to one which is further away? [55]

SIMPLICIO. I understand and agree this, yes; so I don't doubt that, for example, when a piece of hot iron cools it passes from 10 degrees of heat to 9 before it passes from 10 to 6.

SAGREDO. Excellent. So now tell me this: if a cannon ball is fired straight up in the air, does its motion not continually slow down until it reaches its highest point, where it is in a state of rest? And as its speed decreases—or if you like, as its slowness increases—is it not reasonable that it should pass from 10 degrees to 11 rather than from 10 to 12, or from 1000 to 1001 rather than to 1002? In short, that it passes from one degree to another that is closer to it, rather than to one which is further away?

SIMPLICIO. Yes, that's reasonable.

SAGREDO. But what degree of slowness can be so far removed from any degree of motion that the state of rest, which is an infinite degree of slowness, is not even further removed from it? So we can't doubt that before the cannon ball reaches a state of rest it passes through greater and greater degrees of slowness, including that degree where it would take more than 1000 years to travel an inch. This being the case, surely it shouldn't seem improbable to you that when the cannon ball begins to fall and moves away from a state of rest, it regains its velocity by passing through the same degrees of slowness as it did on the way up, and not by missing out those degrees which are closer to a state of rest and jumping straight to another degree which is further away.

SIMPLICIO. This explanation has helped me to understand far better than all those mathematical subtleties; so now Salviati can continue his exposition.]

SALVIATI. So let's return to our main discussion, picking it up at the point where we digressed. If I remember rightly, we were establishing how rectilinear motion cannot serve any useful purpose where the world is properly ordered, and we were going on to say that the same is not true of circular motion, since the motion of a body rotating on its axis keeps it always in the same place, and the motion of a body on the circumference [56] of a circle around a fixed and unmoving centre does not cause any disorder either to itself or to the bodies around it. For such motion, first of all, is finite and defined; more than that, there is no point on its circumference which is not both the beginning and the end of its revolution, so continuing on its allotted circumference, it leaves every other space both inside and outside its circle free for other bodies, without obstructing or disordering them in any way. Because it is a motion which makes the body both continually leave and continually arrive at

Circular motions which are finite and defined do not disrupt the order of the world.

In circular motion any point on the circumference is both a beginning and an end.

the end, it alone can be uniform; for acceleration occurs when a body is moving towards the place to which it is inclined, and its motion is retarded because of its resistance to moving away from the same place. But because in circular motion a body is always both moving away from and returning to its natural place, its resistance and its inclination are always equally balanced, and this equilibrium produces a motion which is neither retarded nor accelerated, but uniform. Being uniform and defined, such motion can continue perpetually, always repeating the same revolution; but this cannot occur naturally in a line which has no defined end or in motion which is continually accelerating and slowing down. I say 'naturally' because when rectilinear motion is slowed down this is due to an external force, which cannot be perpetual. When it accelerates it necessarily reaches its goal, if there is one; if there is not, then there can be no motion towards it, because nature does not move towards a point which cannot be reached.

Circular motion alone is uniform.

Circular motion can continue perpetually.

Rectilinear motion cannot naturally be perpetual.

I conclude therefore that the only natural motion consistent with the natural bodies which make up the universe and are perfectly arranged is circular; and that the most that can be said of rectilinear motion is that it is used by nature when some body or part of a body has moved away from its proper place and is wrongly disposed, and therefore needs to return by the shortest route to its natural state. Hence it seems reasonable to conclude that the only motion which serves to maintain the parts of the world in perfect order is circular, and that if there are any bodies which do not move in circular motion they must necessarily be immovable, since only circular motion and a state of rest serve to maintain order. And I am surprised that Aristotle, who believed that the terrestrial globe was placed in the centre of the universe where it [57] remained immovable, did not state that some natural bodies are mobile by nature and others immovable, especially since he defined nature as being the principle of motion and rest.

Rectilinear motion occurs in nature to restore bodies to a perfect order when they have moved away from it.

Only circular motion or a state of rest can maintain order.

SIMPLICIO. Aristotle, for all his great insight, did not expect more from his reason than it could provide. He considered in his philosophy that the experience of the senses should take precedence over any argument constructed by human reason,

The experience of the senses should precede human reasoning.

Anyone who denies the senses deserves to be deprived of them.

Our senses shows that heavy bodies move towards the centre, and light bodies move towards the concave surface of the Moon's orbit.

and he said that anyone who denied the evidence of one of the senses deserved to be deprived of that sense. Who then is so blind that they cannot see that earth and water, being heavy, move naturally downwards, that is, towards the centre of the universe, this being the goal and end fixed by nature for downward rectilinear motion? And who likewise does not see that fire and air naturally move upwards in a straight line towards the concave surface of the Moon's orbit,* as to the natural goal of upward motion? And if this is manifestly true, and given that we know that *eadem est ratio totius et partium* ('the same reasoning applies to the whole and to the parts'*), is it not self-evidently true that the natural motion of the earth is rectilinear motion towards the centre, and that of fire is rectilinear motion away from the centre?

It is doubtful whether heavy bodies fall in a straight line.

SALVIATI. The most I'll concede in response to your argument is that, if parts of the earth were removed from the whole, i.e. from the place where they naturally belong and so, ultimately, in violation of their place in the natural order, they would spontaneously and therefore naturally return to their place in a straight line, then we can infer that—granted that *eadem sit ratio totius et partium*—if the terrestrial globe itself were to be moved from its natural place by some external force, it would move in a straight line to return there. This, as I say, is as much as can be conceded to you, making all possible allowances. But anyone who examined these arguments rigorously would, first of all, deny that when parts of the earth return to the whole they move in a straight line, rather than with circular or mixed motion. Then you would have your work cut out to argue against them, as you will see from the replies to the arguments and experiments adduced by Ptolemy and Aristotle. Secondly, if it were put to you that parts of the Earth move not in order to be at the centre of the world but rather to rejoin the whole of which they are part, and that their natural motion is therefore towards the centre of the terrestrial globe, [58] then what other whole and what other centre could you give to which the terrestrial globe itself would seek to return, if it were dislodged from it? Yet this would have to happen, if the same reasoning is to apply to the whole and to the parts. Finally, neither Aristotle nor you will ever be able to prove that the

The Earth is a sphere because its parts cohere together around its centre.

Earth is in reality at the centre of the universe. In fact, if any centre can be assigned to the universe, we shall find it is more likely to be where the Sun is, as you will see in due course.

It is more probable that the Sun is the centre of the universe than the Earth.

If all the parts of the Earth come together to form the whole because they are drawn together by an equal inclination from all sides, and form themselves into a sphere so as to be as closely united as possible, is it not likely that the Moon, the Sun and the other heavenly bodies are also spherical for the same reason, simply because all their component parts are naturally and instinctively drawn together? In which case, if any part of them were separated from the whole by some external force, is it not reasonable to suppose that they would spontaneously and instinctively return to them, and would this not mean that rectilinear motion belongs equally to all the heavenly bodies?

The parts of all the heavenly globes have a natural inclination towards their centre.

SIMPLICIO. If you intend to deny not just the principles of science but even the plain evidence of our senses, it's clear that you will never be persuaded or shifted from any of your preconceived opinions. So I shan't object, not because I am convinced by your arguments but because *contra negantes principia non est disputandum* ('there is no arguing with those who deny first principles'*). But staying with what you have just said, since you even cast doubt on whether falling objects move in a straight line, how can you deny that heavy objects, being parts of the earth, fall towards the centre in a straight line? If you drop a stone from a high tower which has straight, perfectly vertical walls, will it not land exactly where a plumb-line suspended from the point from which you dropped it ends, almost brushing against the wall as it falls? Surely this is clear evidence that this is rectilinear motion towards the centre?

Evidence of the senses that falling objects move in a straight line.

Second, you question whether the parts of the Earth move towards the centre of the universe, as Aristotle affirms, even though he proved it conclusively by means of contrary motion,* as follows: the motion of heavy bodies is the contrary of that of light bodies; light bodies can be seen to move directly upwards, i.e. towards the outer circumference of the world. Therefore heavy bodies move directly towards the centre of the world, which coincidentally is also the centre of the Earth, since this is where the Earth happens to be placed. As for asking what would happen if part of the lunar or solar globe were separated from

[59]

Aristotle's argument to prove that falling bodies move towards the centre of the universe.

Falling bodies move towards the centre of the Earth by coincidence.

It is vain to seek what would follow from an impossibility.

Aristotle's view that heavenly bodies are neither heavy nor light.

the whole, this is a vain question because it is asking what would follow from an impossibility. For Aristotle also shows that the heavenly bodies are impassible, impenetrable, and unbreakable, so such a thing could never happen. Even if it did and the separated part did return to the whole, it would not do so by virtue of being either heavy or light, because Aristotle himself proves that heavenly bodies are neither heavy nor light.

SALVIATI. Whether my doubts about heavy bodies moving perpendicularly and in a straight line are reasonable or not will become clear, as I have said, when I come to discuss this question in detail. But as for the second point, I'm surprised that I have to point out the flaw in Aristotle's logic to you, since it seems self-evident. Surely you realize that Aristotle is presupposing the point which he is trying to prove? Note that . . .

Aristotle cannot make logical mistakes, since he was the inventor of logic.

SIMPLICIO. Please, Salviati, don't speak so disrespectfully of Aristotle. No one is going to believe that the first, only and unequalled exponent of the syllogism, the demonstration, the Socratic dialogue, how to recognize sophisms and paralogisms, in short of the whole science of logic, should have made such a basic error as to presuppose as given the point which is under discussion. At least understand him first, before you try to refute him.

SALVIATI. Simplicio, by all means point out my errors and I'll be grateful for the correction, since this is a discussion among friends and we are trying to establish the truth; so please tell me if I have failed to grasp Aristotle's meaning. But in the meantime do allow me to explain the difficulty I have, and first of all to respond to your last point. Logic, as you know, is the organ which we use* as the instrument on which to practise philosophy. But just as there can be excellent organ-builders who have no expertise in playing, so it's possible to be a great logician but have little expertise in knowing how logic is to be used. In the same way, there are many who know all the rules of poetics off by heart, but would struggle to put four verses together; and there are others who know all Leonardo's rules but are unable to paint a chair. We learn to play the organ from an organist, not an organ-builder; we learn poetry by constantly reading the works of poets, and painting by constant practice in

[60]

drawing and painting. And so we learn how to construct a proof by reading books full of proofs, which are written by mathematicians, not logicians.

So to return to the question we were discussing, Aristotle's view of the motion of light bodies is based on his observation of fire, which moves directly upwards from anywhere on the surface of the terrestrial globe. This is indeed motion towards a circumference greater than that of the Earth; Aristotle himself defines it as motion towards the concave surface of the Moon's orbit. But we can only say that this is also the circumference of the world, or concentric with it, and therefore that motion towards it is also motion towards the circumference of the world, if we presuppose that the centre of the Earth, from which we see light bodies rising, is the same as the centre of the world—in other words that the centre of the Earth is located at the centre of the world, which is the point Aristotle is trying to prove. How can you deny that this is a clear logical fallacy?

Aristotle's faulty logic in his proof that the Earth is at the centre of the universe.

SAGREDO. I found this argument of Aristotle's faulty and inconclusive in another respect, even if it were granted that the circumference towards which fire moves is the sphere which encloses the whole world. For if a body moves in a straight line starting from any point inside a circle, not just from its centre, it will undoubtedly be moving towards the circumference, and will reach it provided it keeps going long enough; so this is not in doubt. But it does not follow that by moving along the same straight line in the contrary direction it would necessarily reach the centre. This would only be true if its starting point were the centre itself, or if it were to move along a line which, extended backwards from the starting point, passes through the centre.

[61]

So to say 'fire moves in a straight line towards the circumference of the world, therefore those parts of the Earth which move along the same lines with a contrary motion go towards the centre of the world' is only valid if the lines along which fire moves would, if extended, pass through the centre of the world. We know for certain that a line which is perpendicular to the surface of the terrestrial globe passes through the centre of the globe; so the statement is only valid if we assume that the centre of the Earth is the same as the centre of the world, or at least that fire and earth in their upward and downward motion move

Aristotle's faulty logic exposed from another angle.

only along one line which passes through the centre of the world. This is clearly false and contrary to our experience, which is that fire rises vertically from the surface of the terrestrial globe along any number of lines produced from the centre of the Earth to every part of the world.

SALVIATI. Sagredo, you very cleverly show how Aristotle's argument leads to the same difficulty by exposing its evident equivocation. But another inconsistency follows from this. We see that the Earth is a sphere, and so we can be sure that it has a centre to which all its parts move, as this necessarily follows if their motion is always perpendicular to the surface of the globe. We understand that as they move towards the centre of the Earth they move towards the whole of which they are part, and towards their universal mother; and we are then so simple that we let ourselves be persuaded that their natural instinct draws them not towards the centre of the Earth but to the centre of the universe, even though we do not know where that is, or even if it exists—and if it does exist it is no more than an imaginary point with no physical properties at all.

Proof that it is more reasonable to say that falling objects move towards the centre of the Earth than the centre of the universe.

Simplicio's last point was that it was vain to ask whether parts of the Sun, the Moon or other celestial bodies which became separated from the whole would naturally return to it, because such a thing could not happen, as Aristotle showed that the heavenly bodies are impassible, impenetrable, indivisible, etc. To this my response is that none of the properties which Aristotle cites as differentiating the celestial bodies from the elemental has any foundation other than what he deduces from the diversity of their natural motions. So if we deny that circular motion belongs solely to the heavenly bodies and affirm that, on the contrary, it is equally appropriate to all moving bodies in nature, then we must acknowledge one of two things: either the [62] attributes of being subject or not subject to generation, change, division, etc., apply equally to all bodies, whether heavenly or elemental; or Aristotle was wrong in saying that those which he assigned to the heavenly bodies were the consequence of their circular motion.

The conditions by which Aristotle differentiates the heavenly bodies from the elemental follow from the motions which he assigns to them.

SIMPLICIO. This kind of reasoning subverts the whole of natural philosophy and leaves heaven, Earth and the whole universe in confusion and disorder. But I believe that the foundations of Peripatetic philosophy are sound enough for there to be no danger that new sciences will be built out of their ruins.

SALVIATI. You needn't worry about the heavens or the Earth being subverted, or philosophy for that matter. As far as the heavens are concerned, there is no reason to fear for them, for you yourself say that they are inalterable and impassible; and as for the Earth, we are trying to perfect it and give it a more noble status, making it more like the celestial bodies and, in a way, giving it a place in the heavens, from which your philosophers have banished it. And philosophy itself can only benefit from our disputes, because if our conclusions are correct new truths will have been gained, and if they are wrong, the original positions will have been strengthened by refuting them. Think rather of certain philosophers and try to help and sustain them, for science itself cannot fail to advance. So to return to our theme, please tell us the arguments which you think will maintain Aristotle's fundamental distinction between heavenly bodies and the elemental part of the universe, such that the former are ingenerable, incorruptible, unchanging, etc., whereas the latter are corruptible, changeable, etc.

Philosophy can be enhanced by disputes and differences of opinion between philosophers.

SIMPLICIO. I don't see that Aristotle has any need of support, since he is still standing as firmly as ever, and I don't think he has yet even been attacked, let alone defeated, by you. How are you going to defend yourself against his first assault? Aristotle writes as follows: 'Generation is produced by a contrary in a subject, and similarly corruption in a subject is produced by one contrary being corrupted into another', so that—note this—corruption and generation occur only in contraries. 'But the movements of contraries are contrary to each other. So if celestial bodies have no contrary, because they move in circular motion which has no contrary motion, it is entirely fitting that nature should have made the heavens, which are not subject to generation or corruption, exempt from contraries.'*

Aristotle's argument to prove the incorruptibility of the heavens.

In Aristotle's view, generation and corruption exist only in contraries.

[63] Once this fundamental principle is established, it clearly follows as a consequence that the heavens cannot be increased or altered and are impervious to change, and in short are eternal and a fit habitation for the immortal gods, as is held by all those

There is no motion contrary to circular motion.

The heavens are the habitation of the immortal gods.

Our senses confirm the immutability of the heavens.

Proof that circular motion has no contrary.

who have any concept of the gods. Our senses, too, confirm this, for throughout recorded time no change has been observed in the highest heaven or in any part of the heavens. As for there being no motion contrary to circular motion, Aristotle proves this in numerous ways and I won't rehearse them all. But the clearest proof is this: there are only three simple motions, towards, away from, and around the centre. Of these the two rectilinear motions, upwards and downwards, are clearly contraries; and because one thing has only one contrary, there remains no other motion to be the contrary of circular motion. And there you have Aristotle's brilliantly conclusive proof that the heavens are incorruptible.

SALVIATI. This is indeed exactly the step in Aristotle's argument which I've already alluded to; and if I affirm that the motion which you attribute to the celestial bodies applies also to the Earth, his argument proves nothing. I say therefore that the circular motion which you claim belongs exclusively to celestial bodies belongs also to the Earth. If the rest of your argument holds, then one of three things follows. I've said this already and will repeat it now: either the Earth, like the celestial bodies, is unaffected by generation and corruption; or the celestial bodies, like those composed of the elements, are subject to generation, change, etc.; or this difference of motion has nothing to do with generation and corruption at all. Aristotle's argument, which you reproduce, contains many propositions which cannot be taken for granted, so to examine it more closely it will be as well to reduce it as far as possible to its essentials—and I ask Sagredo's indulgence if I bore him by repeating things that have been said more than once already. Perhaps he can imagine that he is hearing the arguments rehearsed in a public debate. So your argument runs as follows: 'Generation and corruption are produced only where there are contraries; contraries are found only among simple natural bodies which move with contrary motion. The only contrary motions are rectilinear motion [64] between contrary goals, of which there are only two, towards the centre and away from the centre. These movements occur only among natural bodies composed of earth, fire and the other two elements; therefore generation and corruption are produced only among these elements. And since the third simple motion, which is circular motion around the centre, has no contrary (the other two being contrary to each other, and each can have only one contrary), a natural body to which such motion belongs also has no contrary; since it has no contrary, it is not subject to generation and corruption, etc., because generation and corruption, etc., are produced only where there are contraries. But circular motion belongs only to celestial bodies; therefore these alone are not subject to generation, corruption, etc.'

My first response to this is to say that I think it is much easier to establish whether the Earth, a vast body which is easy for us to study because it is so close at hand, moves at such a speed that it turns on its axis in twenty-four hours, than it is to understand whether generation and corruption are the product of contraries, or for that matter whether corruption, generation and contraries exist in nature at all. And if, Simplicio, you can explain to me how nature is able to generate thousands of flies in almost no time from a few fumes of wine must, and show me what are the contraries in this process, what it is that is corrupted and how, then my estimation of you will be even greater than it is, because these are things which I cannot understand at all. I would dearly like to know, too, why these corrupting contraries are so benign in their treatment of crows and so harsh towards doves, or why they are so tolerant of deer and so impatient with horses, for the former are granted more years of life—i.e. of incorruptibility—than the latter are of weeks. Or again, peach trees and olive trees both have their roots in the same soil, are exposed to the same cold and heat, the same wind and rain, in short to all the same contraries, and yet peach trees die in a short time whereas olive trees live for centuries. What's more, I have never been able to understand this idea of transmutation of substance (still speaking in purely natural terms) whereby matter is so totally transformed that its original being is completely destroyed so that nothing of it remains, and another completely different body is produced from it. For a body to appear first under one aspect and then under another quite different one a short time later, I don't find it impossible that this can come about through a simple transposition of parts, without corruption or generation of anything new, for we

It is easier to establish whether the Earth moves than whether corruption is the product of contraries.

[65]

Bodies can appear under different aspects through a simple transposition of parts.

see such metamorphoses happening all the time. So I repeat, you will have an uphill task if you want to persuade me that the Earth cannot move in circular motion because of the principle of corruption and generation, because I will prove the opposite to you with arguments which are no less conclusive, though they may be more difficult.

SAGREDO. Salviati, forgive me if I interrupt your exposition, fascinating as I find it, for I find myself embroiled in exactly the same difficulties. But I fear we could only get to the end of it by setting aside our main subject altogether; so I propose that, in order to continue with our original discussion, we keep this whole debate about generation and corruption as a topic for a separate dialogue in its own right. And if you and Simplicio agree, I will keep a note of this and any other specific questions which come up in the course of our discussions, so that we can devote another day to examining them in detail. So returning to the matter in hand, you say that if we deny Aristotle's claim that circular movement does not affect the Earth as it does the other celestial bodies, it follows that whatever happens on Earth as regards being subject to generation, change, etc., also affects the heavens. So, leaving aside whether or not generation and corruption are to be found in nature, let us return to investigating what the terrestrial globe does.

SIMPLICIO. I can't bring myself to listen to someone denying that generation and corruption exist in nature, when the evidence for them is continually before our eyes, and Aristotle has written two whole books* on the subject. If you start denying the principles of science and casting doubt on things which are perfectly plain to see, then of course you can prove whatever you want and maintain any paradox at all. If you don't see the constant generation and corruption of grasses, plants and animals all around you, what do you see? How can you fail to see the perpetual clash of contraries, and earth turning into water, water into air, air into fire, and then condensing again into clouds, rain, hail and tempests?

Any paradox can be maintained by denying the principles of science.

SAGREDO. We do see all these things, of course, and so we grant you Aristotle's argument as far as these examples of generation and corruption produced by contraries are concerned. But if I were to prove to you, on the basis of these same

[66]

propositions of Aristotle's, that the celestial bodies are subject to generation and corruption just as much as those made up of the elements, what would you reply?

SIMPLICIO. I would reply that you have done the impossible.

SAGREDO. Then tell me, Simplicio, are not these qualities contraries?

SIMPLICIO. Which qualities?

SAGREDO. The qualities of being alterable or inalterable, passible or impassible, generable or ingenerable, corruptible or incorruptible.

SIMPLICIO. Of course they are contraries.

SAGREDO. In that case, if the celestial bodies are ingenerable and incorruptible, I will prove to you that they must necessarily be generable and corruptible.

SIMPLICIO. This can only be pure sophistry.

SAGREDO. Listen to the argument, then criticize it and resolve it. Celestial bodies, being ingenerable and incorruptible, have contraries in nature, namely bodies which are generable and corruptible; but generation and corruption are found where there are contraries; therefore the celestial bodies are generable and corruptible.

Celestial bodies are generable and corruptible, because they are ingenerable and incorruptible.

SIMPLICIO. I said this was nothing but sophistry. This is one of those horned arguments known as sorites,* like the example of the Cretan who said that all Cretans were liars. If he was a Cretan he must have been lying when he said that all Cretans were liars. Therefore Cretans must tell the truth, and he, as a Cretan, must have been telling the truth when he said that Cretans were liars, including himself; and therefore he must have been lying. This kind of sophism simply goes round in circles without ever reaching any conclusion.

A horned argument, otherwise known as a sorites.

SAGREDO. So you've identified the argument; now resolve it, and show where the fallacy lies.

SIMPLICIO. Resolve it and demonstrate the fallacy? For a start, surely you can see the evident contradiction—'the celestial bodies are ingenerable and are incorruptible; therefore they are generable and corruptible'? In any case, the contraries [67] are not in the celestial bodies but in the elements, which are contraries in that they are light and heavy and that they move upwards and downwards. The heavens, on the other hand,

There are no contraries in the celestial bodies.

move in a circle, to which there is no contrary motion. Therefore they have no contraries, and hence they are incorruptible, etc.

SAGREDO. Just a moment, Simplicio. This contrariety which you say causes some simple bodies to be corruptible: does it reside in the body which is corrupted, or in some other body? For example, does the humidity which causes corruption in some part of earth reside in the earth itself, or is it in some other body such as air or water? I think you will say that, just as upward and downward motion, weight and lightness, which you call the primary contraries, cannot exist in the same subject, so it is with the contraries of moist and dry, hot and cold. It follows then that when a body is corrupted this is the result of a quality found in another body which is contrary to its own. Therefore a heavenly body is corruptible if there are bodies in nature which have qualities contrary to it; and that is exactly what the elements are, if it is true that corruptibility is the contrary of incorruptibility.

The contraries which are the cause of corruption are not inherent in the body which is corrupted.

SIMPLICIO. That's not a sufficient proof. The elements change and are corrupted because they touch and mingle with each other, and so exercise their contrary qualities; but the celestial bodies are separated from the elements and are not so much as touched by them, although they do have an influence on the elements. To prove that generation and corruption exist in celestial bodies you would have to show that contrariety resides among them.

Heavenly bodies touch the elements, but are not touched by them.

SAGREDO. I'll demonstrate this as follows. The contrary qualities in the elements derive in the first place, in your view, from their contrary upward and downward motion; therefore the principles from which these motions derive must also be contraries. Now upward motion is the result of lightness, and downward motion the result of heaviness; therefore these must also be contraries, as too must be the qualities which make one body light and another heavy. But your own school of philosophy maintains that lightness and weight are the consequence of rarity and density; so rarity and density must also be contraries. And these are such pervasive qualities among the celestial bodies that you hold the stars to be simply the densest parts of the sphere in which they rotate.* If this is correct, then the density of the stars must be almost infinitely greater than that

Heaviness and lightness, rarity and density, are contrary qualities.

[68]

of the rest of the heavens. This is clear from the fact that the heavens are extremely transparent and the stars are extremely opaque, and the fact that varying degrees of density and rarity are the only qualities found in the heavens which could be the cause of these varying degrees of transparency. Therefore, since these contraries exist among the celestial bodies, the celestial bodies must also be subject to generation and corruption, in the same way as elemental bodies; or else corruptibility, etc., are not caused by contraries.

The stars are infinitely more dense that the substance of the rest of the heaven.

SIMPLICIO. Neither of these conclusions follows. Density and rarity in the celestial bodies are not contraries to each other as they are in elemental bodies, because they do not depend on the prime qualities of heat and cold, which are contraries, but on the greater or lesser amount of matter in proportion to their quantity. The opposition between more and less is what is known as a relative opposition, which is the least significant kind, and has nothing to do with generation and corruption.

*Rarity and density in the celestial bodies are not the same as they are in the elements (Cremonini).**

SAGREDO. So for density and rarity to be the cause of heaviness and lightness in the elements, which in turn can cause the contrary upward and downward motion on which the contraries underlying generation and corruption depend, it is not enough for these elements to be dense or rare because of varying amounts of matter contained in the same quantity or mass of a body. Their density and rarity must derive from the primary qualities of heat and cold; otherwise they would have no effect. But if this is the case, Aristotle has deceived us: he should have said straight away that generation and corruption occur in simple bodies which are subject to simple upward and downward motion, depending on lightness or heaviness, which are caused by the rarity or density produced by a greater or lesser amount of matter as a result of heat or cold; instead of which he just refers to upward and downward motion, without any other qualification. Because I can assure you that any kind of density or rarity is enough to make bodies heavy or light, and therefore subject to contrary motion, whether as a result of heat and cold or anything else. Heat and cold have nothing to do with this effect, as is clear from the fact that a piece of iron which has been in the fire, and so can be called hot, has the same weight and moves in the same way as when it is cold. But

Aristotle is diminished by his explanation of the causes of generation and corruption in the elements.

leaving this aside, how do you know that density and rarity in the heavens are not dependent on heat or cold?

[69] SIMPLICIO. Because these qualities do not occur among the celestial bodies, which are neither hot nor cold.

SALVIATI. I can see that we're adrift again in a boundless sea from which there is no escape, since we are navigating without compass, stars, oars or rudder, so that all we can do is go from rock to rock, or run aground, or else sail aimlessly for ever. So if, as you suggest, we are to press ahead with our main topic, we must leave aside for now this general discussion of whether rectilinear motion is necessary in nature and belongs to some bodies and not others, and come to consider the specific demonstrations, observations and experiments which have been put forward, first by Aristotle, Ptolemy, and others to prove that the Earth is fixed, and try to resolve them; and then those which have convinced others that the Earth, no less than the Moon or any other planet, is to be numbered among the natural bodies endowed with circular motion.

SAGREDO. I gladly agree to this, all the more so because I find your general architectural exposition much more satisfactory than Aristotle's. Yours resolves all my difficulties, whereas with Aristotle's I find obstacles in my path at every step. In fact I don't see how Simplicio can fail to be convinced by your argument proving that rectilinear motion can have no place in nature if we assume that the parts of the universe are optimally disposed in perfect order.

SALVIATI. Allow me to interrupt you, Sagredo, since I have just thought of a way to convince even Simplicio, as long as he isn't so wedded to every word of Aristotle's that he considers any departure from him to be sacrilege. There is no doubt that maintaining the optimal disposition of every part of the universe in perfect order, in terms of their location, requires circular motion and a state of rest. As for rectilinear motion, I cannot see that it serves any function apart from restoring to its natural place some fragment of an integral body which has been accidentally separated from it, as we have seen. So now let us consider what is needed for the terrestrial globe, given that it, like all the other bodies in [70] the universe, must be kept in its optimal natural disposition.

There are three possibilities: that it rests and remains immobile in its proper place in perpetuity; or that it turns on its axis while remaining always in the same place; or that it moves around a centre, moving along the circumference of a circle.

Of these, Aristotle, Ptolemy and their followers say that its natural disposition has always been the first, and that it will remain so for all eternity, i.e. that it is perpetually at rest in the same place. So why, I ask, do we not say that the Earth's natural condition is to remain motionless, instead of saying that it naturally moves downwards, since it never has moved downwards and there is no reason to think that it ever will? And as for rectilinear motion, let us allow that nature uses it to restore to their proper place all those fragments of earth, water, air, fire or any other integral body which for any reason have become separated from the whole, and so removed from their properly ordered place—unless even such a restoration might not be achieved more effectively by a circular motion of some kind. So it seems to me to fit much better with all the other consequences, even in Aristotle's own terms, to say that the Earth's natural condition is to be in a state of rest, instead of making rectilinear motion the intrinsic natural principle of all the elements. This is clearly the case: the Peripatetics believe that the celestial bodies are incorruptible and eternal; but if I were to ask one of them if he believes that the terrestrial globe is corruptible and mortal, and therefore that there will come a time when the Sun, the Moon, and the other stars continue in their courses but the Earth no longer exists, having been destroyed and annihilated along with all the other elements, I am sure that he would say no. So corruption and generation must belong to the parts, not to the whole, and what's more to very small and superficial parts, so small that they are almost imperceptible in comparison to the whole. Therefore, since Aristotle attributes generation and corruption to the contraries of rectilinear motion, let us leave these motions to the parts, as it is only they that undergo change and corruption, and conclude that the whole globe and sphere of the elements either moves with circular motion or is permanently motionless in the same place, since only these two contribute to the maintenance of perfect

Aristotle and Ptolemy suppose that the terrestrial globe is motionless.

The natural condition of the terrestrial globe should be defined as a state of rest rather than downward rectilinear motion.

There are more grounds for attributing rectilinear motion to parts of the elements than to the elements as a whole.

order. What we have concluded about the Earth applies equally to fire and to the greater part of the air. The Peripatetics end up saying that the intrinsic and natural motion of these two elements is one with which they never have moved and never will, while they define as contrary to nature the motion with which [71] they do move, as they always have and always will. They say that air and fire naturally move upwards, even though neither of these elements has ever moved upwards but only fragments of them, simply so as to return to their properly ordered position from which they had been displaced. They say that circular motion, by which these elements are constantly moved, is unnatural, forgetting that Aristotle said on many occasions that nothing violent can last for very long.

The Peripatetics groundlessly claim that the natural motions of the elements are ones with which they never move, and that the motions with which they move all the time are unnatural.

SIMPLICIO. We have conclusive replies to all these points, but I shall leave these for the moment to concentrate on more specific arguments and the experience of the senses, which Aristotle rightly says should take precedence over what can be worked out by human reason.

The experience of the senses should be preferred to human reasoning.

SAGREDO. Good; so let everything that has been said so far serve to set out two general accounts and to help us consider which appears the more probable. Aristotle's view, based on the different kinds of simple motion, would persuade us that the nature of sublunary bodies, being generable, corruptible, etc., is entirely different from the nature of the celestial bodies which are impassible, ingenerable, incorruptible, etc. Salviati, on the other hand, assumes that the integral parts of the universe are optimally disposed, and consequently that rectilinear motion cannot belong intrinsically to simple natural bodies, because it would serve no purpose in nature. So he considers the Earth to be itself one of the celestial bodies, endowed with the same prerogatives as all the others. I must admit that this latter seems to me so far to be much the more satisfactory of the two. So let me now invite Simplicio to put forward the detailed arguments, experiments and observations, both natural and astronomical, to persuade us that the Earth is different from the celestial bodies, and that it is immobile, fixed at the centre of the universe, and whatever else excludes it from moving like Jupiter, the Moon or any of the other planets; and Salviati will be so good as to reply to each point in turn.

SIMPLICIO. To start with, here are two powerful demonstrations to prove that the Earth is completely different from the celestial bodies. First, bodies which are generable, corruptible, mutable, etc., are completely different from those which are ingenerable, incorruptible, immutable, etc. The Earth is generable, corruptible, mutable, etc., and the celestial bodies are ingenerable, incorruptible, immutable, etc.; therefore the Earth is completely different from the celestial bodies. [72]

SAGREDO. Your first argument is simply reintroducing the one we've just disposed of.

SIMPLICIO. Wait a minute; listen to what follows and you'll see how different it is. The previous argument proved the minor premise *a priori*; now I shall prove it *a posteriori*,* and you'll see that it is not the same at all. I shall prove the minor premise, since the major premise is self-evident. Experience shows us that processes of generation, corruption, change, etc., are continually going on here on Earth, whereas we have never seen them in the heavens, and none of the traditions or writings of our forebears has any record of them; therefore the heavens are unchangeable, etc., and the Earth is changeable, etc., and hence different from the heavens. The second argument is based on a principal and essential property, namely this: a body which is naturally dark and devoid of light is different from bodies which shine with their own light. The Earth is dark and devoid of light; the heavenly bodies shine brilliantly with their own light; therefore, etc. Perhaps you would reply to these arguments before I go on to others, so as not to accumulate too many at a time.

The heavens are immutable because no change has ever been seen in them.

Bodies which are naturally sources of light are different from those which are dark.

SALVIATI. As regards the first argument, which you derive from experience, I would like you to tell me more specifically what are the changes you see taking place on Earth and not in the heavens, on the basis of which you say the Earth is mutable and the heavens are not.

SIMPLICIO. On the Earth I see a continuous process of generation and corruption in grasses, plants, and animals; I see wind, rain, storms, and tempests arising; in short, this Earth which we see is in a constant state of metamorphosis. And I see none of these changes in the celestial bodies, whose position and configuration remain exactly as they have always been recorded, without the generation of anything new or the corruption of anything old.

SALVIATI. But if your belief rests on these visible phenomena—or rather, ones which you have seen—then you must consider China and America to be celestial bodies, since you have certainly not seen in them the changes which you see here in Italy; so in your view they must be unchangeable.

[73] SIMPLICIO. I may not have physically seen these changes in those places, but there are reliable accounts of them. In any case, *cum eadem sit ratio totius et partium*—since the same reasoning applies to the parts and to the whole—these countries are parts of the Earth the same as ours, and so they must be subject to change just as our country is.

SALVIATI. And why have you not seen and observed them with your own eyes, so that you have to rely on accounts provided by other people?

SIMPLICIO. Because these countries are not accessible to our eyes, and in any case they are so far away that even if we could see them our eyes would not be able to perceive changes of this kind.

SALVIATI. See how you have now yourself exposed the fallacy in your argument. If you say that the changes which we see before our eyes here on Earth would not be visible to you in America because it is so far away, how much less would you be able to see them in the Moon, which is hundreds of times further away? And if you believe in changes in Mexico on the basis of reports which have come from there, what reports have you received from the Moon to tell you that there are no changes there? So while you quite rightly argue for the existence of changes on Earth because we can see and recognize them, you can't argue that no changes take place in the heavens because you don't see them, since if there were any you wouldn't be able to see them because they are so far away, or because you have no reports of them, since there is no way for such reports to reach us.

SIMPLICIO. I can give you examples of changes on Earth such that, if anything comparable were to happen on the Moon, it could easily be seen from here on Earth. There is a very ancient memory that Abyla and Calpe* were once joined together at what is now the strait of Gibraltar, and that together with other smaller mountains they held back the western ocean. Then, for whatever reason, the mountains separated, opening the way for the sea water to rush in so that it formed the whole of the Mediterranean sea. Now if we consider the size of the Mediterranean, and the difference in appearance that there must be between the surface of the sea and the land when seen from a great distance, we can be sure that this change could easily have been visible if anyone had been looking at it from the Moon; so a similar change on the Moon's surface would be visible to us from Earth. But there is no record that any such thing has ever been seen; so there are no grounds for saying that the celestial bodies are mutable, etc.

The Mediterranean formed by the separation of Abyla and Calpe.

SALVIATI. Well, I don't presume to know whether changes of such magnitude have ever happened on the Moon, nor can I say categorically that they could not have happened. And since in any case all we would see of such a change would be [74] some variation in the lighter and darker areas on the Moon's surface, I don't know whether anyone on Earth has mapped the Moon accurately enough over a sufficiently long period of years for us to be certain that there never has been any such change on its surface. The best we can do in describing it is that some people have said it is like a human face, others that it resembles a lion's muzzle, and others that it is Cain carrying a bundle of thorns* on his back. So it doesn't prove anything to say 'The heavens are unchangeable because there are no alterations in the Moon or in any other celestial body which can be seen from the Earth'.

SAGREDO. This first argument of Simplicio's has raised another doubt in my mind which I would like to have resolved. So let me ask him this: was the Earth generable and corruptible before the flood which formed the Mediterranean, or did it only become so afterwards?

SIMPLICIO. There's no doubt that it was already generable and corruptible before that; but the flood was such a huge mutation that it could have been seen even from the Moon.

SAGREDO. But if the Earth was generable and corruptible even before this flood, why can the same not be true of the Moon even if there has never been such a mutation there? Why should something that is irrelevant for the Earth be necessary in the case of the Moon?

SALVIATI. A very telling example. But I wonder if Simplicio hasn't slightly changed the meaning of what Aristotle and the other Peripatetic philosophers wrote. They say that they consider the heavens to be immutable because no one has ever seen the generation or corruption of a star, which is a very small part of the heavens, possibly less than a city would be on Earth. And yet innumerable cities have been destroyed so that not a trace of them remains.

SAGREDO. I understood what Simplicio said differently; I thought that he was deliberately obscuring the meaning of this text so as to spare the Master and his disciples from an even greater absurdity. What kind of empty argument is it to say 'The heavens are unchangeable because stars are not subject to generation and corruption'? Has anyone ever seen a terrestrial globe decompose and a new one generated from it? And yet don't all philosophers grant that there are very few stars in the heavens which are smaller than the Earth, and that many of them are very many times bigger? So for a star to become [75] corrupted in the heavens would be at least the equivalent of the whole terrestrial globe being destroyed. Well, if introducing the principle of generation and corruption into the universe requires bodies as vast as a star to be corrupted and regenerated, we might as well forget the principle altogether, because I can assure you that seeing the terrestrial globe or any other integral body in the universe dissolve so that no trace of it is left, after it has been seen to be there for centuries, is simply not going to happen.

It is no less impossible for a star to be corrupted than it would be for the whole terrestrial globe.

SALVIATI. But to be generous to Simplicio and rescue him from error, if we can, we should acknowledge that there have been such new discoveries and observations in our own time that I'm quite sure Aristotle, if he were alive today, would change his opinion. This is clear from his own method of reasoning. Since he says that he considers the heavens to be immutable, etc., because no one has seen any new bodies being generated or old ones dissolving there, he implies that if he ever did see such a thing he would believe the contrary. He would quite rightly have rated the evidence of his senses above that of human reason, since it was the lack of any sense evidence of change in the heavens that had convinced him of their immutability.

Aristotle would change his opinion if he saw the discoveries that have been made in our time.

SIMPLICIO. Aristotle based his conclusions primarily on arguing *a priori*, and he demonstrated the necessity of the immutability of the heavens by means of clear natural principles; and then he confirmed this *a posteriori*, drawing on the evidence of the senses and the traditions of the ancients.

SALVIATI. He did indeed write his works in this way, but I don't think that was how he arrived at his conclusions. I'm quite sure that he tried first to assure himself as far as he could of the conclusion on the basis of observation and the evidence of the senses, and that he then tried to find the means to demonstrate it. This is the usual way of proceeding in the demonstrative sciences, because if the conclusion is correct, then the analytical method will readily lead to some proposition which has already been proved or to some generally accepted principle. But if the conclusion is false, the argument can proceed *ad infinitum* without ever reaching any recognized truth, if indeed it does not lead to some impossibility or manifest absurdity. There can be no doubt that Pythagoras, long before he arrived at the proof for which he sacrificed a hundred oxen, was already confident in his own mind that the square on the hypotenuse of a right-angled triangle is equal to the sum of the squares on the other two sides; for in the demonstrative sciences, certainty of the conclusion is a great help in finding the proof. But whichever way Aristotle proceeded, whether the argument *a priori* came before the evidence *a posteriori* or vice versa, the important thing is that, as we have already said more than once, Aristotle rated the evidence of the senses more highly than argument of any kind.

Being certain of the conclusion helps to find proof by the analytical method.

[76]

Pythagoras sacrificed a hundred oxen when he discovered a geometrical proof.*

So to return to our subject, I say that the things which have been discovered in the heavens in our own time are, and have been, enough to satisfy any philosopher. For both in particular bodies and in the great expanse of the heavens as a whole we have seen, and continue to see, occurrences similar to what we describe as generation and corruption here on Earth. Eminent astronomers have observed many comets generated and then disintegrated in regions beyond the orbit of the Moon, as well as the two new stars* which appeared, undoubtedly far beyond all the planets, in the years 1572 and 1604. And now, thanks to the telescope, we can see dense dark spots forming and

New stars have appeared in the heavens.

Spots which form and dissolve on the face of the Sun. Sunspots bigger than the whole of Africa and Asia.

dissolving on the face of the Sun itself, very similar in appearance to the clouds around the Earth, many of them so vast that they are not just bigger than the Mediterranean sea but the whole of Africa and Asia as well. What do you think Aristotle would have said and done, Simplicio, if he had seen these things?

SIMPLICIO. Aristotle was the master of the sciences, and I don't know what he would have said or done, but I do know something of what his followers say and do—quite rightly, if they are not to be left without any kind of guide or leader in philosophy. As regards the comets, surely those modern astronomers who claimed that they were celestial bodies have been proved wrong by the *Anti-Tycho*?* Proved wrong, what's more, by their own arguments, by means of parallaxes* and a hundred kinds of calculations, all pointing to the conclusion that Aristotle was right to say that the comets are all elemental bodies. And once that claim has been disproved, since it was the foundation of all their other innovations, I don't see that they've got a leg to stand on.

Astronomers refuted by the Anti-Tycho.

SALVIATI. Just a minute, Simplicio. What does this modern author say about the new stars of 1572 and 1604, and about the sunspots?* Because as far as the comets are concerned, I'm [77] quite ready to accept that they are generated either below or above the sphere of the Moon, and I've never set much store by Tycho's verbosity. And I don't have any problem in believing that their matter is elemental, or that they can ascend freely in the heavens without encountering any obstacle in passing through the Peripatetic heavens, which I consider to be much thinner, finer and more yielding than our atmosphere. As for the calculation of parallaxes, I'm equally sceptical of both sets of opinions, first because I doubt whether comets are subject to such accidental factors, and secondly because the observations on which their computations are based are so inconsistent – especially since the *Anti-Tycho* seems to me to adapt them at will, or to dismiss as erroneous any which don't fit into its scheme.

The Anti-Tycho adapts astronomical observations to its own purposes.

SIMPLICIO. The *Anti-Tycho* thoroughly disposes of the new stars in a few words, saying there is no certainty that these new modern stars are celestial bodies at all, and that if his adversaries want to prove that change and generation are found in the heavens they will have to demonstrate changes in the stars which are indisputably celestial, having been recorded there over many years; and this they will never be able to do. As for the materials which some say form and disintegrate on the surface of the Sun, he makes no mention of them; which I take to mean that he regards them as a fiction, or as illusions produced by the telescope, or at most as minor disturbances in the atmosphere—anything, in short, but celestial matter.

SALVIATI. And you yourself, Simplicio, what response have you come up with to these importunate sunspots which have come along to disrupt the heavens, and even more, the Peripatetic philosophy? You must surely have found some reply and solution, intrepid defender of Aristotle as you are, so please give us the benefit of your thoughts.

SIMPLICIO. I have heard a range of opinions on this particular matter. Some say that they are stars which revolve in their own orbits around the Sun in the same way as Venus and Mercury; that they appear dark to us as they pass between us and the Sun; and that because they are so numerous it often happens that some of them cluster together and then separate. Others believe that they are impressions in the atmosphere; others that they are optical illusions produced by the lenses; and others offer different explanations. For myself, I am inclined to believe—in fact I am convinced—that they are a conglomeration of many different opaque bodies which come together almost at random, so that when we look at a sunspot we can often count ten or more of these minute, irregularly shaped bodies, which appear to us like snowflakes or tufts of wool or flying insects. Their relative positions change, and they join together and separate again, especially below the sphere of the Sun, around which they revolve. But it does not follow from this that they are subject to generation and corruption; rather, they are sometimes hidden behind the body of the Sun, and at other times they are invisible because of their proximity to the dazzling light of the Sun, although they are not on the Sun's surface. For the Sun's eccentric sphere* has layers like an onion, one inside the other, each of which moves and each of which is studded with a number of small spots; and although their motion appears at first inconstant and irregular, nonetheless

*Various opinions on the sunspots.**

[78]

I understand that recent observations have shown that the same spots recur within a fixed time period. This seems to me to be the best explanation that has been put forward so far to account for these appearances, while at the same time maintaining the incorruptibility and ingenerability of the heavens. If this explanation should be found wanting, other more elevated minds will be able to find better ones.

SALVIATI. If we were discussing some point of law or some other humanistic discipline, where there is no final truth or falsehood, then we could indeed have confidence in a writer's subtlety of mind and greater skill and experience in speaking, and hope that someone who excelled in these abilities would also be able to make their own argument prevail. But conclusions in the natural sciences are necessarily true, and have nothing to do with human choice; so we must be careful not to commit ourselves to defending what is false, because then all the skill of Demosthenes* and Aristotle would be left standing by any run-of-the-mill thinker who happened to have latched on to the truth. So, Simplicio, you should stop thinking and hoping that there might be men so much more learned and erudite than us that they could defy nature and make what is false become true. You have concluded that, of all the opinions put forward so far to account for the nature of these sunspots, the one you have just expounded is true; so it follows that, if you are correct, all the others must be false. Now in order to show you that this opinion, too, is completely fanciful and mistaken, I shall leave aside all its other improbabilities, and cite just two pieces of evidence against it.

The art of oratory is ineffective in the natural sciences.

[79]

The first is that many of these spots are seen to arise in the middle of the Sun's disc, and many similarly disintegrate and disappear when they are a long way from its rim. This must necessarily show that they generate and disintegrate, for if they were simply the product of local motion we would see them all enter and leave the Sun's disc at the very edge. The second observation concerns the apparent change in their shape and the velocity of their motion, and conclusively shows, to anyone who is not completely ignorant of perspective, that the sunspots must be contiguous with the body of the Sun, and that, touching its surface, they move either with it or upon it; they cannot possibly revolve in an orbit separate from it. This is

Argument proving that the sunspots are generated and dissolve.

Conclusive demonstration that the sunspots are on the surface of the Sun.

shown by their motion, since they appear to move very slowly as they approach the circumference of the solar disc and faster towards the middle. This is shown by their shape, since towards the circumference they appear much narrower than when they are in the middle. This is because in the middle they appear in all their glory as they really are, whereas near the circumference they appear foreshortened because of the curvature of the Sun's globe. What is more, it is clear to those who have been able to observe and calculate their motions systematically that both these changes, in shape and in motion, correspond exactly to what we would expect if they were contiguous with the Sun, and is completely incompatible with their moving in circles separated from the solar body, even by only a small distance. This has all been demonstrated at length by our friend in his *Letters on the Sunspots* to signor Mark Welser.*

The motion of the sunspots appears slower towards the Sun's circumference.

The sunspots are narrow in shape when they approach the Sun's circumference, and why.

The fact that they change shape in this way also shows that none of them are stars or any other spherical body, for the sphere is the only shape which never appears foreshortened, or anything but perfectly round. So if any one of these spots was a spherical body, as we assume all stars to be, then its shape would always appear equally round, whether it was in the middle of the Sun's disc or at the edge; whereas the fact that they are so foreshortened at the edge, and by contrast so expansive and wide at the middle, clearly shows that they are like flakes, with very little depth or thickness in relation to their length and breadth. Finally, Simplicio, you should not believe anyone who claims that recent observations have shown the sunspots reappearing unchanged after a fixed period of time. Whoever told you that is deceiving you. Why else would they have said nothing to you about the spots which form and disintegrate on the face of the Sun, well away from its circumference, or about their apparent foreshortening, which necessarily proves that they are contiguous with the Sun's surface? As for the spots reappearing unchanged, all this means is that some of them may occasionally last longer than it takes for the Sun to rotate once on its axis, which is less than a month, before they disintegrate. Our friend has made this clear in the *Letters* we have already mentioned.

The sunspots are not spherical in shape, but stretched out like fine flakes.

[80]

SIMPLICIO. To tell the truth, I haven't observed them for long enough or systematically enough to have mastered the *quod est* of this matter; but I certainly shall do so, and try to see for myself whether I can harmonize what we learn from experience with what is demonstrated by Aristotle, because it's clear that two truths cannot contradict each other.

SALVIATI. As long as you aim to harmonize what you perceive with your senses with the most solid teachings of Aristotle, you will have no trouble at all. After all, doesn't Aristotle say that we cannot have complete certainty in matters concerning the heavens, because of their great distance from us?

We cannot speak with certainty about the heavens because of their great distance, according to Aristotle.

SIMPLICIO. He does indeed.

SALVIATI. And doesn't he also say, with great emphasis and without reservation, that what we learn from our senses should take precedence over any reasoned argument, however well founded it may seem?

The senses prevail over reason, according to Aristotle.

SIMPLICIO. Yes, he does.

SALVIATI. So of these two propositions, both of them Aristotle's, the second, saying that the senses should be given precedence over reason, is much more firmly established than the one which says that the heavens are immutable. Hence it is more faithful to Aristotle to say 'The heavens can change, because this is what the evidence of my senses shows', than to say 'The heavens cannot change, because this was what reason led Aristotle to conclude'. What's more, we are in a much better position to speak about the heavens than Aristotle was. He admitted that such knowledge was difficult for him because the heavens were so far removed from his senses, and he conceded that someone whose senses gave them a better representation of the heavens would have a more secure basis for speculating about them. We, thanks to the telescope, have made the heavens [81] thirty or forty times closer to us than they were to Aristotle, so that we can see many things there which he could not, including the sunspots, which were completely invisible to him. So we have a more secure basis for speculating about the Sun and the heavens than Aristotle.

It is more in conformity with Aristotle to say that the heavens can change than to say that they are immutable.

Thanks to the telescope, we are better able to speak about the heavens than Aristotle.

SAGREDO. I sympathize with Simplicio. I can see that he is moved by the strength of these all too conclusive arguments, and yet on the other hand he sees the universal authority in which Aristotle is held, and remembers all the famous commentators who have laboured to expound his meaning. He sees the other sciences, so necessary to the common good, which base such a large part of their reputation on the standing of Aristotle, and he is confused and alarmed.

Simplicio's declamation.

It is as if I hear him say: 'Who are we to turn to for a resolution of our disputes if Aristotle is unseated from his throne? What other author should we follow in schools, academies and universities? What other philosopher has written on every part of natural philosophy, in such a systematic way that not a single conclusion is omitted? Are we to lay waste the building which gives shelter to so many travellers? Must we destroy that refuge, the Prytaneum,* which enables so many scholars to shelter in comfort and, without being exposed to the outside air, to acquire knowledge of every aspect of nature simply by turning a few pages? Are we going to demolish that rampart within which we are safe from every enemy attack?' I feel sorry for him, as I would for a landowner who has spent years building a magnificent palace, sparing no expense and employing hundreds of craftsmen, only to find it starting to crumble because its foundations are unsound. Rather than have the grief of seeing the walls decorated with so many fine paintings collapse, the pillars supporting the splendid loggias fall down, the gilded balconies, the doorways, the cornices, and the marble mouldings all in ruins, he would do all he could with chains, props, buttresses, embankments, and supports to stave off collapse.

The Peripatetic philosophy is immutable.

SALVIATI. Oh, I don't think Simplicio need fear any such collapse; I would undertake to insure him against loss for far less expense than that. There's no danger of such a large number of wise and prudent philosophers being overcome by one or two individuals creating a bit of commotion. They won't even need to sharpen their pens against them; simply meeting them with silence will expose them to general scorn and derision. [82] It's vain to think that anyone could introduce a new philosophy simply by criticizing this or that author; they would need first to remake men's minds and make them able to distinguish true from false, and that's something that only God can do.

But how did we get on to this? I'll need the help of your memory to get me back onto the right track.

SIMPLICIO. I remember exactly where we were. We were discussing the response of the *Anti-Tycho* to the arguments against the immutability of the heavens, and you added the question of the sunspots, which the *Anti-Tycho* doesn't mention. I think you were going to consider his response to the example of the new stars.

SALVIATI. Yes, now I remember the rest. So to carry on where we left off, there are several points in the *Anti-Tycho*'s response which seem to me to deserve criticism. First of all, he says that the two new stars, which he has no choice but to place in the highest part of heaven, and which lasted a long time before they finally disappeared, don't undermine his belief in the immutability of the heavens, because it is not certain that they are celestial bodies at all, and in any case they are not changes in the stars which have been observed since antiquity. But in that case, why does he go to such trouble over the comets, to exclude them at all costs from the celestial regions? Surely he could just have said the same thing about them as he did about the new stars—that it is not certain that they are part of the heavens and they do not involve changes to any of the stars, and so they don't affect either the heavens or the teaching of Aristotle. Secondly, I can't follow his line of thought when he admits that any changes in the stars would destroy the prerogatives of the heavens, of being incorruptible, etc., because everyone agrees that the stars are clearly celestial bodies; and yet it doesn't seem to trouble him that the same changes might affect the rest of the expanse of the heavens, away from the stars themselves. So does he not consider the heavens to be part of the celestial regions? I always thought that the stars were called celestial bodies by virtue of their being in the heavens or made of celestial material, and that therefore the heavens were more celestial than the stars—in the same way as there can be nothing more terrestrial than the Earth, and nothing more fiery than fire itself. And as for his not mentioning the sunspots, which have been shown conclusively to form and disintegrate, to be [83] close to the body of the Sun, and to rotate with it or around it, this seems to me to show that this author was writing to please others rather than for his own satisfaction. I say this because he clearly shows that he understands mathematics, and so he cannot possibly fail to be convinced by the demonstrations that the sunspots must be contiguous with the body of the Sun, and that they are examples of generation and corruption on a far larger scale than any that ever occur on Earth. And if they are found so frequently and on such a scale in the globe of the Sun itself, which must surely be considered among the noblest parts of the heavens, then what grounds remain for denying that other examples can occur in the other globes?

SAGREDO. I am astonished when I hear it said that this quality of impassibility, immutability, unchangeableness, etc., is a source of great nobility and perfection in the natural and integral bodies of the universe, whereas being changeable, generable, mutable, etc., is considered a great imperfection. In fact I find such an idea repugnant, for I consider the Earth to be noble and marvellous because of the many different changes, mutations, generations, etc., which constantly occur in it. If it were not subject to any kind of change but was just a vast solitude of sand or a mass of jasper-hard rock, or if the water which covered the Earth at the time of the flood had frozen to form an immense globe of crystal where there was never any change or mutation of any kind, I would consider that to be a vile body which served no useful purpose but was simply superfluous, as if it had no existence in nature. It would be like the difference between a living creature and a dead one; and the same goes for the Moon, for Jupiter and all the other globes in the universe.

To be subject to generation and change is a greater perfection in the bodies of the universe than the opposite.

The Earth is noble because of the many mutations which occur in it.

The Earth would be useless and idle if it did not undergo change.

In fact, the more I reflect on the vacuous way the common people reason, the more shallow and insubstantial it appears. What could be more foolish than to say that gold, silver, and gems are precious, and that earth and mud are merely vile? Does it not occur to them that if earth was as scarce as jewels or precious metals, there would not be a prince who would not willingly give a sackful of diamonds and rubies and four cart-loads of gold simply to have enough earth to plant a jasmine in a little pot, or to grow an orange tree from seed and see it sprout, grow, and produce such lovely foliage, such fragrant blossom, [84] and such noble fruit? So it is scarcity and abundance which make the common people esteem or despise things. They will say that a diamond is beautiful because it is like pure water, and yet they would not exchange it for ten barrels of water. I think

The Earth is more noble than gold or jewels.

Scarcity or abundance make things costly or cheap.

that those who exalt incorruptibility, immutability, etc., so highly say these things because they are so anxious to have a long life and are so terrified of death; they don't realize that if men were immortal they would never have been brought into the world. It would serve them right if they were to encounter a Gorgon's head which would turn them into jasper or diamond statues, so that they could become more perfect than they are.

The common people extol incorruptibility because of their fear of death.

Those who despise corruptibility deserve to be turned into statues.

SALVIATI. It might even be a change for the better, because I think not talking at all is preferable to talking nonsense.

SIMPLICIO. There's no doubt that the Earth is far more perfect as it is, mutable, subject to change, etc., than it would be if it were a mass of rock, even if it were a single diamond, hard and impassible. But these qualities which confer nobility on the Earth would make the celestial bodies more imperfect, because they would be superfluous for them. The celestial bodies—the Sun, the Moon, and the other stars—are ordained solely for the benefit of the Earth, and to achieve this end they need only two things, motion and light.

The celestial bodies are ordained so as to serve the Earth, for which they need only motion and light.

SAGREDO. Are you saying, then, that nature has produced and ordained all these vast, perfect, and noble celestial bodies, impassible, immortal, and divine, for no other purpose than to serve the needs of the Earth, which is changeable, transient, and mortal? Purely for the benefit of what you call the dregs of the universe, the bilge where all the filth accumulates? What would be the point of making the celestial bodies immortal, etc., simply to serve one which is transient, etc.? The whole vast array of celestial bodies would be completely useless and superfluous if it were not for serving the needs of the Earth, given that they are all immutable and impassible and so can't possibly have any reciprocal effect on each other. If the Moon, for example, is impassible, what effect can the Sun or any other star have on it? Surely it would be less than that of someone who tried to liquefy a large mass of gold by looking at it or thinking about it. Indeed, it seems to me that if the celestial bodies work together to produce generation and change on Earth, they must themselves be subject to change. Otherwise, expecting the Moon or the Sun to produce generation on Earth would be like putting a marble statue alongside a bride and expecting the union to produce offspring.

The celestial bodies have no reciprocal effect on each other.

[85]

SIMPLICIO. Corruptibility, change, mutability, etc., don't apply to the entire terrestrial globe, which as a whole is no less eternal than the Sun or the Moon, but its external parts are subject to generation and corruption. In these parts, however, generation and corruption are perpetual, and as such they need the operation of eternal effects from the heavens. Hence it is necessary for the celestial bodies to be eternal.

Mutability applies not to the whole terrestrial globe, but to some parts of it.

SAGREDO. So far so good. But if the eternity of the terrestrial globe as a whole is not undermined by the corruptibility of its external parts, and if indeed this susceptibility to generation, corruption, change, etc., is an adornment to it and enhances its perfection, could you not—indeed, should you not—apply the same argument to the celestial spheres? Why not allow that change, generation, etc., in their external parts are an adornment to them, without in any way diminishing their perfection or limiting their effects? Indeed, their effects would be increased, because as well as acting on the Earth they would also produce reciprocal effects on each other, and the Earth on them.

Celestial bodies are subject to change in their external parts.

SIMPLICIO. That's not possible, because the generation, mutation, etc., that was produced, for example, on the Moon, would be vain and would serve no useful purpose, and *natura nihil frustra facit*: nature does nothing in vain.*

SAGREDO. Why would they be vain and serve no useful purpose?

SIMPLICIO. Because it's plain to see that all the generations, mutations, etc., that are produced on Earth are designed, directly or indirectly, to serve the use, convenience, and benefit of humankind. Horses are born for the convenience of humans; the Earth produces hay to feed the horses, and the clouds water it. Grass, grain, fruits, animals, birds, fish, are all produced for the convenience and nourishment of humans. In short, if we carefully examine and resolve all these things we shall find that the end to which they are all directed is the need, the utility, the convenience, and the delight of humankind. Now, what use to the human race would be any generation that was produced on the Moon or another planet? Assuming, that is, that you don't claim that there are also humans on the Moon to enjoy its fruits, an idea that is either fantastic or impious.

The generations and mutations on Earth are all for the benefit of humankind.

[86] SAGREDO. I have no way of knowing whether herbs, plants or animals similar to ours are generated on the Moon or any other planet, or whether they have rain, wind, or thunderstorms like those around the Earth. I don't believe so, and even less do I believe that the Moon is inhabited by humans.* But I don't see how it follows that just because they do not have species similar to ours, there cannot be change or alteration of any kind, or other things which change, are generated and dissolve which are not just different from ours, but beyond our imagination and wholly unknowable to us. I'm quite sure that a man who was born and brought up in a vast forest, among wild animals and birds, who had no knowledge of the element of water, could never imagine that there was another natural world quite different from dry land, full of creatures which move swiftly without legs or wings. Such creatures move not only on its surface as animals do on land, but anywhere in its depths as well; and they can also remain motionless wherever they wish, something which is impossible for birds in the air. Humans, too, live in this element and build palaces and cities there, and can travel so easily that they can transport their whole families, households, and entire cities to distant countries. So if such a man, however fertile his imagination, could never conceive of fish, the ocean, ships, fleets, and navies, how much less can we conceive what substances and effects there might be on the Moon, which is such a great distance away from us and which could, for all we know, be made of material quite different from the Earth? They might well be not just remote from our experience but completely beyond our imagination, having no resemblance to anything we know and hence being inconceivable to us. For the products of our imagination are necessarily either things we have already seen or a combination of things or parts of things we have seen before; this explains such things as sphinxes, mermaids, chimeras, centaurs, etc.

The Moon does not have species similar to ours; it is uninhabited by humans.

The Moon may have species which are different from ours.

Anyone who had no knowledge of the element of water would be unable to imagine ships or fish.

SALVIATI. I have often exercised my imagination about these matters, and I think I am now able to identify some of the things which are not and cannot be on the Moon, but not any of the things which I believe are or could be there, except in the most general terms. Of the latter, I can imagine only that they are an adornment to the Moon, that they act, move, and live and, in ways that may be completely different from ours, contemplate and marvel at the greatness and beauty of the universe and of its Creator and Ruler, continually singing His [87] glory. In short, I imagine them doing what is so frequently affirmed in Holy Scripture, namely that all creatures are perpetually occupied in praising God.

The Moon may have substances which are different from ours.

SAGREDO. If these are, in the most general terms, what might be found on the Moon, I would be glad to hear what you think are the things which are not and cannot be there. You must be able to identify these more specifically.

SALVIATI. Sagredo, this will be the third time that our conversation has led us away from the main topic which we set ourselves, without our realizing it. If we keep on digressing we'll never get to the end of our discussion, so I think it would be a good idea if we set this to one side, together with the other matters which we've agreed to come back to on a separate occasion.

SAGREDO. Please, since we've reached the Moon, let's deal with the questions concerning it, so as not to have to make such a long journey again.

SALVIATI. Very well, if you wish. Starting, then, at the most general level, I consider that the lunar globe is very different from that of the Earth, although we can also see some similarities; I'll describe the similarities first, and then the differences. We can be sure that the Moon is similar to the Earth in shape, as there is no doubt that it is spherical. This is proved by the fact that we see it as a perfectly circular disc, and by the way in which it receives light from the Sun. If its surface were flat, then it would all be bathed in light at once, and similarly it would all simultaneously become dark; whereas in fact the parts facing the Sun are illuminated first, followed by the rest, so that it is only at full Moon, when it is in opposition to the Sun, that the whole disc appears light. Conversely, the opposite would happen if its visible surface were concave; in that case the parts opposite the Sun would be illuminated first.

First similarity between the Moon and the Earth: its shape, as is proved by way it is illuminated by the Sun.

Secondly, the Moon is dark in itself and opaque, like the Earth. Its opacity means that it can receive and reflect the light of the Sun, as it could not do otherwise. Third, I consider the matter of which it is made to be dense and solid, no less than the

The second similarity is that the Moon is dark, like the Earth.

Third, the Moon is made of dense matter, and is mountainous, like the Earth.

Earth. I find clear evidence of this in the fact that the greater part of its surface is uneven, with many peaks and hollows which can be seen with the aid of a telescope. There are very many such peaks, in every respect like the steepest and most [88] rugged of our mountains, some of which are in ranges which stretch for hundreds of miles; others are in more compact groups, and there are also many isolated rocks which are very steep and precipitous. But the features which occur most often are high banks (this is the best word I can find to describe them) which surround and encircle plains of different sizes. These plains form various shapes, but the majority of them are circular; many of them have a mountain which stands out prominently in the middle, and a few are filled with darkish matter, similar to that of the spots which we can see with the naked eye. These are the largest of these flat areas; added to which there are a great number of smaller ones, nearly all of them circular.

Fourth, the Moon has distinct light and dark areas, like the sea and the land surface on Earth.

The surface of the sea, seen from a distance, would appear darker than that of the land.

Fifth, the Earth changes shape in the same way as the Moon, and with the same periodicity.

Fourth, in the same way as the surface of our globe is divided into two large areas, the land and the sea, so too on the surface of the Moon we can see a marked difference between some large areas which shine brightly and others less. I think that anyone who could observe the Earth from the Moon or from some other point a similar distance away would see the Sun illuminate it in the same way, with the seas appearing as darker areas and the land lighter. Fifth, we see the Moon from the Earth sometimes wholly illuminated, sometimes half, sometimes more or less. Sometimes we see it as a crescent, and sometimes it is completely invisible to us, as happens when it is directly below the Sun's rays, so that the part of it facing the Earth is in darkness. The Sun's light on the face of the Earth would appear exactly the same from the Moon, with the same time period and with the same alterations in shape. Sixth, . . .

SAGREDO. Just a moment, Salviati. I can quite well see that the Earth would appear illuminated in the same changing shapes to someone observing it from the Moon as the Moon does to us. But I don't see how this could follow the same time period, since what the Sun's illumination does on the surface of the Moon in the course of a month is the same as it does on the surface of the Earth in twenty-four hours.

SALVIATI. It's true that when the Sun shines on these two bodies and illuminates their whole surface, it completes this on the Earth in a natural day and on the Moon it takes a month. But this is not the only factor affecting the different shapes in which the illuminated part of the Earth would appear if it were observed from the Moon. This depends also on the varying positions which the Moon has in relation to the Sun. If, for [89] instance, the Moon exactly followed the motion of the Sun, and if it always stood in a direct line between the Sun and the Earth—what we mean when we say it is in conjunction with the Sun—then the hemisphere of the Earth facing the Sun would also be facing the Moon, and it would always appear fully illuminated. If, on the other hand, the Moon were always in opposition to the Sun, then an observer on the Moon would never see the Earth, because the Earth's dark side would always be facing the Moon and so would be invisible. But when the Moon is at its first or last quarter, then, of the hemisphere of the Earth visible from the Moon, the half which is facing the Sun would be shining and the half which is facing away from the Sun would be dark; and so the illuminated part of the Earth would appear to the Moon as a semicircle.

SAGREDO. That's all quite clear to me now. I can see how none of the illuminated part of the Earth's surface is visible from the Moon when it is in opposition to the Sun, but then as it moves day by day towards the Sun it gradually begins to see a small part of the Earth, which appears as a fine crescent because the Earth is round. Then, as the Moon's motion brings it daily closer to the Sun, more and more of the illuminated hemisphere of the Earth is revealed, so that when the Moon reaches first quarter it sees exactly half of it, just as we see half of the Moon. As it continues towards its conjunction with the Sun, more and more of the Earth's bright surface appears, until finally at conjunction the whole hemisphere is shining brightly. And I can see how the experience of Earth-dwellers in watching the phases of the Moon would be replicated for anyone observing the Earth from the Moon, but in reverse: when we see the full Moon when it is in opposition to the Sun, for someone on the Moon the Earth would be in conjunction with the Sun and so would be wholly dark and invisible; and what for us is the conjunction of the Moon with the Sun, when the Moon appears absent and invisible, for them the Earth would be in opposition to the Sun and it would be, as it were, 'full Earth', i.e. fully illuminated. Finally, at any given time whatever proportion of the Moon's surface appears illuminated to us, a corresponding proportion of the Earth's surface would appear dark to the Moon, and however much of the Moon appears dark to us, the [90] equivalent part of the Earth would appear light to the Moon. Only at the first and last quarter, when we see a semicircle of the Moon illuminated, would a Moon-dweller see the Earth in the same way. There's just one respect in which, as I see it, these reciprocal effects differ: assuming for the sake of argument that there was someone on the Moon in a position to observe the Earth, they would see the whole of the Earth's surface every day, because of the Moon's motion around the Earth every twenty-four or twenty-five hours; whereas we never see more than half of the Moon, because it does not turn on its axis, as it would have to do for us to be able to see it all.

SALVIATI. Unless the opposite is true—that it's because the Moon turns on its axis that we never see the other side, as would have to be the case if the Moon had an epicycle.* But aren't you overlooking another difference, which compensates for the one you've already noted?

SAGREDO. What difference is that? For the moment I can't think of any others.

The whole of the Earth can see only half of the Moon, but only half of the Moon can see the whole of the Earth.

SALVIATI. It's this: that while, as you've rightly pointed out, only half of the Moon is visible from the Earth, but the whole of the Earth is visible from the Moon, it's also the case that the Moon is visible from the whole of the Earth, but the Earth is only visible from half of the Moon. The inhabitants, so to speak, of the upper hemisphere of the Moon, which is invisible to us, cannot see the Earth: perhaps these are the Antichthons.*

More than half of the Moon's globe is visible from the Earth.

But I have just remembered a particular detail which has recently been observed in the Moon by our friend the Academician, from which two consequences follow: one is that we actually see rather more than half of the Moon, and the other is that the Moon's motion has a precise relation to the centre of the Earth. His observation is as follows. If it is indeed the case that the Moon has a natural correspondence and agreement with the

Earth, and that a specific part of it always faces towards the Earth, then it follows that a straight line joining the centres of the two bodies must always pass through the same point on the surface of the Moon. This means that anyone looking at the Moon from the centre of the Earth would always see the same lunar disc, bounded by exactly the same circumference. But when someone standing on the surface of the Earth looks at the Moon, the ray from their eye* to the centre of the Moon would only pass through the same point on the Moon's surface as the line joining the centres of the two bodies if the Moon were directly overhead. If the Moon is either to the east or to [91] the west, the observer's line of vision will strike the Moon's surface at a higher point than the line joining the two centres, and therefore part of the hemisphere above the circumference will appear, and a corresponding part below the circumference will be hidden—in relation, that is, to the hemisphere which would be visible from the true centre of the Earth. And since the upper part of the Moon's circumference when it rises is the lower part when it sets, there should be a quite noticeable variation in the appearance of these parts, which should be visible by the presence or absence of their spots or other distinctive features. The same should be true of the northern and southern extremities of the Moon's disc, depending on the Moon's location in relation to the ecliptic: when it is in the north, part of the area towards the northern circumference should be hidden, and part of the southern area should be revealed, and vice versa.

Two spots on the Moon which show that its motion is related to the centre of the Earth.

Now the telescope enables us to confirm that these consequences do in fact follow. There are two distinctive spots on the Moon, one which is towards the north-west when the Moon is on the meridian,* and the other almost diametrically opposite it. The first is visible even without a telescope, but not the second. The one to the north-west has an oval shape, and is separate from the other great spots; the one opposite it is smaller, and is also separate from the great spots, and is situated in a light-coloured field. Both of them clearly reveal the variations mentioned above; they appear opposite each other, at one time close to the edge of the Moon's disc and at another time further away. In the case of the north-western spot, its distance from the circumference more than doubles; the other is closer to the circumference so the variation is even greater, and is more than three times as much at one time than another. So it is clear that one side of the Moon is constantly facing the terrestrial globe, as if held there by magnetic force, and never diverges from it.

SAGREDO. Is there no end to the new observations and discoveries which can be made with this wonderful instrument?

SALVIATI. If its progress follows the course of other great inventions, we can hope that as time passes we shall be able to see things which now we cannot even imagine. But to come back to our earlier discussion, the sixth similarity between the Moon and the Earth is that, just as the Moon makes up for our lack of sunlight much of the time by reflecting the Sun's rays, making our nights relatively light, so the Earth repays the debt [92] when the Moon needs it most, reflecting the Sun's rays to give it a very strong light, which I reckon must exceed the light which the Earth receives from the Moon by the same amount as the Earth's surface is larger than the Moon's.

Sixth, the Earth and the Moon illuminate each other reciprocally.

SAGREDO. Don't say any more, Salviati; allow me the pleasure of showing you how what you have just said has enabled me to penetrate the reason for a phenomenon which I have thought about a hundred times without ever being able to fathom it. You're saying that the faint light which is sometimes visible in the Moon, especially at new Moon, is the light of the Sun reflected from the surface of the Earth and the sea; and the thinner the crescent Moon, the brighter this light shines, because that's when the illuminated area of the Earth's surface visible from the Moon is largest. This follows the principle you stated just now, that the illuminated part of the Earth which faces the Moon is as large as the dark area of the Moon facing the Earth. So when the crescent Moon is thinnest, and hence its dark area is largest, the illuminated area of the Earth seen from the Moon is largest, and its reflection of the Sun's rays is at its most powerful.

Light from the Earth is reflected in the Moon.

SALVIATI. That was exactly my meaning. What a pleasure it is to talk to people of sound judgement who learn quickly, especially when we are exploring and discussing truths in this way! I've lost count of the number of times I have encountered such obtuse minds that, however many times I repeated to them what you have just worked out for yourself, they have never been able to grasp it.

SIMPLICIO. I'm very surprised if you mean that you've never managed to make them understand it, because I find your explanation perfectly clear, so if they can't understand when you explain it to them I'm sure they will never understand it from anyone else. But if you mean that you haven't been able to persuade them to believe it, that doesn't surprise me at all; I must confess that I myself am one of those who understand your reasoning, but still am not persuaded by it. In fact, I have many difficulties with this and with some of the other six similarities you have listed; and I'll put these forward when you have finished your exposition of them all.

SALVIATI. I'll deal with the rest very briefly, because I'm always eager to discover new truths, and the objections voiced by an intelligent person like yourself can be a great help. The seventh similarity, then, is that the reciprocity between the Earth and the Moon extends to offences as well as favours: so as the Moon, when it's at its brightest, is often deprived of light and eclipsed by the Earth coming between it and the Sun, so it avenges itself by coming between the Earth and the Sun and casting its shadow on the Earth. It's true that the revenge is not equal to the offence, because the Moon is often totally immersed in darkness by the Earth's shadow for quite a considerable time, whereas it never happens that the whole Earth is completely overshadowed by the Moon, nor is it eclipsed for long periods of time. Nonetheless, considering how much smaller the Moon's body is than the Earth's, it can't be denied that the Moon puts up a spirited defence. So much for the similarities between the two. I could go on to discuss their differences; but since Simplicio is going to favour us with his doubts about the similarities, I think we should hear and consider these before we go on.

[93]

Seventh, the Earth and the Moon eclipse each other reciprocally.

SAGREDO. Indeed, because I dare say he won't object to the disparities and differences between the Earth and the Moon, since he considers their substances to be utterly different.

SIMPLICIO. Of the resemblances you've listed in establishing your parallel between the Earth and the Moon, the only ones I can accept without reservation are the first and a couple of others. I accept the first, namely its spherical shape; although even here I don't entirely agree with you, because I consider the Moon to be absolutely clear and smooth, like a mirror, whereas we can see that the Earth is very broken up and rugged. But the unevenness of its surface comes under one of the other similarities you have mentioned, so I shall leave what I have to say about this until we come to that point. As for your second resemblance, that the Moon is opaque and dark in itself like the Earth, I accept only the first attribute, its opacity. Solar eclipses are proof enough of this: if the Moon were transparent, the air would not be as dark as it is when the Sun is eclipsed, because some refracted light would still pass through the Moon's transparent body, as it does through very thick clouds.

Secondary light considered to be the Moon's own light.

The Earth is incapable of reflecting the Sun's rays.

The material of the heavens is impenetrable, according to Aristotle.

As for its being dark, I don't believe the Moon is completely lacking in its own light, as the Earth is. In fact I think that the faint light visible in the rest of the Moon's disc when a thin crescent is [94] illuminated by the Sun, is the Moon's own natural light, not light reflected from the Earth. I consider the Earth to be incapable of reflecting the Sun's rays because it is so uneven and dark. As for your third parallel, I agree with you on one count, and I dispute the other. I agree that the Moon is solid and hard, like the Earth—indeed much more so, because we learn from Aristotle that the heavens are impenetrably hard,* and the stars are the densest part of their heavens, so they must be truly solid and impenetrable.

SAGREDO. What wonderful material the heavens would be for building a palace, if you could get hold of any—so hard and so transparent!

SALVIATI. On the contrary, it would be quite useless: because it's so transparent as to be invisible, you couldn't walk through its rooms without risking colliding with the door-jambs and breaking your head.

Heavenly matter is intangible.

SAGREDO. There wouldn't be any risk of that if, as some of the Peripatetics say, heavenly matter is intangible; because if you can't touch it, you certainly couldn't collide with it.

SALVIATI. That wouldn't help at all. It may be true that you can't touch heavenly matter because it lacks the quality of tangibility, but it can still touch elemental bodies. And it would hurt just as much, if not more, for it to collide with us as for us to collide with it. But enough of these palaces, or rather castles in the air; we mustn't stand in Simplicio's way.

SIMPLICIO. The question which you've raised in passing is one of the most difficult in philosophy, and I've heard some wonderful thoughts about it from a distinguished professor in Padua;* but there isn't time for us to go into this now. So to return to our subject, I reply that I consider the Moon to be even more solid than the Earth; but I deduce this, not as you do from the ruggedness and unevenness of its surface, but rather from the opposite. I see it as being, like the hardest gemstones, capable of being polished to a lustre greater than that of the most highly polished mirror, as its surface must be for it to reflect the Sun's rays so brightly. All the appearances which you describe—mountains, rocks, banks, valleys, etc.—are illusions. I've heard the case made convincingly in public debates against those who put forward these novelties, that these appearances derive simply from varying degrees of opacity and clarity in the body of the Moon and on its surface. We can often observe this in crystals, amber and other highly polished precious stones, in which some parts are opaque and others transparent, so that they appear to have various concave and protruding parts.

The Moon's surface is smoother than a mirror.

[95]

Peaks and hollows on the Moon are illusions produced by opacity and clarity.

As regards your fourth similarity, I accept that the surface of the terrestrial globe, seen from a distance, would present two different appearances, some parts being brighter and others darker, but I think that these would be the opposite way round from what you say: I believe that the surface of the water would shine more brightly because it is smooth and transparent, and the dry land would be darker because it is opaque and uneven, and so less capable of reflecting the Sun's rays. The fifth comparison I accept entirely, and I can quite well see that if the Earth were to shine like the Moon it would appear, to someone observing it from there, as having the same shapes as those we see on the Moon. I can see, too, that the period of its illumination and variation of shape would be one month, even though the Sun circles it in twenty-four hours. Finally I have no difficulty in agreeing that only half of the Moon can see the whole of the Earth, while the whole of the Earth can see only half of the Moon. On the sixth point, I regard as totally false the idea that the Moon can receive light from the Earth, since the Earth is dark, opaque, and quite unsuited to reflecting the rays of the Sun, as the Moon reflects them to us. And as I have said, I consider that the light which can be seen in the rest of the Moon's surface when just its crescent is brightly illuminated by the Sun is the Moon's own natural light, and it will take a great deal to persuade me otherwise. The seventh comparison, about reciprocal eclipses, is acceptable, although what you choose to call an eclipse of the Earth is normally more properly called an eclipse of the Sun. These, I think, are all the points I have to make in refuting your seven similarities, and if you have anything to say in reply I shall be glad to hear it.

SALVIATI. If I've understood your response correctly, we differ on some of the features which I said were common to the Moon and the Earth, as follows. You consider the Moon to be smooth and polished like a mirror, and therefore able to reflect the Sun's rays, whereas you don't believe that the Earth is capable of reflecting them because its surface is so rough. You agree that the Moon is hard and solid, but you deduce this from [96] its being polished and smooth, not from its mountainous surface; and you attribute its mountainous appearance to its having different parts which are more and less opaque and clear. Finally, you consider that the Moon's secondary light is its own and not reflected from the Earth, although you do agree that the sea could reflect light because of its smooth surface. I don't hold out much hope of disabusing you of your erroneous belief that the Moon reflects light like a mirror, since it's clear that you've learnt nothing from what our mutual friend wrote about this in *The Assayer* and his *Letters on the Sunspots**—assuming that you have read what he says about this carefully.

SIMPLICIO. I've skimmed it rather superficially, as much as the time I could spare from more weighty studies allowed. So if you think you can resolve my difficulties by reproducing some of his arguments, or by introducing new ones, I'll give them my best attention.

SALVIATI. I'll give you my thoughts as they occur to me now, so they may be a mixture of my own ideas and those I read in these two books. I remember being entirely convinced by them, although when I first read them I found them highly paradoxical. So, Simplicio, we're trying to establish whether, in order to produce reflected light such as we receive from the Moon, the reflecting surface needs to be smooth and polished like a mirror, or whether a surface which is uneven and unpolished would be more effective. Suppose, then, we had light reflected from two different surfaces, one light brighter than the other, which surface do you think would shine more brightly and which would be darker?

SIMPLICIO. I have no doubt that the one which reflected the light more vividly would appear brighter, and the other would be darker.

SALVIATI. Be so good as to take that mirror that's attached to the wall, and let's go out into the courtyard; you come too, Sagredo. Now let's fix the mirror to the wall here where it's in the Sun, and we shall retire over here into the shade. So there we have two surfaces both exposed to the Sun, the wall and the mirror. Now tell me: which of the two shines more brightly, the wall or the mirror? What, no answer?

Detailed proof that the Moon has a rough surface.

SAGREDO. I'll let Simplicio reply, since he's the one who has the difficulty. For myself, just this small experiment is enough to convince me that the Moon's surface must be very unpolished.

SALVIATI. Simplicio, tell me whether, if you had to paint a picture of that wall with the mirror on it, where would you use the darker colours: to paint the wall or the mirror? [97]

SIMPLICIO. The paint for the mirror would be much darker.

SALVIATI. Then if it's true that the surface which reflects the light more strongly is the one which appears brighter, the wall must reflect the Sun's rays more strongly than the mirror.

SIMPLICIO. Really, is that the best experiment you can come up with? You've made us stand away from the reflection of the mirror, but if you come a little this way—no, just come over here.

SAGREDO. Are you looking for the place where the mirror casts its reflection?

SIMPLICIO. That's right.

SAGREDO. It's there on the opposite wall, exactly the same size as the mirror, and almost as bright as if the Sun were shining directly on it.

SIMPLICIO. Come here then, and look at the surface of the mirror over there, and then tell me whether it's darker than the wall.

SAGREDO. You can look at it if you want to; I don't want to be blinded, and I know very well without needing to look that it's as bright and clear as the Sun, or only slightly less so.

SIMPLICIO. How can you say, then, that a mirror reflects light less strongly than a wall? Looking at this wall opposite, onto which light is reflected both from the sunlit wall and from the mirror, the reflection of the mirror is much the brighter of the two; and when I look at the mirror itself it, too, is much brighter than the wall.

SALVIATI. You've cleverly anticipated what I was going to say next: I was going to use precisely this observation to point out what follows. You see the difference, then, between the reflections produced by the two different surfaces, the wall and the mirror, when they are both identically exposed to the Sun's rays. You can see that the reflection from the wall is evenly diffused across the whole of the surface opposite, whereas the mirror is reflected in just one place which is no bigger than the mirror itself. You can see, too, how the surface of the wall has the same brightness from wherever you look at [98] it, and is much brighter than the mirror except from that one small place where the mirror's rays fall; and from there the mirror appears much brighter than the wall. Now I think it's quite easy to work out from this tangible evidence of our senses whether the light of the Moon comes to us as from a mirror or as from a wall, in other words from a smooth or a rough surface.

SAGREDO. I don't think I could better grasp the roughness of the Moon's surface if I was there and could reach out and touch it than I have from listening to your explanation. Whatever the Moon's position in relation to the Sun and to us, the part of its surface illuminated by the Sun always appears equally bright. This is exactly the same as the wall, which appears equally bright from wherever you look at it, and is quite different from the mirror, which only appears bright from one position and is dark from everywhere else. And also, the light reflected off the wall is weak and not dazzling at all, compared to the reflection of the mirror which is so strong that it hurts the eyes almost as much as looking at the Sun itself. In the same way, we can easily look at the face of the Moon, but if it were a mirror its brightness would be absolutely intolerable and it would be like looking at another Sun, especially as its closeness to us would make it appear as big as the Sun itself.

SALVIATI. Please, Sagredo, don't make my demonstration carry more weight than it can bear. Let me raise an objection which may not be very easy to resolve. You make it a point of great difference between the Moon and the mirror that the Moon sheds its light equally in all directions, like the wall, whereas the mirror is reflected only in a single specific place; and hence you conclude that the Moon is like the wall and not like the mirror. But the reason the mirror is reflected in only one place is that its surface is flat; and since the angle of reflection of the rays must be equal to their angle of incidence, the rays reflected off a flat surface must necessarily all go to the same place. But the Moon's surface is spherical, not flat; and rays striking a spherical surface are reflected at the same angle as their angle of incidence in all directions, because the surface of a sphere has an infinite number of inclinations. Hence the Moon can shed its reflected light in all directions, and the light doesn't all have to be reflected from the same place, as it does from the mirror which is flat.

Plane mirrors reflect in only one direction, but spherical mirrors in every direction.

SIMPLICIO. This is precisely one of the objections I was going to raise. [99]

SAGREDO. If this was one of your objections, that means you must have others. Let's hear them, because I think that this first point may count against you rather than in your favour.

SIMPLICIO. You stated it as a self-evident fact that the light reflected off that wall was as bright as the light that comes to us from the Moon, but I consider it incomparably less. For in this matter of illumination it is important to distinguish the sphere of activity; and no one doubts that the celestial bodies have a greater sphere of activity than our elemental, transient, and mortal bodies. And what is that wall, in the last resort, but a small piece of earth, dark and quite incapable of shedding light?

The celestial bodies have a greater sphere of activity than elemental bodies.

SAGREDO. Here too I think you are much mistaken. But let me come to Salviati's first point. For an object to appear illuminated to us, it's not enough for rays from the source of light to fall on it; they must also be reflected to our eyes. This is clear from the example of the mirror, which is clearly exposed to the sun's rays even though it didn't appear illuminated to us unless we were looking from the precise spot where the reflected rays fell. Now let's consider what would happen if the mirror were spherical: we would undoubtedly find that only a tiny part of the light reflected from its illuminated surface would fall on our eyes at any particular point, because the reflection at that point would come from only a tiny part of the spherical surface. So only a very small part of the sphere's surface would appear bright to us, and all the rest would remain dark. If, then, the Moon were smooth like a mirror, our eyes would only perceive the light coming from a very small part of its surface, even if its whole hemisphere was exposed to the Sun. The rest of it would not appear illuminated and so would be invisible. The final result would be to make the Moon itself invisible, since the part reflecting the light would be so small and so far away that it [100] would be lost to our sight; and just as the Moon would remain invisible to our eyes, so we would receive no illumination from it, because it's impossible for a luminous body to shed light in the darkness without being visible.

If the Moon were like a spherical mirror, it would be invisible.

SALVIATI. Stop there, Sagredo, because I can see from Simplicio's expression that either he hasn't grasped what you've so evidently and correctly said, or he isn't convinced by it. I've just had an idea for another experiment which will remove all his doubts. I noticed a big spherical mirror in one of the rooms upstairs; let's have it brought down here, and while we're waiting for it to be brought, perhaps Simplicio would like to look again at the brightness of the light reflected from the plane mirror on the wall here under the loggia.

SIMPLICIO. I can see that it's scarcely less bright than if it were in direct sunlight.

SALVIATI. It is indeed. Now tell me, if we were to take away the small plane mirror and put the big spherical one in its place, what effect do you think its reflection would have on the same wall?

SIMPLICIO. I think the light on it would be much greater and more widely spread.

SALVIATI. What would you say if there proved to be no light at all, or so little you hardly noticed it?

SIMPLICIO. I'll wait until I see the effect, and then I'll consider how to reply.

SALVIATI. Here's the mirror; I'd like it to be placed alongside the other one. But first let's look closely at the reflection from the plane mirror, and note its brightness: look how bright it is here where it strikes the wall, and how clearly you can make out every detail of the wall.

SIMPLICIO. I've seen and observed it carefully. Now have the other mirror placed next to the first one.

SALVIATI. It's there already. It was put there as soon as you began looking at the detail of the wall, and you didn't notice it, because the increase of light on the rest of the wall was just as great. Now let's take away the plane mirror. You see how all the reflection has gone, even though the big convex mirror is still there. You can take it away and put it back as many times as you like, but you won't see any change in the light on the wall. So there you have the evidence of your senses to show that sunlight [101] reflected from a convex spherical mirror does not cast any appreciable light on its surroundings. Now how do you respond to this experiment?

SIMPLICIO. I fear you may have introduced some sleight of hand. But when I look at that mirror I see it shedding a brilliant light, so bright it almost blinds me; and what's more, I see it from wherever I look, and the light comes from a different place on the mirror's surface depending on where I am standing to look at it. This must mean that the light is reflected brightly in every direction, and so it must shine as strongly on the whole of that wall as it does to my eyes.

SALVIATI. That shows how cautious you must be in assenting to a conclusion which is arrived at solely by means of argument. There's no doubt that what you say seems very plausible; and yet you can see that the evidence of your senses tells you otherwise.

SIMPLICIO. So how do we make any progress in resolving this question?

SALVIATI. I'll tell you what I think, though I don't know how far you will be convinced by it. To start with, that brilliance which you see when you look at the mirror, and which seems to you to occupy quite a large part of it, is not actually as large as it looks; in fact it comes only from a very small part. But its brightness has an effect on your eyes, because it is reflected in the moisture at the edge of your eyelids covering the surface of the pupil. It looks like a kind of extra irradiation, rather like the apparent halo around a candle seen from a distance. Or you could compare it to the extra brightness which you see surrounding a star. If you compare the size, for example, of the Dog Star when you see it through a telescope in the daytime, without irradiation, with its appearance at night seen with the naked eye, you will be left in no doubt that the irradiation makes it appear more than a thousand times larger than the bare body of the star itself. The image of the Sun that you see in that mirror is enlarged in the same way—in fact more, because the sun's light is so much stronger than that of a star, as is clear from the fact that looking at a star is much less dazzling to the eyes than looking at this reflection in the mirror. Therefore, the reflected light which is shared over the whole surface of this wall comes from only a small part of that mirror, whereas the reflection from the whole of the plane mirror was restricted to only a small part of the wall. So it's not surprising that the first reflection shone very brightly, while the second remained barely perceptible.

Irradiation around the body of a star makes it appear many times larger than it really is.

[102]

SIMPLICIO. I'm more confused than ever, and now I have another difficulty. How can it be that that wall, which is made of such dark material and has such an unpolished surface, reflects light more strongly and vividly than a smooth, polished mirror?

SALVIATI. Not more vividly, but diffused more widely. As for being vivid, you can see that the reflection from that plane mirror where it strikes the wall there under the loggia is very bright, but the rest of the surface, which receives reflected light from the wall with the mirror on it, is far less brightly lit than the small area where the mirror is reflected. If you want to understand the principle behind this, think of the rough surface of that wall as having innumerable tiny surfaces all at different angles, of which many will necessarily reflect rays in one direction and many in another. So there will not be anywhere which does not receive many rays reflected from many tiny surfaces scattered over the whole of the rough surface which is exposed to the light. Hence every part of any surface facing the one on which the primary rays fall will receive some reflected rays, and consequently will be illuminated. It follows, too, that the object on which the illuminating rays fall will appear wholly illuminated and bright, from whatever point one looks at it.

Light reflected from rough surfaces is more diffused than that from smooth surfaces, and why.

That is why the Moon, since its surface is rough not smooth, reflects the Sun's light in all directions, and appears equally bright to anyone who looks at it. But since it is a sphere, if its surface was smooth like a mirror it would be invisible, because the very small part of its surface which could reflect the Sun's image to any one observer would be invisible from such a great distance, as we have said.

If the Moon were smooth and polished, it would be invisible.

SIMPLICIO. I understand your explanation perfectly; but I think I can answer it very easily, and maintain that the Moon is round and polished and that it reflects the Sun's light to us like a mirror. Nor does this mean that we should be able to see the reflected image of the Sun in the middle of the Moon; for 'we should not expect to see the small image of the Sun in the form of the Sun itself at such a great distance; rather we should understand the illumination of the whole lunar body as being the light produced by the Sun. We can observe a similar phenomenon in a well-burnished gilded plate, which when it is struck by rays from a luminous body, appears when seen from a distance to be all shining, and it is only when it is seen close at hand that we discern in the middle the small image of the luminous body.'*

[103]

SALVIATI. I must confess my obtuseness, and admit that I don't understand anything of what you have said apart from the bit about the gilded plate. In fact, if you'll allow me to speak freely, I have an idea that you don't understand it either, but have simply learnt by heart what someone has written in order to contradict and appear more intelligent than his opponent. But this only impresses those who, wanting to appear intelligent themselves, applaud what they don't understand, and whose opinion is highest of those they understand least. That's always assuming that the writer himself is not one of those many who write things which they don't understand, so that no one else can understand what they write either.

Some write of what they do not understand, and therefore what they write cannot be understood.

So leaving the rest aside and responding to what you say about the gilded plate, if it is flat and not very large it could appear from a distance to be very bright when a strong light strikes it. But that only happens when you are looking at it directly from the point to which the light is reflected. And it will appear more fiery than if it were made, for example, of silver, because of its colour and because the greater density of the metal makes it better suited to being highly burnished. If its surface were not only highly polished but also not entirely flat, but had variously inclined surfaces, then its splendour would be visible from even more places, as many as caught the reflections from its various surfaces. This is why diamonds are worked into many facets, so that their delightful brilliance can be seen from many different places. If the plate were very large, however, it would not appear from a distance as if its whole surface were shining, even if it were completely flat. To show what I mean more clearly, imagine a very large flat gilded plate which is exposed to sunlight. When looked at from a distance the image of the Sun will appear to occupy only part of the plate—the part from which the incident rays of the Sun are reflected—but the brilliance of the light will make the image appear to be surrounded with many rays, so that it will look as if it occupies a much larger part of the plate than it really does. To confirm this, note the particular point on the plate where the reflection comes from, and also estimate how large the bright area appears to be; then cover up most of this area, leaving just a small space around the centre exposed. You will see that, when looked at from a distance, the bright area will not appear diminished in the least; in fact it will spread out over the cloth or whatever you used to cover it up. So if anyone who sees a small gilded plate appear from a distance as if its whole surface shone, and then imagines that the same thing would happen with a plate as big as the Moon, they are as much mistaken as if they imagined the Moon to be no bigger than the bottom of a barrel. If the surface of the plate were spherical, then the strong reflection would be seen in only one small part of it, but its vividness would make it appear to be surrounded by many glittering rays. The rest of the sphere would appear as coloured, as long as it was not highly polished; if it was highly burnished, it would appear dark.

Diamonds are worked into many facets; the reason for this.

[104]

We can see an everyday example of this in silver vessels which, if they are simply boiled in bleach, have a white frosting on them, like snow, and they don't reflect any image at all; but if any part of them is polished, that part immediately appears dark, and reflects images like a mirror. The reason for

Burnished silver appears darker than silver which has not been burnished; the reason for this.

its appearing dark is simply that polishing it has smoothed off a very thin patina which made the surface of the silver uneven, and therefore reflect light in all directions, so that it appeared illuminated from wherever you looked at it. Burnishing it then smooths away these tiny irregularities so that the incident rays are all directed to the same point. This means that when seen from that point the burnished part looks much brighter and clearer than the rest, which was just bleached, but it looks dark when seen from anywhere else. It's well known that because the appearance of burnished surfaces varies so much when seen from different viewpoints, if a painter wants to represent something like a piece of burnished armour in a picture, he has to use black and white alongside each other for places where the armour is in fact equally lit.

Burnished steel appears very bright when seen from some angles, and very dark from others.

SAGREDO. So supposing these philosophers were willing to concede that the surface of the Moon, Venus and the other planets is not as polished and smooth as a mirror, but just marginally less—like a silver plate which has been bleached but not burnished—would that be enough to make them visible and able to reflect the Sun's rays to us?

[105]

SALVIATI. It would be partly enough, but it would not give a light as strong as the Moon does thanks to its mountainous surface, with its high peaks and deep hollows. But these philosophers will never concede that the Moon is less polished than a mirror; in fact they insist that it is far more so, if such a thing can be imagined. They believe that perfect bodies must have a perfect shape, and therefore the celestial bodies must be absolutely perfect spheres; and in any case, if they conceded even the tiniest irregularity I would have no scruples about claiming much greater ones, because perfection is indivisible and so if it falls short by a hair's breadth it might as well fall short by the height of a mountain.

SAGREDO. This raises two questions in my mind: first, how it is that a more irregular surface produces a more powerful reflection of light; and second, why these Peripatetics are so insistent on this perfect shape.

SALVIATI. I'll reply to the first question, and I'll leave Simplicio to respond to the second. You need to know, then,

A rough surface produces a greater reflection of light than one which is less rough.

that the same surface can be illuminated to a greater or lesser extent by the same rays, depending on whether the rays strike it more or less obliquely; the greatest illumination is produced when they fall vertically. Here's a demonstration of what I mean. I'll fold this piece of paper so that one part of it is at an angle to the rest. If I hold it up to the light reflected from that wall over there, you can see how this surface of the paper, which the rays strike obliquely, is less bright than the other where they strike it at right angles; and see how the illumination diminishes as I turn the paper to receive the light more and more obliquely.

Perpendicular rays give more illumination than those which strike obliquely; the reason for this.

SAGREDO. I can see the effect, but I don't understand the cause.

SALVIATI. I'm sure you would if you thought about it for a moment; but to save time, here is a drawing to demonstrate it.

[106] SAGREDO. Yes, I see now just by looking at the drawing—but carry on.

SIMPLICIO. Please explain the rest to me; I'm not so quick to understand.

The reason why oblique rays illuminate less.

SALVIATI. Let these parallel lines starting from points A, B be rays of light which strike the line CD at right angles. Now incline the line CD so that it is in a new position, DO. Don't you see how many of the rays which struck the line CD pass over DO without touching it? If DO is illuminated by fewer rays, it follows that the light it receives must be weaker. Now consider the Moon. As it is a sphere, if its surface were as smooth as this sheet of paper then the parts near the edge of the hemisphere illuminated by the Sun would receive very much less light than those in the centre, since in the centre the Sun's rays would strike the surface at right angles and at the edges at a very oblique angle. This would mean that at full Moon, when almost the full hemisphere is illuminated, the central parts ought to appear much brighter to us than those at the circumference; and yet this doesn't happen. But if you now picture the Moon's surface with very

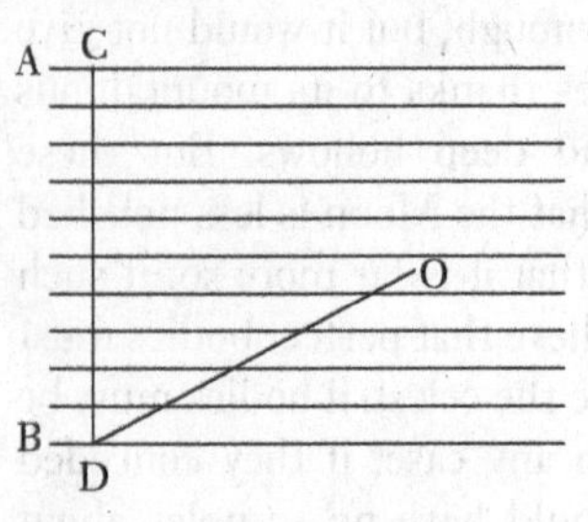

high mountains, you can see how their peaks and ridges, standing out from the perfectly spherical surface, are exposed to the Sun and in a position to receive its rays at a much less oblique angle, and therefore to appear as brightly illuminated as the rest:

SAGREDO. I see all that. If there are such mountains near the circumference, it's certainly true that the Sun's rays will fall more directly on them than on the inclination of a polished surface. But it's also true that the valleys between these mountains would all be dark, because of the very long shadows the mountains would cast at such a time. But the parts near the centre, even if they were also full of mountains and valleys, would not have shadows because the Sun would be much higher in the sky. So the parts in the centre should shine much more brightly than those at the edge which would have as much shadow as light; and yet no such difference is apparent.

SIMPLICIO. I was just thinking of the same difficulty.

SALVIATI. Isn't it remarkable how much more readily Simplicio grasps the difficulties which support Aristotle's view than he does their solutions? But I think he sometimes deliberately keeps quiet about them; and in the present case, since he was able to see the objection by himself—which is [107] quite an ingenious one—I can't believe he hasn't also spotted the reply. So let me try to winkle it out of him, so to speak. Tell me then, Simplicio: do you think it is possible for there to be shadow in a place exposed to the Sun's rays?

SIMPLICIO. I think not; in fact I'm sure of it. The Sun is the greatest luminary which scatters the darkness with its rays, so anywhere touched by the Sun's rays cannot possibly remain dark. Besides, we have the definition that *tenebrae sunt privatio luminis*: darkness is the absence of light.*

SALVIATI. Therefore, when the Sun beholds the Earth, the Moon, or any other opaque body, it never sees any of their parts that are in shadow, since it has no eyes to see apart from its light-giving rays. Hence any observer who was on the Sun would never see anything that was in shadow, because his visual rays* would always accompany the Sun's rays of light.

SIMPLICIO. That's very true, beyond contradiction.

SALVIATI. But when the Moon is in opposition to the Sun, what difference is there between the line of your visual rays and that of the Sun's rays?

SIMPLICIO. Now I understand; you mean that since the rays of our vision always follow the same lines as the Sun's rays, we can't see any of the valleys of the Moon that are in shadow. Please don't think that I am pretending or dissembling with you; I give you my word as a gentleman that this reply had not occurred to me, and I doubt whether I would have found it without your help, or at least without thinking about it at length.

SAGREDO. The solution to this latest difficulty which you have arrived at between you has convinced me as well. But at the same time, this notion of visual rays accompanying the Sun's rays has raised another doubt in my mind on the other side of the argument. I don't know if I'll be able to explain it, because it has only just occurred to me and I still haven't clarified my thoughts about it, but let's see whether we can resolve it between us. We've established that the areas near the circumference of a hemisphere which is smooth but not burnished, when the Sun shines on it, are struck by the Sun's rays obliquely and therefore receive fewer of them than the areas in the middle, which the rays strike directly. So it could be that a band which is, say, twenty degrees wide near the edge of [108] the hemisphere receives no more rays than another band near the centre which is only four degrees wide; and therefore the band at the edge will appear much darker than the one at the centre, to anyone looking straight at them—seeing them in all their glory, as it were. But if the observer's eye were so placed that the dark, twenty-degree band appeared no wider than the four-degree band at the centre, then it's quite possible that the two might appear to shine equally brightly. They would both appear as a band four degrees wide, and an equal number of rays would reach the eye from them both—both the band in the middle which really was four degrees wide, and the other which was twenty degrees wide but appeared to be only four degrees because of foreshortening. This will be the case when we are between the illuminated hemisphere and the source from which the light is coming, because then the lines of sight and of the rays of light are the same. Hence, it seems not impossible that the Moon's surface is really quite even, and that it nonetheless appears at full Moon to shine as brightly at the edges as it does in the centre.

SALVIATI. Your doubt is an ingenious one, and it deserves to be taken seriously. And as it has only just occurred to you, I too will answer with the thoughts that come to mind immediately; maybe when I have thought more about it I will be able to give a better response. First, though, before I introduce any new arguments, it would be a good idea to see experimentally whether your objection is borne out in practice. So let's take this piece of paper again, and fold it so that a small part of it is at an angle to the rest; and then let's hold it in the light so that the rays fall directly on the smaller part and obliquely on the rest. You see, this clearly shows that the smaller part is noticeably brighter. Now, to see whether your objection is conclusive, we need to lower our eyes so that we look obliquely at the larger, less bright part until it appears no wider than the other, more brightly illuminated part, and we see the same visual arc for both. If you're right, then the light from the wider part will increase so that it appears to be as bright as the other. I'm looking at it now, so obliquely that the wider part appears narrower than the other, but it still remains as dark as ever. Now you look and see if you find the same thing.

SAGREDO. I've looked, and however much I lower my eyes [109] I don't see that surface as any brighter or more illuminated; if anything it seems to get darker.

SALVIATI. We've established, then, that the objection is invalid. As for the explanation, I think that because this paper is not completely smooth, only a few rays are reflected back in the direction of the incident rays compared with the many which are reflected in opposite directions; and of these few, more are lost the more closely our visual rays approach the incident light rays. And since the brightness of the object comes not from the incident rays but from those which are reflected to our eyes, it follows that in lowering our eyes we lose more rays than we gain, and this is borne out by your observation that the paper appeared to become darker.

SAGREDO. I'm convinced, both by the experiment and by your argument. So it remains now for Simplicio to reply to my other question, and to tell me what moves the Peripatetics to require such perfect rotundity in the celestial bodies.

Why the Peripatetics assume a perfectly spherical shape for the celestial bodies.

SIMPLICIO. The fact that the celestial bodies are ingenerable, incorruptible, unchangeable, impassible, immortal, etc., means that their perfection is absolute. A necessary consequence of this perfection is that they should be perfect in every respect, and hence that their shape should also be perfect, that is, spherical, and absolutely and perfectly spherical, not rough or irregular.

SALVIATI. And from what do you derive this quality of incorruptibility?

SIMPLICIO. Directly, from their lack of contraries; indirectly, from their simple circular motion.

SALVIATI. So, it appears from what you say that if the essential quality of the celestial bodies is that they are incorruptible, immutable, etc., then rotundity is not a cause or necessary condition. If it were the cause of immutability, then we could confer incorruptibility on a piece of wood, or wax, or any other elemental material, simply by giving it a spherical shape.

Shape is a cause not of incorruptibility, but of greater durability.

SIMPLICIO. But isn't it clear that a ball made of wood will last longer and be better preserved than a spire or some other angular shape made from a piece of the same wood?

The corruptible can vary in degree, but the incorruptible cannot.

SALVIATI. That's very true, but this doesn't mean that it will no longer be corruptible and become incorruptible: rather, it will still be corruptible, but more durable. The point to note is that there can be degrees of corruptibility, so we can say 'This [110] is less corruptible than that'—for instance, that jasper is less corruptible than sandstone. But incorruptibility doesn't admit of different degrees; it's not possible to say 'This is more incorruptible than that', if both are incorruptible and eternal.

Perfection of shape affects corruptible bodies, but not those which are eternal.

Differences of shape, therefore, can only affect materials which are capable of being more or less durable; but materials which are eternal must all be equally eternal, and so shape no longer has any effect on them. So, since celestial material is not incorruptible on account of its shape but for some other reason, there is no need to be so anxious about its being perfectly spherical. If it is incorruptible it will remain so, whatever shape it may have.

SAGREDO. I have been thinking about this a little further. If it were granted that a spherical shape had the capacity to confer incorruptibility, then all bodies, of whatever shape, would be eternal and incorruptible. For if a spherical body was incorruptible, then corruptibility would be confined to those parts of the body which make it less than perfectly round. A dice, for instance, contains within it a perfectly round ball, which as such is incorruptible, and so the corruptible parts must be the corners which cover and conceal its rotundity; so at the most it would be these corners and, so to speak, excrescences which would be subject to corruption. But if we take the argument a stage further, those parts which form the corners have within them other smaller balls of the same material, and so they too, being round, are incorruptible. Similarly in the residue which remains around each of these eight smaller spheres it is possible to imagine yet others; so that finally the entire dice resolves itself into innumerable balls, and we have to admit that it is incorruptible. And you could pursue the same argument to a similar conclusion with any other shape.

If spherical shape conferred eternity, all bodies would be eternal.

SALVIATI. The argument works very well. So if, for example, a crystal sphere were incorruptible, i.e. able to resist any alteration either internally or externally, because of its shape, is it not clear that adding more crystal to it in order to make it, say, into a cube, would change it both internally and externally? It would then be less able to resist a new enclosure of the same material than others of a different material—especially since Aristotle says that corruption is produced by contraries. For what could surround that ball of crystal that would be less contrary to it than crystal itself?

But we're losing track of the time, and we shall be very late concluding our discussions if we pursue every detail at such length. What's more, my memory is so confused by all these different topics that I have difficulty in recalling all the propositions which Simplicio set out so logically for us to consider.

[111]

SIMPLICIO. I remember where we were. On this question of the mountainous surface of the Moon, you haven't yet dealt with the explanation I put forward which I think saves this appearance very well: that it is an illusion deriving from the fact that different parts of the Moon are more and less opaque and clear.

SAGREDO. When Simplicio was speaking earlier about the apparent unevenness in the Moon's surface and, following the opinion of a Peripatetic philosopher friend of his, attributed it to the varying degrees of opacity and clarity in its different parts, he compared this to the similar illusions which we see in crystals and various kinds of precious stones. This reminded me of another material which is much more suitable to replicate this kind of effect, so much so that I think this philosopher would pay any price for it: mother-of-pearl, which can be worked into various shapes, and even when it is polished to a high degree of smoothness, still gives the optical illusion of having such a variety of hollows and protrusions that even touching it is hardly enough to convince you that it's smooth.

Mother-of-pearl can replicate the apparent unevenness in the Moon's surface.

SALVIATI. That's a charming idea; no doubt there will be more which haven't been thought of yet, and if there are any other gems and crystals which have nothing to do with the illusions produced by mother-of-pearl, I dare say these will be excellent ideas as well. Meanwhile, as I don't want to stand in anyone's way, I shall suppress the answer which I could give to this, and just concentrate on answering Simplicio's objections.

I say, then, that your argument is too general, and since you haven't applied it to all the varying appearances which can be observed in the Moon, which have prompted me and others to believe it to be mountainous, I don't think you will find anyone who is convinced by it. Indeed, I don't think that either you or the author himself will find it any more satisfactory than any other explanation that has nothing to do with the case. If you fashioned a ball with a smooth surface and parts which had varying degrees of opacity and clarity, you wouldn't be able to rep- [112] licate a single one of the countless different appearances which are observed in the course of a lunar month. But you could take any solid and non-transparent material you like and make it into balls which, solely by means of peaks and hollows and varying kinds of illumination, would replicate exactly the same appearances and mutations which can be observed hour by hour in the Moon. You would see the ridges of the high peaks catching the sunlight and shining brightly, and the dark shadows which they cast behind them. You would see them appear larger or smaller depending on their distance from the boundary which divides the bright part of the Moon from its dark part. You would see that this boundary or dividing line is not evenly spread, as it

The observed unevenness in the Moon's surface cannot be replicated by means of parts which are more and less opaque and clear.

The changing appearance of the Moon can be replicated with any opaque material.

would be if the ball's surface was smooth, but jagged and crenellated. You would see numerous peaks catching the light on the dark side of this boundary, detached from the rest of the area which was already illuminated. You would see the shadows mentioned above diminishing and, as the light grew higher, eventually disappearing altogether, until when the whole hemisphere was illuminated none would remain. Then, as the light moved towards the Moon's other hemisphere, you would recognize the same peaks which you had observed before, and this time you would see their shadows on the other side, and growing. And I repeat that you would not be able to replicate a single one of these things with your variations in opacity and clarity.

Various appearances from which can be deduced the mountainous surface of the Moon.

SAGREDO. Actually, you could replicate one of them, which would be the full Moon, when none of these shadows or other variations produced by the peaks and hollows are visible because the whole surface is illuminated. But please, Salviati, don't waste any more time on this point, because anyone who had had the patience to observe the Moon over one or two months and still could not grasp these plainly evident truths, would have to be considered lacking in any kind of judgement. And there is no point in wasting time and words with such people.

SIMPLICIO. To tell the truth, I haven't made such observations, because I haven't had sufficient curiosity, or a suitable instrument for the purpose, but I would very much like to do so. So let's leave this question pending and move on to our next point. Please give me your reasons for believing that the Earth can reflect the Sun's light no less powerfully than the Moon; because it seems to me to be so dark and opaque that I find such an effect quite impossible.

SALVIATI. Simplicio, your reason for believing the Earth incapable of shedding light is equally impossible. Wouldn't it be a fine thing if I were able to understand your argument better than you can yourself? [113]

SIMPLICIO. Whether my argument is sound or not is something that I dare say you can judge better than me; but sound or unsound, I'll never believe that you can understand my argument better than I can.

SALVIATI. I think I can make you believe it right now. Tell me: when the Moon is nearly full, so that it can be seen both by day and in the middle of the night, when do you think it shines more brightly, by day or by night?

The Moon appears brighter by night than by day.

SIMPLICIO. By night; there's no comparison. I think the Moon is like the pillar of cloud and fire which accompanied the children of Israel, which was like a cloud when the Sun was shining but was brilliantly bright by night.* I've sometimes observed the Moon among clouds in the daytime, and it appeared white just like one of the clouds, but then at night it shines resplendently.

The Moon seen in the daytime is like a little cloud.

SALVIATI. So if you had only ever seen the Moon in the daytime, you would have thought it was no brighter than a cloud.

SIMPLICIO. I believe so, yes.

SALVIATI. So now tell me this: do you think the Moon is really brighter at night than it is by day, or is there some other factor that makes it appear so?

SIMPLICIO. I think its brightness is intrinsically the same both by day and by night, but its light appears greater at night because we see it against the dark background of the sky. In the daytime, because everything around it is light, it hardly stands out at all, and so it appears much less bright to us.

SALVIATI. Tell me now, have you ever seen the terrestrial globe illuminated by the Sun in the middle of the night?

SIMPLICIO. That's either a trick question, or one which you only ask someone who is known to have no sense at all.

SALVIATI. No, no: I know you to be a man of very good sense, and I'm asking the question quite seriously. So don't hesitate to answer; if then you think I'm talking nonsense, I'm happy to be considered the one who has no sense. After all, if someone asks a stupid question it's they who are stupid, not the person they are asking.

[114] SIMPLICIO. Very well: if you don't take me for a complete simpleton, assume that I have answered and said it is impossible for someone who is on the Earth, as we are, to see at night that part of the Earth where it is day, i.e. where the Sun is shining.

SALVIATI. So you have only ever seen the Earth illuminated in the daytime, whereas you see the Moon shining in the sky even on the darkest night; that's why you do not believe that the Earth shines like the Moon. If you were able to see the Earth with the Sun shining on it when you were in darkness similar to our night, you would see it shining more brightly than the Moon. Now, to make a valid comparison, you must compare the light of the Earth with the light of the Moon when you see it in the daytime, not with the Moon at night, because we never see the Earth illuminated except by day. Isn't that the case?

SIMPLICIO. Yes, it must be.

SALVIATI. And since you have said yourself that you have seen the Moon by day among little white clouds and similar in appearance to one of them, you have already acknowledged that the clouds, although they are elemental matter, are capable of receiving illumination from the Sun just as much as the Moon is—indeed more, if you think of times when you have seen very large clouds which are as white as snow. We can be sure that if such a cloud could shine as brightly as that in the middle of the night, it would light up the area around it more than a hundred Moons. If, then, we could be certain that the Earth receives light from the Sun as readily as one of these clouds, then we could no longer doubt that it shines no less brightly than the Moon. And we can be sure of this, because we can see these same clouds at night, in the absence of the Sun, appear as dark as the Earth. More than this, we have all had the experience of seeing clouds low on the horizon, and not being sure whether they were clouds or mountains. This clearly shows that the mountains are no less luminous than these clouds.

Clouds, no less than the Moon, are capable of being illuminated by the Sun.

SAGREDO. That's enough arguments! Look there at the Moon, which is more than half full, and over there at that high wall with the Sun shining on it. Come over here so you can see the Moon and the wall alongside each other, and look: which of them is brighter? Isn't it clear that, if anything, the wall is brighter? The Sun's rays are striking it; from there, they are reflected onto the walls of this room, and from them into that [115] other small room over there. So they are reflected three times to reach that small room, and yet I'm sure it has more light than it would if it were lit directly by moonlight.

A wall illuminated by the Sun, compared to the Moon, reflects no less light.

The third reflection of light from a wall is stronger than the first reflection from the Moon.

SIMPLICIO. No, I don't agree about that; moonlight is very bright, especially when the Moon is full.

SAGREDO. It seems very bright because the area all around it is dark and in shadow, but in absolute terms it's not; in fact

Moonlight is weaker than twilight.

it's less than the light at twilight half an hour after sunset. You can see that, because only then do objects lit by the Moon cast a shadow on the Earth. If you want to prove whether the third reflection in that little room gives more light than the first reflection from the Moon, you can go in there now and read a book, and then try this evening to read by moonlight, and see which is easier to read by. I'm sure that it will be harder by moonlight.

SALVIATI. So now, Simplicio, if you're convinced, you can see how you really knew all along that the Earth shines no less than the Moon; to convince you I had only to remind you of things you knew already, without needing to learn them from me. I didn't teach you that the Moon shines more brightly by night than in the daytime; you knew it for yourself, just as you knew that a cloud can appear as bright as the Moon. You knew, too, that the Earth's illumination can't be seen at night; in short, you knew everything without knowing that you knew it. So it shouldn't be hard for you to grant that the dark part of the Moon can be illuminated by the reflection of the Earth with no less light than the Moon gives when it lights up the night—in fact more, for the Earth's light exceeds the Moon's by as much as the Earth is bigger than the Moon, which is forty times more.*

SIMPLICIO. But I did believe that the Moon's secondary light was intrinsic to the Moon itself.

SALVIATI. And yet you knew this too, without realizing it. You knew for yourself, surely, that the Moon shines more brightly by night than in the daytime, in contrast to the darkness of the area around it. So did you not know in consequence the general principle that every bright body appears brighter when its surroundings are darker?

Illuminated bodies appear brighter in dark surroundings.

SIMPLICIO. Yes, I know that very well.

SALVIATI. When the Moon is new and its secondary light is brightest, is it not always close to the Sun, and hence in twilight?

[116] SIMPLICIO. Yes, it is; and I've often wished that the sky would get darker so that I could see this light more clearly, but the Moon has set before it got completely dark.

SALVIATI. So you must have known, then, that the secondary light would appear more clearly when it was completely dark.

SIMPLICIO. Yes, of course; and it would appear more brightly still if it were not for the brightness of the crescent which is illuminated by the Sun, which detracts from the secondary light.

SALVIATI. But doesn't it sometimes happen that you can see the whole of the Moon's face without the Sun shining on it at all?

SIMPLICIO. As far as I know, the only time that happens is when there is a total eclipse of the Moon.

SALVIATI. Then at such a time its light ought to shine very brightly indeed, since its surroundings are completely dark and there is no illuminated crescent to detract from it. But how bright have you seen it under those conditions?

SIMPLICIO. I've sometimes seen it appear copper-coloured with a touch of whiteness, but at other times it has been so faint that I've lost sight of it altogether.

SALVIATI. In that case how can that be its own intrinsic light, since you can see it quite brightly at twilight despite the much greater brightness of the crescent alongside it, but when the night is completely dark and all other light is absent, it's not visible at all?

SIMPLICIO. I understand some people have said that the light is shed on it from the other stars, and particularly from Venus, which is nearest to it.

SALVIATI. This is another vain idea, because in that case it ought to shine more brightly than ever when it is totally eclipsed, since no one supposes that the Earth's shadow conceals it from Venus or the other stars. But in fact it's completely devoid of light at an eclipse, because the hemisphere of the Earth which is facing the Moon at that time is in darkness, so there is a complete absence of light from the Sun. And if you observe carefully, you will see clearly that when the Moon is new it gives very little light to the Earth, but as the area illuminated by the Sun grows so the light which it reflects to us becomes brighter. At the same time, the new Moon appears very bright to us, because it is [117] between the Sun and the Earth and therefore is exposed to a large part of the Earth's sunlit hemisphere. As it approaches first quarter and moves away from the Sun, its brightness diminishes, and once it has passed its first quarter it becomes quite dim, because its exposure to the illuminated part of the Earth is continually decreasing. But if the Moon's light were intrinsic to it or derived from the stars, the opposite would happen, because then we would be able to see it in the darkest night and in completely dark surroundings.

SIMPLICIO. Please stop there. I've just recalled reading a recent booklet of conclusions* full of new ideas, which explained this secondary light as follows. 'It is produced neither by the stars nor by the Moon itself; even less is it communicated to the Moon from the Earth. Rather, it comes from the light of the Sun itself, which penetrates through the body of the Moon, the substance of which is somewhat transparent, although it shines most brightly on the surface of the Moon's hemisphere which is exposed directly to its rays. But the interior of the Moon absorbs and soaks up, as it were, the light like a cloud or crystal, and transmits it so that it is visibly illuminated. The author proves this, if I remember rightly, by means of authority, experience and reason: he cites Cleomedes, Vitellio, Macrobius,* and some other modern author,* and he adds that it can be seen by experience that the Moon shines very brightly when it is close to conjunction with the Sun, i.e. when it is crescent, and also that it shines most brightly around the edge. He also says that in a solar eclipse, when the Moon is under the Sun's disc, light shines through it, especially at the outer edges. His arguments, I think, are that since this light cannot derive from the Earth, the stars, or the Moon itself, it must necessarily come from the Sun, and moreover that this assumption gives a coherent explanation for all the observed phenomena. The fact that this secondary light appears brighter around the outer edge is explained by the fact that the Sun's rays only have to penetrate a shorter distance, since the longest line through a circle is the one which passes through its centre, and those further away from this line are always shorter than those close to it. The same principle, he says, explains why this light does not diminish very much. Finally, it explains why it is that in a solar eclipse the brighter ring around the outer edge of the Moon is visible [118] around the part which is under the Sun's disc but not around the part outside the disc. The reason for this is that the Sun's rays pass through the part of the Moon placed under the Sun

Some say that the Moon's secondary light is caused by the Sun.

directly to our eyes, but those which go through the parts outside fall outside our line of sight.'

SALVIATI. If this philosopher had been the first author to put forward this view, it would have been understandable for him to be so attached to it that he embraced it as true. But since he learnt it from others, there is no excuse for his failure to see the fallacies in it, especially if he had heard the true cause of this effect and had the opportunity to confirm it by any number of experiments and manifest proofs: that it is reflected light from the Earth, and nothing else. Of those authors who find this explanation unconvincing, I can excuse the ancients who did not hear it expounded and who did not think of it themselves; I'm sure that if they heard it now they would have no hesitation in accepting it. To speak quite frankly, I can't believe that this modern author doesn't assent to it privately, but I suspect the fact that he can't claim to be its originator prompts him to try to suppress it or discredit it at least among the ignorant, of whom as we know there are a great number. For there are many who have much more pleasure in the applause of the common crowd than in the approval of the select few.

SAGREDO. I don't think you're really getting to the heart of the matter, Salviati. These people are always out to set traps for the unwary, and they're quite capable of making themselves authors of other people's ideas, as long as the ideas haven't been common currency for so long that everybody knows them.

SALVIATI. Oh, I'm more cynical than you are. Never mind about ideas being common currency; does it make any difference whether the ideas and discoveries are new to men, or the men are new to the discoveries? If you wanted nothing more than the esteem of those who are beginners in science, you could claim to be the inventor of the alphabet and they would all wonder at you; and even if in time your deception was discovered, that wouldn't do you any serious damage, because there would always be newcomers to keep up the number of your supporters.

It makes no difference whether ideas are new to men, or the men new to the ideas.

[119] But let's return to our discussion, and show Simplicio how his modern author's arguments are undermined by various errors, non sequiturs, and untenable opinions. First of all, it's not correct to say that this secondary light is brighter around the Moon's outer edge than in the middle, so as to form

The Moon's secondary light appears in the form of a ring, bright at the outer circumference and not in the middle; the reason for this.

How to observe the Moon's secondary light.

a brighter ring or circle around the rest of the field. It is true that the Moon appears at first sight to be surrounded by such a circle when it is observed at twilight, but this is an illusion caused by the different borders around its surface when it shines with this secondary light. The side which is towards the Sun is bordered by the Moon's brilliantly bright crescent, whereas on the other side it borders on the darkness of twilight. The contrast makes this side of the Moon's disc appear brighter, while the other appears darker against the greater brightness of the crescent. Your modern author could have seen this if he had tried blocking out the crescent by looking at it with a rooftop or some other obstacle in the way. Then he would have realized that the rest of the Moon's surface, apart from the crescent, is in fact equally illuminated.

SIMPLICIO. I seem to remember he does say that he used some such device to obscure the bright crescent.

SALVIATI. Well, in that case what I thought was just carelessness on his part turns out to be a lie, and a barefaced lie at that, since anyone can easily put it to the test for themselves. As for the Moon being visible in a solar eclipse, I doubt very much whether this can be for any reason other than the lack of light shining on it—especially when it is not a total eclipse, as must be the case with any which this author has observed. But even if it did appear to shed some light, that would support our argument rather than undermining it, since in a solar eclipse the whole of the Earth's sunlit hemisphere is facing the Moon, and even if the Moon itself obscures part of it this is only a very small part compared to what remains illuminated. The author goes on to say that in an eclipse the edge of the Moon shines brightly around the part which overlaps with the Sun, but not around the rest, and that this is because the Sun's rays pass through the overlapping part of the Moon directly to our eyes, but not through the rest. This is one of those fantasies that show up the other fictitious ideas he is propounding; for does the poor fellow not realize that if the Moon's secondary light is only visible to us when the Sun's rays pass directly through it to [120] our eyes, then the only time we would ever see this secondary light would be in a solar eclipse? And if a part of the Moon which is less than half a degree away from the Sun is enough to

In a solar eclipse the Moon's disc is visible only through lack of light.

deflect the Sun's rays away from our sight, what will happen when the Moon is twenty or thirty degrees away from the Sun, as it is when it first rises? And in any case, how will the Sun's rays pass through the body of the Moon to reach our eyes?

This man proceeds by imagining facts as his theory requires them to be, rather than accommodating his theories to fit the facts as they are. So, in order for the Sun's rays to penetrate the substance of the Moon he says that the Moon is partly transparent, in the same way as, for example, a cloud or a crystal; but I don't know how he conceives of a transparency which allows the Sun's rays to penetrate a cloud more than two thousand miles thick. Let's assume that he confidently replies that this is perfectly possible in celestial bodies, which are quite different from the elemental bodies we know, impure and tainted as they are; and let's refute his error by means which don't allow any riposte, or rather subterfuge. So if he wants to maintain that the Moon's substance is transparent, he will have to explain how its transparency allows the Sun's rays to pass right through its thickness, a distance of more than two thousand miles, but that when they are blocked by a thickness of only a mile or less they are no more able to penetrate it than they could a mountain here on Earth.

The author of the booklet of conclusions adapts the facts to his ideas, not his ideas to the facts.

SAGREDO. This reminds me of a man who wanted to sell me a secret device* which would enable me to talk to someone two or three thousand miles away by means of some kind of attraction of magnetic needles. I said I would certainly like to buy it, but first I would like to see it working, and that it would be enough for him to go into another room and talk to me from there. He replied that its operation couldn't be seen properly over such a short distance. So I sent him away, and said that I wasn't inclined to go to Cairo or Moscow to try the experiment, but he was welcome to go and then he could talk to me from there while I stayed here in Venice. But let's hear what follows from this author's argument, and why he must admit that the Moon is permeable to the Sun's rays to a depth of two thousand miles, yet more opaque than one of our mountains at a thickness of only one mile.

Joke at the expense of a man who wanted to sell a secret device for talking to someone a thousand miles away.

SALVIATI. The mountains on the Moon themselves are evidence of this. When the Sun strikes them on one side, they [121] cast dark shadows on the other, sharper and more clearly defined than the shadows of our mountains; whereas if they were transparent, we would never have discerned any irregularity on the Moon's surface, or seen those illuminated peaks which stand out detached from the boundary between its light and dark areas. In fact, this boundary itself would be much less distinct if the sunlight really did penetrate through the body of the Moon. It follows from the author's own words that the dividing line between the part which was exposed to the Sun's rays and the part which was not would have to appear blurred and a mixture of light and dark; for a material which lets the Sun's rays pass through a depth of two thousand miles must be so transparent that they would hardly be obstructed at all by a thickness of only a hundredth part of that or less. But in fact, the dividing line between the light and dark areas is as sharp and distinct as the contrast between black and white, especially where it crosses the part of the Moon which is naturally brighter and more uneven. Where it cuts across the ancient spots,* which are plains, it is not so clear-cut because the light is fainter, as the inclination of the Moon's spherical surface makes the Sun's rays strike them more obliquely. Finally, what he says about the secondary light not diminishing or becoming fainter as the Moon becomes more full, is simply wrong. In fact, this light is hardly seen at all when it is at its first or last quarter, when it would be expected to be most clearly visible because then it is surrounded by complete darkness, well away from the twilight.

We conclude, therefore, that the Earth's reflection has a very powerful effect on the Moon. Even more important, another marvellous congruity can be deduced from this: that if it is true that the planets affect the Earth with their light and motion, the Earth may be no less able to affect the planets reciprocally, with its light and possibly with its motion as well. In fact, even if the Earth did not move it could still affect them in the same way, for as we have seen, the action of the light is the same in either case, since it is the reflected light of the Sun. Motion simply produces variations in aspect which follow in the same way whether the Earth moves and the Sun is fixed or the other way round.

The Earth can reciprocally affect the celestial bodies by means of light.

[122] SIMPLICIO. No philosopher has said that these inferior bodies affect the celestial bodies, and Aristotle clearly states the contrary.

SALVIATI. This is excusable in Aristotle and the others who did not know that the Earth and the Moon reflect light to each other; but there would be no excuse for philosophers whom we have shown how the Earth gives light to the Moon if they insisted on denying that the Earth can affect the Moon, while expecting us to agree that the Moon's light affects the Earth.

SIMPLICIO. I have to say that I am extremely reluctant to accept this relationship between the Earth and the Moon that you want to persuade me of, placing the Earth, so to speak, among the stars. Apart from anything else, it seems to me that the great distance separating the Earth from the celestial bodies must necessarily mean that there is a huge difference between them.

SALVIATI. That just goes to show, Simplicio, how powerful a long-standing attachment and a deeply-rooted opinion can be. It is so tenacious that it makes you cite arguments in support of your belief which actually undermine it. If you are convinced that distance and separation are a mark of great divergence in nature, then the converse must also be true, and proximity and contiguity must indicate similarity; and how much closer is the Moon to the Earth than to any of the other celestial globes? So by your own admission you must concede that there is a great affinity between the Earth and the Moon—and you will have other philosophers for company if you do. Now let's proceed: tell me if any of the other objections which you raised against the congruity between these two bodies remain to be discussed.

Affinity between the Earth and the Moon on account of their proximity.

SIMPLICIO. I still have some uncertainty about the solidity of the Moon, which I deduced from its being supremely polished and smooth, and you from its being mountainous. And I also had a difficulty arising from my belief that the sea, because of the smoothness of its surface, must reflect light more strongly than dry land, which is so uneven and opaque.

SALVIATI. I will respond to your first doubt by reference to the Earth, all of whose parts are drawn by their natural gravity to come together as close to its centre as they can, though some do remain further away than others, the mountains being further from the centre than the plains. We attribute this to their hardness and solidity, since if they were made of more [123] fluid matter they would level out. In the same way, the fact that some parts of the Moon remain elevated above the spherical surface of its plains is evidence of their hardness, since it is reasonable to suppose that lunar matter too has formed itself into a sphere because its parts are all drawn together to the same centre.

Solidity of the Moon's sphere deduced from its being mountainous.

As regards your other doubt, I think it is clear from our investigation of what happens with mirrors that the light reflected from the sea is much less than from dry land, at least as far as its overall reflection is concerned. There's no doubt that a calm sea reflects light very strongly to a particular place, and anyone observing from that spot would see a brilliant reflection from the water, but from anywhere else the surface of the water would appear darker than that of the land. We can demonstrate this experimentally if we go into the main room here and pour a little water onto the floor. There: doesn't that brick which is wet look much darker than the dry ones around it? Clearly it does, and it will appear the same from everywhere except one place, which is where it reflects the light coming from that window. Just move back slowly . . .

Light is reflected more weakly from the sea than from dry land.

Experiment to demonstrate that water reflects less brightly than dry ground.

SIMPLICIO. From here the wet area appears lighter than the rest of the floor, and I can see that this is because the light coming in through the window is reflected towards me.

SALVIATI. The water has simply filled up all the little cavities in the brick so that it has a perfectly smooth surface, so that the reflected rays all converge on the same place; whereas the rest of the floor which is dry still has all its irregularities, and hence an innumerable variety of tiny inclined surfaces, so that the light is reflected in all directions, but more weakly than if the rays all converged together. This means that its appearance hardly changes at all from wherever you look at it; it looks the same from any angle, but much less bright than the reflection from the part which is wet. I conclude from this that since the surface of the sea, seen from the Moon, would appear completely smooth—apart from any islands or rocks—it would appear [124] darker than the land, which is mountainous and uneven. And at the risk of appearing to be pushing my luck, so to speak, I will add that I have observed the Moon's secondary light, which

I say is reflected from the Earth, to be noticeably brighter two or three days before the Moon's conjunction with the Sun than after it—when, that is, we see this light before dawn in the east, rather than after sunset in the west. This difference is caused by the fact that when the Moon is facing the Earth's eastern hemisphere it is exposed to much more land, since it includes the whole of Asia, and very little sea, whereas when it is facing west it looks down on a great expanse of sea, all the way across the Atlantic Ocean to the Americas. This makes it highly probable that the surface of the sea reflects light less brightly than the land.

The Moon's secondary light is brighter before its conjunction with the Sun than after.

SIMPLICIO. So in your view the Earth would have a similar appearance to the two main areas that we see in the Moon. Do you think, then, that those great spots that are visible on the Moon are seas, and that the other lighter areas are land or something similar?

SALVIATI. This is the first of the differences that I believe exist between the Earth and the Moon—and it's time we dealt with these quickly, as I think we've lingered too long on the Moon already. I say, then, that if nature had only one way of making two surfaces illuminated by the Sun appear one darker than the other, namely by making one of dry land and the other of water, then we would necessarily have to conclude that the Moon's surface was partly dry land and partly water. But since we know of many ways of producing the same effect, and there may well be others that are unknown to us, I do not presume to say that one rather than another occurs on the Moon. We have already seen how a touch with a burnishing brush changes a bleached silver plate from white to dark; and the parts of the Earth which are moist are darker than those which are arid. On a mountainside, the areas which are forested appear much darker than those which are bare and barren; this is because there is a great deal of shade among the trees, while the open [125] spaces are illuminated by the Sun. We can see the same effect of light and shade in a piece of patterned velvet, where the cut silk appears much darker than where it is uncut, because of the shadow between one thread and another; similarly plain velvet appears much darker than a smooth lining made from the same silk. So, if there were anything on the Moon resembling our great forests, this could appear to us as the dark spots that we can see; the same effect would be apparent if they were seas; and finally, it is not impossible that these spots could in reality be a darker colour than the rest, rather as snow makes mountains appear lighter. What is quite clear on the Moon is that the darker areas are all plains, with a small number of rocks and banks contained in them. The rest, which is brighter, is full of rocks, mountains, and small banks, some round and some in other shapes. In particular, the dark spots are surrounded by very long ranges of mountains. We can see that the surface of the spots is smooth because the boundary line dividing the light and dark parts of the Moon cuts across them evenly, whereas in the lighter areas it appears jagged and crenellated. I don't know, however, whether the smoothness of the surface is sufficient on its own to explain the darkness of these areas; I rather think not.

The darker parts of the Moon are plains, and the lighter parts mountainous.

The Moon's spots are surrounded by long ranges of mountains.

I also consider that the Moon differs greatly from the Earth in other respects. I imagine it is not completely dead and sterile, but nor do I affirm that movement or life are to be found there. Even less do I believe that plants, animals and other things similar to ours are generated there. If any generation does occur there it will be of things totally unlike what we know and beyond anything we can imagine. My first reason for believing this is that I do not think the Moon is made up of earth and water, and this alone means that it cannot have generation and change as we know them. But even if water and earth did exist there it would still not produce plants and animals similar to ours, for two main reasons.

There is no generation on the Moon of things similar to ours; if there is generation, it is of things which are completely different.

The Moon is not made up of earth and water.

First, generation on Earth would not be possible without the varying aspects of the Sun, which behaves quite differently towards the Earth than it does towards the Moon. For the greater part of the Earth, the alternating cycle of night and day is completed every twenty-four hours, but on the Moon this is spread over a month. The annual rise and fall in the Sun's altitude which brings us the changing seasons and the varying [126] length of day and night is also completed in a month on the Moon. And whereas for us the variation in the Sun's altitude is approximately forty-seven degrees, this being the distance from one tropic to the other, on the Moon the variation is only ten degrees or a little more, this being the maximum extent of the Sun's movement in relation to the ecliptic. Consider now the effect which the Sun would have in the tropics if its rays beat down

The aspects of the Sun, which are necessary for generation on Earth, are not the same on the Moon.

Each natural day on the Moon lasts for a month.

continually for fifteen days, and it will be clear that all the plants, grasses and animals would be scattered; so even if any species were generated there, they would be grasses, plants and animals completely different from ours. Second, I am firmly convinced that there is no rain on the Moon, because if clouds ever formed over any part of its surface as they do above the Earth, they would block out some of the things which we can see through a telescope, and we would see some variation in its appearance. In all my long and diligent observations I have never seen any such change, but I have always seen the Moon under a perfectly clear sky.

The Sun's altitude rises and falls by 10 degrees on the Moon, and by 47 degrees on Earth.

There is no rain on the Moon.

SAGREDO. To this one could reply either that the Moon could have very heavy dew, or that it rained during their night, when it is not illuminated by the Sun.

SALVIATI. If other considerations gave us grounds for believing that there might be generation similar to ours on the Moon, and all that was lacking was evidence of rain, then we might find some mitigating condition to fulfil the same function, like the flooding of the Nile in Egypt. But since we have not found any conditions similar to ours of the many that it would take to produce similar effects, there's no point in going to great lengths to introduce just one, especially since there is no positive evidence of it but simply because it is not inconceivable. And in any case, if I were asked what my initial understanding and pure natural reason tell me about the likelihood of the Moon producing things which are similar to or different from ours, I would always say, utterly different and completely beyond our imagination; for this is what the riches of nature and the omnipotence of its Creator and Ruler seem to me to require.

SAGREDO. It has always struck me as the height of temerity to make the limits of human understanding the measure of what nature can do, when there is not a single effect in nature, however small, which even the most penetrating human intelligence can fully understand. This vain presumption of understanding everything can only come from never having understood anything; for anyone who has just once experienced a perfect understanding of just one thing, and has truly tasted what knowledge is made of, will recognize that there is an infinity of other truths of which they understand nothing. [127]

Never having fully understood anything makes some people believe that they understand everything.

SALVIATI. Your argument is irrefutable; and it is confirmed by the experience of those who have reached some understanding, who, the wiser they become, the more they recognize and freely admit how little they know. And the wisest of the Greeks, who was judged to be so by the oracles, openly confessed that he knew nothing.

SIMPLICIO. Then we will have to say that either the oracle or Socrates himself was lying, when the oracle said he was the wisest of men, and he said that he knew himself to be the most ignorant.

The oracle spoke the truth in declaring Socrates the wisest of men.

SALVIATI. Neither of these follows, since both statements can be true. The oracle judges Socrates the wisest of men, and men's wisdom is limited; Socrates knows that he knows nothing in comparison to absolute wisdom, which is infinite. And since in relation to infinity much is no more than little or nothing—for to arrive at an infinite number it makes no difference whether you multiply thousands, tens, or zeros—Socrates knew that the finite wisdom which he possessed was nothing compared to the infinity which he lacked. Since, however, there is some wisdom among men, and it is not equally shared among all, it was possible for Socrates to have a greater share than anyone else, and so for the oracle's statement to be true.

SAGREDO. I think I understand the point here very well. In the same way, Simplicio, there exists among men the power to act, but not all share it equally. The power of an emperor is clearly much greater than that of a private person, but both are as nothing compared to the omnipotence of God. Or among men, there are some who understand agriculture better than others; but what is knowing how to plant a vine-shoot in [128] a trench, compared to being able to make it take root, draw nourishment from the earth, and select the appropriate parts to form leaves, tendrils, bunches, grapes, and pips, all of which are the work of nature in her great wisdom? This is just one small example of the innumerable works of nature, and this alone shows infinite wisdom, so we can conclude that divine knowledge is infinitely infinite.

Divine knowledge is infinitely infinite.

The sublime genius of Michelangelo.

SALVIATI. Here's another example. Would we not all agree that Michelangelo's* ability to reveal a beautiful statue in a block of marble has raised his genius to a sublime level far above that of ordinary men? And yet he only imitated just one external, superficial arrangement of the limbs of a motionless man. What is this in comparison with a man made by nature, composed of so many external and internal members, and muscles, tendons, nerves, and bones, which allow so many and such diverse movements? Not to speak of man's senses, his imaginative faculties, and finally his understanding. Surely we can justifiably say that the creation of a statue falls infinitely short of the creation of a living man, or even of the vilest worm?

SAGREDO. Or what difference do we think there was between the dove of Archytas* and one produced by nature?

SIMPLICIO. Either I am not one of those men endowed with understanding, or there is a plain contradiction in your argument. You identify as one of the most praiseworthy attributes of man made by nature, indeed the greatest of all his attributes, this quality of understanding; and yet a moment ago you agreed with Socrates that his understanding was as nothing. So you must conclude that not even nature has understood how to make an intellect which can understand.

SALVIATI. A very telling objection; and to answer it we must make a philosophical distinction between *intensive* and *extensive* understanding. Extensively—that is, in relation to things that can be understood, of which there is an infinite number—human understanding is as nothing, even if we understand a thousand propositions, because a thousand is the same as zero in relation to infinity. But if we speak of understanding intensively—meaning that we understand a proposition perfectly—then it is possible for the human intellect to understand some propositions as perfectly, and with as much certainty, as nature itself. The pure mathematical sciences, geometry and arithmetic, are of this kind; and while the divine intellect knows infinitely more propositions in these sciences, nonetheless I believe that in those few which the human mind can understand, we can achieve the same objective certainty as the divine mind itself. For we can understand that they are necessarily true, and there can be no greater certainty than this.

Man understands much intensively, but little extensively.

[129]

SIMPLICIO. This strikes me as a very bold and daring statement.

SALVIATI. These are generally accepted propositions* which are not bold or presumptuous at all, nor do they detract in any way from the majesty of divine wisdom, any more than it diminishes God's omnipotence to say that He cannot undo what has been done. But I suspect, Simplicio, that you may have taken my words amiss because you have misunderstood them; so let me try to explain myself better. The truth which we arrive at by means of mathematical proofs is the same truth that is known to divine wisdom; but I readily concede that the way in which God knows the infinite number of propositions of which we know only a few, is vastly superior to ours. We proceed by arguing logically from one conclusion to another, whereas God's knowledge is a simple intuition. So, for example, we gain an understanding of some of the properties of the circle, of which there are an infinite number, by starting from one of the simplest and, taking this as our definition, argue step by step to the second, the third, the fourth and so on; but the divine mind comprehends them all instantly, in their infinity, by a simple act of understanding. These properties are in any case present implicitly in the definitions of all things, and perhaps, being infinite, they are ultimately all one in their essence and in the divine mind. Indeed, this is not entirely unknown to the human mind, but we perceive it through a deep and dense fog, which may be partly dispersed and penetrated when we have mastered some firmly-established conclusions which we grasp so readily that we can run over them very quickly. So, for example, to say that the square on the hypotenuse of a right-angled triangle is equal to the sum of the squares on the other two sides is the same as saying that parallelograms on a common base and between parallel lines are equal; and this is ultimately the same as saying that two plane figures which coincide exactly when they are superimposed are equal. These deductions which our [130] intellect makes step by step in time and space, the divine intellect makes instantaneously, with the speed of light, which is the same as saying that they are all constantly present in the divine mind. I conclude, therefore, that both in its manner of understanding and in the number of things it understands, our intellect is infinitely surpassed by the divine; but I do not therefore despise human understanding or consider it to be

God's way of understanding is different from that of men.

Human understanding is gained by reasoning.

Definitions contain implicitly all the properties of the things defined.

The infinite number of properties may perhaps be only one.

Deductions which human reason makes in time, the divine intellect makes instantaneously, which is to say that they are always present in the divine intellect.

worthless. Indeed, when I consider how many and how marvellous are the things which men have been able to do and understand, I perceive all too clearly that the human mind is among the most excellent works of God.

SAGREDO. As I have reflected on these matters, I have often thought how acute the human mind is. As I consider all the many and marvellous discoveries which men have made, both in the arts and in letters, and then think of what I myself know, and how far I am from even learning everything that has been discovered, much less discovering anything new, I am filled with bewilderment and despair, and count myself little more than a wretch. I look at one of these excellent statues, and I say to myself, 'When could you ever remove the excess from a block of marble to reveal such a beautiful figure concealed within it? Or mix and paint different colours on a canvas or a wall, and with them depict every kind of visible object, like a Michelangelo, a Raphael, or a Titian?' If I consider what composers have discovered about the distribution of musical intervals, and how they have established precepts and rules for arranging them to produce delightful sounds for the ear, how can I fail to be amazed? And what am I to say about the number and diversity of musical instruments? Can anyone read the most excellent poets attentively and not wonder at the inventiveness of their conceits and how they are developed? What are we to say of architecture, or the art of navigation?

Human genius is wonderful in its acuteness.

But perhaps the most stupendous inventor of all was the one who dreamt of finding a way to communicate his innermost thoughts to all, however far away they might be in time and place; of speaking to those who are in the Indies, or to those who are not yet born and who will not be born for another thousand or ten thousand years. And how? By means of varied arrangements of twenty little characters on a sheet of paper. So let this set the seal on all humankind's marvellous inventions, and let it [131] be the conclusion to our discussions for today. The hottest hours of the day have now passed, and I think Salviati may like to enjoy the cool of the evening in a gondola; and I shall expect you both to continue our discussion tomorrow.

The invention of writing is the most stupendous of all.

[132] SECOND DAY

SALVIATI. We digressed so often and so far from the main theme of our discussions yesterday that I think I will need your help to find the thread again before we can proceed.

SAGREDO. I'm not surprised you are confused, since your memory is crowded and encumbered with everything we've said as well as with the things that still remain to be said. But since I am simply a listener and only have to retain what I've already heard, perhaps I can recapitulate them briefly and so put our discourse back on the right track.

As I recall, then, the gist of our discussions yesterday was to examine from first principles which of two opinions was more plausible and reasonable. On one hand, there is the view that the substance of the heavenly bodies is ingenerable, incorruptible, unchangeable, impassible, and in short immune from any kind of change, apart from change of location; and that therefore it is a fifth essence,* totally different from the essence of our terrestrial bodies, which are subject to generation, corruption, alteration, etc. The other view is that the universe has no such difference between its parts, and that the Earth enjoys the same perfection as its other integral bodies; in short that it is a wandering globe just as much as the Moon, Jupiter, Venus, or any other planet. Lastly we established many comparisons between the Earth and the Moon—more with the Moon than any other planet, perhaps because our senses allow us to have more knowledge about the Moon as it is not so far away from us. Having concluded in the end that this latter view was the more probable of the two, we should now, I think, go on to con- [133] sider whether the Earth should be regarded as immovable, as the majority view has held up to now, or mobile, as some ancient philosophers believed and as others in recent times assert; and if it is mobile, what the nature of its motion might be.

SALVIATI. Yes, now I remember and recognize the path we have taken. But before we go on I must say something in response to your last remark. You said we had concluded that it was more plausible to regard the Earth as being endowed with the same qualities as the celestial bodies than to take the contrary view. I haven't reached any such conclusion, just as I don't intend to come to a conclusion about any other of these controversial propositions. My purpose has been simply to put forward the arguments and responses, the objections and solutions for both points of view, which have so far been put forward by others, together with some other considerations which have occurred to me as I have thought about them at length; and to leave it to the judgement of others to decide.

SAGREDO. I let myself be carried away by my own sentiments. I thought that others must have the same feeling as I did, and so I claimed as universal a conclusion which I should have made clear was only my own. In this I was wrong, especially as I don't know the view of Simplicio who is here with us.

SIMPLICIO. I have to confess that all last night I have been turning over in my mind what we discussed yesterday, and indeed I find many of the arguments new and compelling. And yet I am held back even more by the authority of so many great writers, in particular . . . You're shaking your head, Sagredo, and smiling as if I had said something quite outrageous.

SAGREDO. I may only be smiling, but believe me, it's all I can do to stop myself from laughing out loud, because you've reminded me of a splendid occasion that I witnessed a few years ago, together with some of my noble friends, whom indeed I could name to you.

SALVIATI. You'd better tell us what it was, otherwise Simplicio might go on thinking you were laughing at him.

SAGREDO. Gladly. One day I was at the house of a highly esteemed doctor in Venice* where people sometimes gathered, in the course of their studies or simply out of curiosity, to witness anatomical dissections by an anatomist who was as learned as he was careful and skilful. On this occasion they were investigating the source and origin of the nerves, a topic on [134] which there is a well-known controversy between the Galenist doctors and the Aristotelians.* The anatomist showed how the great mass of nerves, starting in the brain and passing down the neck, extends down the spinal column and branches out to every part of the body, and that only a single strand, as fine as a thread, reached the heart. Turning to a gentleman whom he

The ridiculous reply of a philosopher in an investigation of the origin of the nerves.

knew to be an Aristotelian philosopher,* and for whose benefit he had taken great pains to expose and demonstrate all this, he asked him if he was satisfied and convinced that the nerves originate in the brain and not the heart. To which the philosopher replied, after some reflection, 'You have made this so clear and evident to the senses that I would be compelled to admit its truth, were it not for the fact that Aristotle's text states the contrary.'

Views of Aristotle and of medical doctors on the origin of the nerves.

SIMPLICIO. I'll have you know, gentlemen, that this debate on the origin of the nerves is by no means as conclusively settled as some people think.

SAGREDO. It never will be, as long as there are men like that to contradict it. But what you say doesn't make the Aristotelian's reply any less outrageous, since he didn't offer any experiments or arguments from Aristotle to counter this demonstration of the evidence of the senses, but simply his authority, his *ipse dixit*: 'Aristotle says so'.

SIMPLICIO. Aristotle gained his great authority because of the strength of his proofs and the profundity of his arguments. But you have to learn to understand him—and not just to understand him, but to become so familiar with his works that you have a perfect grasp of his thought, so that all his sayings are constantly present in your mind. He didn't write for the common crowd, or undertake to string his syllogisms together by the trivial ordinary method; rather he made use of the perturbed method,* and sometimes placed the proof of a proposition in among texts that seem to be about quite different matters. So you need to grasp the whole grand scheme of his thought, and be able to combine one text with another, sometimes bringing together passages which are far removed from each other. There's no doubt that anyone who has this skill will be able to derive from his books the proof of all that can be known, because everything is contained in them.

Requirements for the proper practice of philosophy based on Aristotle.

SAGREDO. Well, my dear Simplicio, since you don't mind things being scattered far and wide and you believe you can extract the meaning from a variety of fragments by juxtaposing them and combining them together, I'll use the verses of Virgil [135] and Ovid to do what you and other skilled philosophers do with the texts of Aristotle: I'll put them together into a patchwork of

A witty device for learning philosophy from any book one chooses.

quotations so as to explain the whole of human affairs and the secrets of nature. Or for that matter, why look as far as Virgil or any other poet? I have a little book here that's much shorter than Aristotle or Ovid, which contains all the sciences and which can be perfectly mastered with a minimum of study. It's called the alphabet, and anyone who can form the right combinations of this or that vowel with these or those consonants will undoubtedly be able to draw out from it the correct answer to any doubtful question, or the instruction needed in all the sciences and all the arts. In the same way a painter uses the different simple colours which he has separately on his palette, taking a bit of this colour and putting it alongside a bit of that and a bit of another to represent men, plants, buildings, birds, fishes—in fact any kind of visible object. But his palette doesn't have any eyes, or feathers, or scales, or leaves, or stones on it; in fact, none of the objects to be depicted, or any part of them, should be found among the colours, since they have to be able to represent anything at all, whereas if the painter did have, say, feathers, they could only be used to depict birds or plumes.

SALVIATI. There are some gentlemen still alive who were present when a doctor who taught at a famous university, on hearing a description of the telescope which he had never seen, said that its invention was derived from Aristotle.* Calling for a copy of the text, he found where it explains why it is possible to see the stars in the sky in the daytime from the bottom of a very dark well. 'Here', he said to those standing around, 'is the well, which represents the tube; here are the thick vapours, from which is taken the invention of the lenses; and here, finally, is how the sight is strengthened when the rays pass through a denser and darker transparent medium.'

The invention of the telescope derived from Aristotle.

SAGREDO. Saying that every kind of knowledge is 'contained' in this way is similar to the way a block of marble contains within it a beautiful statue, or rather a thousand beautiful statues: the key is being able to reveal them. Or it's like the prophecies of Joachim* or the replies given by the pagan oracles, which can only be understood after the events they prophesy have happened.

SALVIATI. And what about the predictions of astrologers, which are so clear to see in their horoscopes, or rather in the disposition of the heavens, after the event? [136]

Alchemists interpret the fables of poets as secrets for making gold.

SAGREDO. It's the same with alchemists. Guided by their melancholy humour,* they find that all the world's greatest geniuses really wrote about just one thing, which was how to make gold; but that they used the most extravagant ways of veiling it under the appearance of various other subjects, so as not to reveal it to the uninitiated. It's endlessly entertaining to listen to their commentaries on the ancient poets and to see how they find the most profound mysteries concealed beneath their fables, and the meanings they attribute to the amorous adventures of the Moon goddess, her descent to Earth for Endymion and her anger at Actaeon, or to Jove changing himself into a shower of gold or into burning flames. And what great secrets of their art they find in Mercury as the interpreter of the Gods, or in Pluto's abductions, or in the golden bough!

SIMPLICIO. I believe there is indeed no lack of people with weird ideas in the world. In fact I know some of them; but their follies shouldn't be held against Aristotle, whom I think you sometimes treat with too little respect. His antiquity alone, and the great reputation he has gained among many distinguished men, should make him someone to be reckoned with by all educated people.

Some followers of Aristotle diminish his reputation by trying too hard to enhance it.

SALVIATI. That's not how things stand, Simplicio. Some of his excessively timid followers diminish our esteem for him, or they would if we took their shallow contributions seriously. Surely you are not so naïve that you can't see that, if Aristotle had been there when the doctor tried to present him as the inventor of the telescope, he would have been far more indignant with the doctor than with those who laughed at his claims? Can you doubt that if Aristotle had seen the new discoveries in the heavens he would have changed his opinion and amended his books to reflect the evidence of his senses? He would surely have rejected those who are so feeble-minded that they timidly feel obliged to defend his every word. They don't realize that Aristotle as they imagine him would have been someone with an unyielding, stubborn mind, a boorish spirit, and a tyrannical will—someone who regarded everyone else as so many silly sheep, and would have insisted that his dictates should take precedence over the senses, over [137] experience, and over nature itself. Aristotle's authority has been wished on him by his followers, not something that he usurped or claimed for himself. Since it's easier to hide behind someone else's shield than to stand up for oneself, they are too scared to take a single step on their own, and they would rather deny what they see in the heavens of nature than change anything in the heavens of Aristotle.

SAGREDO. They remind me of the sculptor who carved a great block of marble into an image of Hercules, or Jove with his thunderbolt—I don't remember which—and such was his skill in making it lifelike and fierce that it inspired fear in everyone who looked at it. He ended up being afraid of it himself, even though all its vivacity and ferocity were his own handiwork, and such was his terror that he no longer dared to approach it with his hammer and chisels.

The ridiculous case of a certain sculptor.

SALVIATI. I've often been amazed how these scrupulous defenders of every saying of Aristotle's fail to realize how much harm they do to his credibility and reputation, and how far their attempts to enhance his authority actually detract from it. When I see them stubbornly upholding propositions which are self-evidently false and trying to persuade me that this is what a true philosopher should do and what Aristotle himself would do, this greatly diminishes my confidence in their reasoning on other conclusions which I don't know about. But if I saw them concede a point and change their view in the face of manifest truths, then I would believe that when they did stand firm they had good grounds for doing so which were outside my knowledge or competence.

SAGREDO. On the other hand, if they thought they were damaging their own and Aristotle's reputation by confessing their ignorance of this or that conclusion discovered by someone else, wouldn't they do better to discover it in his writings by bringing different texts together, using the method Simplicio has described to us? Because if all knowledge is contained in his works, then it must surely be possible to find it there.

SALVIATI. Sagredo, don't make fun of this procedure, because I detect a note of sarcasm in your advocacy of it. Not long ago a well-known philosopher* wrote a book on the soul, in which he cited many texts to illustrate Aristotle's view on whether or not it is immortal. But he didn't give any of the texts quoted by Alexander,* because he said that in them Aristotle

[138] wasn't even writing about the soul, much less stating any conclusion about it; instead he gave others which he had found in various out-of-the-way places, which tended in a dangerous direction. When a friend pointed out to him that he would have trouble getting permission for the book to be published, he replied that that was no reason not to press ahead with it, because if that was the only obstacle he would have no difficulty in changing Aristotle's teaching, and finding and expounding other texts to support the contrary view and to show that it too reflected the mind of Aristotle.

An opportune decision by a Peripatetic philosopher.

SAGREDO. Now there's a doctor I can follow—one who isn't bamboozled by Aristotle but is prepared to lead him by the nose and make him say what serves his purpose! That shows how important it is to seize the moment, and confront Hercules when he's telling stories to Omphale's maids, not when his fury is aroused.* But to be willingly subservient, to accept judgements as unchallengeable, to declare oneself convinced by arguments so 'powerful' and 'conclusive' that they can't decide whether they prove their point or not or even whether they refer to the matter in hand—this is just abject intellectual cowardice. The crowning absurdity is that they themselves don't know whether their author was in favour of their view or against it. This is tantamount to setting up a wooden statue as an oracle and running to it for a response, and to fearing, revering, even worshipping it.

Faint-heartedness of some followers of Aristotle.

SIMPLICIO. But if you abandon Aristotle, who is going to be our guide in philosophy? Perhaps you'd like to suggest someone.

SALVIATI. We need a guide in wild and unknown territory, but in flat open country only the blind need an escort. Such people would do better to stay at home, but anyone who has eyes in their head and an open mind can act as a guide. I'm not saying that we shouldn't pay attention to Aristotle; on the contrary, I think it's right to read him and study him carefully. My criticism is of those who submit to him so that they blindly assent to everything he says and take his judgements as unchallengeable, without seeking any other reason for doing so. This abuse leads to another harmful practice, which is that people no longer apply themselves to understanding the validity of Aristotle's proofs. There is nothing more shameful in a public

It is wrong to adhere too closely to Aristotle.

Those who never practise philosophy should not claim the title of philosopher.

debate on demonstrable proofs than to hear someone interrupt [139] with a text, often written for a quite different purpose, in order to silence their opponent. If you want to go on studying in this way, you should stop claiming the name of philosophers and call yourselves historians, or practitioners of memory, because those who never practise philosophy have no right to usurp the honourable title of philosopher.

But we must return to the shore, before we find ourselves in a boundless sea from which we wouldn't emerge all day. So, Simplicio, please bring forward your arguments and proofs—your own or Aristotle's, but not just texts and bare authorities, because our discussions concern the world of the senses, not a world of paper. In our discussion yesterday we brought the Earth out of darkness and exposed it to the open heavens, and we showed that it was not a wholly unreasonable or hopeless proposition to count it among what we call the heavenly bodies. It follows that we should now examine how plausible it is to consider the Earth—that is, the globe in its entirety—as wholly fixed and immobile, or what probability there may be in considering it mobile, and if so, with what kind of motion. Since I am undecided on this point, and since Simplicio is with Aristotle in being firmly on the side of immobility, let him explain step by step the reasons for their position, and I will give the responses and arguments for the opposite view; and Sagredo will give us his reactions and tell us to which side he finds himself drawn.

SAGREDO. I'm very happy to do so, on condition that I remain free to put forward whatever seems to me to be dictated by plain simple reasoning.

SALVIATI. I would particularly ask you to do so. I think there are very few of the easier and, so to speak, material considerations that writers have not already discussed, so that only some of the more subtle and elusive points remain; and who better than Sagredo, with his exceptionally acute and perceptive intelligence, to investigate these?

SAGREDO. Describe me as you like, Salviati, but please let us not be diverted into elaborate courtesies; I'm here as a philosopher, in the classroom and not in the piazza.

SALVIATI. We should begin our investigation, then, by noting that any motion that is attributed to the Earth will

The motions of the Earth are imperceptible to its inhabitants.

necessarily be imperceptible to us who inhabit it, since it is a motion that we share, and as long as we look only at what is on [140] the Earth it will be as if it did not exist. On the other hand, such a motion will necessarily appear to us to be common to all other bodies and visible objects which are separate from the Earth and so do not share it. Hence, the correct way to investigate whether any motion can be attributed to the Earth and, if so, what kind of motion it is, is to observe whether there is any apparent motion in bodies separate from the Earth which is common to all of them—for a motion which was apparent only in, say, the Moon, without affecting Venus or Jupiter or any other stars, could clearly not belong to the Earth or indeed to any body except the Moon. Now there is a motion which is the most general and supreme over all, and that is the motion whereby the Sun, the Moon, the other planets and the fixed stars—in short, the whole universe, with the sole exception of the Earth—appear to us to move together from east to west in the space of twenty-four hours. At first sight this motion could equally well belong to the Earth alone, or to the rest of the universe excluding the Earth, because in either case the appearance would be the same. Aristotle and Ptolemy understood this perfectly well, so when they wanted to prove that the Earth was immobile, it was only the diurnal motion that they sought to refute. Aristotle did in fact touch on some other motion which an ancient writer* had attributed to the Earth, which we shall discuss in due course.

The only possible motions of the Earth are those which appear to us to be common to the rest of the universe apart from the Earth.

The diurnal motion is common to the whole universe, apart from the terrestrial globe.

Aristotle and Ptolemy argue against attributing the diurnal motion to the Earth.

SAGREDO. I'm entirely convinced by the necessary conclusion of your argument, but I have a doubt which I can't resolve. Copernicus attributes another motion to the Earth besides the diurnal one, and according to the rule we've just established this ought to be imperceptible to us on the Earth but apparent in the rest of the universe. So I can't escape the conclusion, either that Copernicus was manifestly in error when he assigned a motion to the Earth which has no corresponding appearance in the heavens, or if such a correspondence does exist, that Ptolemy was equally at fault for not arguing against this motion as he did against the other.

[141] SALVIATI. Your doubt is very reasonable, and when we come to discuss the other motion you will see how much more

sharp-witted and perceptive Copernicus was than Ptolemy. He saw what Ptolemy failed to see, which is the marvellous way this motion is reflected in all the other heavenly bodies. But let's put this to one side for now and return to our first argument. I shall put forward the reasons which seem to weigh in favour of the mobility of the Earth, starting with the most general, and then we shall hear Simplicio's reasons against it.

Why it is more likely that the diurnal motion belongs to the Earth than to the rest of the universe.

First, then, if we consider simply the enormous size of the sphere of the fixed stars, in comparison with the smallness of the terrestrial globe which it contains millions of times over, and then think of the speed at which the celestial sphere has to travel in order to complete a revolution in the space of a day and a night, it's hard to imagine that anyone would think it more reasonable and credible that it's the celestial sphere that moves and the Earth which remains at rest.

SAGREDO. Even if all the effects in nature which follow from such motions were exactly the same in either case, my first and overall impression would still be that it was quite unreasonable to expect the entire universe to be in motion so that the Earth could remain at rest. It would be even more unreasonable than someone who climbed to the top of the cupola of your cathedral in Florence simply to see the view of the city and the surrounding countryside, and then expected the whole scene to revolve around him so as to save him the trouble of turning his head. The advantages of the new theory would have to be many and great indeed to outweigh such an absurdity and to persuade me that this was the more credible alternative. But perhaps Aristotle, Ptolemy, and Simplicio do see such advantages, in which case it would be good for us to hear them; if not, let us acknowledge that no such advantages exist or ever could.

SALVIATI. For all the thought I've given to this, I haven't been able to find that it would make any difference at all. So I've concluded that there can be no difference, and that there is no [142] point in looking any further for one. The point to note is that motion is motion and acts as such in so far as it relates to things which have no part in it; among things which have an equal share in it, motion has no effect and it might as well not exist. Consider the cargo on a ship: it moves when the ship leaves Venice and travels via Corfu, Crete, and Cyprus to Aleppo,

Motion is not perceived among things that are equally moved by it. It is effective only in relation to things which have no part in it.

while Venice, Corfu, Crete, etc., stay where they are and don't move with the ship. But for the bales, chests, and other packages loaded on the ship, in relation to each other and to the ship itself, the motion from Venice to Syria is as nothing and has no effect at all, because it's common to them all and shared equally by them all. If one of the bales on the ship should shift by as much as an inch from the chest next to it, that would be a greater movement in relation to the chest than the two-thousand-mile journey they have all made together.

SIMPLICIO. This is good sound Aristotelian doctrine.

A proposition taken by Aristotle from the ancients, but altered by him.

SALVIATI. I think it's older than Aristotle, and I suspect that when Aristotle took it from a good school of thought* he didn't entirely understand it. In the altered version in which he wrote it down, it has been a source of confusion thanks to those who want to defend his every word. When he wrote that 'everything that moves, moves over some unmoving object' I suspect that this was a confused rendering of 'everything that moves, moves in relation to some unmoving object',* a proposition which raises no difficulties at all, whereas Aristotle's version raises a great many.

SAGREDO. Please let's not lose the thread; carry on with the argument you began.

First argument to prove that the diurnal motion belongs to the Earth.

SALVIATI. It's clear, then, that motion which is common to many moving objects is idle and ineffective as regards the relation between the objects themselves, because nothing changes between them. Such motion produces an effect only in the relation between these moving objects and others which have no part in it, since their relative positions are changed. We have, further, divided the universe into two parts, of which one is necessarily in motion and the other motionless. As far as anything which depends on this motion is concerned, it makes no difference whether it is the Earth alone that moves or everything else in the universe, because the effect of this motion is confined to the relation between the celestial bodies and the Earth, this being the only relation that is changed. Now since nature can produce precisely the same effect whether the Earth alone moves while the rest of the universe is at rest, or the Earth alone is at rest while everything else in the universe shares the same motion; [143] and since it is generally agreed that nature does not work by intervening in many things when it can achieve the same effect by means of a few; is it credible that it should have chosen to set in motion a vast number of enormous bodies, moving at unimaginable speed, to achieve what could have been done by the moderate motion of one single body turning on its own axis?

Nature does not act through many things when it can act through few.

SIMPLICIO. I don't see how this great motion can be considered as nothing as regards the Sun, the Moon, the other planets, and the innumerable array of fixed stars. The Sun moves from one meridian to the other, rising above one horizon and setting below the other and bringing alternately day and night. The Moon, the other planets, and the fixed stars show similar variations. How can you call all these motions 'nothing'?

SALVIATI. None of these variations that you mention is anything except in relation to the Earth. To prove that this is so, imagine that there was no Earth. There would then be no more rising or setting of the Sun or the Moon, no horizons or meridians, no day or night. In short, this motion produces no change whatsoever between the Sun, the Moon, or any other planets or fixed stars; the changes are all in relation to the Earth. All it means, in fact, is that the Sun shines now in China, then in Persia, and then in Egypt, Greece, France, Spain, America, etc., and that the Moon and the other celestial bodies do the same. And this would happen in precisely the same way simply as a result of the terrestrial globe turning on its axis, without disturbing so many other parts of the universe.

The diurnal motion produces no changes among all the celestial bodies; the changes all relate to the Earth.

There is another very great obstacle which redoubles the first, namely this: by attributing this tremendous motion to the heavens, we necessarily have to make it contrary to the individual motions of all the planets and spheres, since no one disputes that they all have their own particular motion, very moderate and pleasing, from west to east. This would involve forcing them to move in the opposite direction, from east to west, in this very rapid diurnal motion; whereas if we assume that the Earth turns on its own axis we eliminate this contrariety of motion, and the west–east motion on its own is enough to accommodate and fully satisfy all the appearances.

Second argument confirming that the diurnal motion belongs to the Earth.

SIMPLICIO. Contrariety of motion isn't a problem, because Aristotle shows that circular motions are not contrary to each other, so it cannot truly be called contrariety.

Circular motions are not contraries, according to Aristotle.

[144] SALVIATI. Does Aristotle show this, or does he simply assert it because it suits his purpose? If contraries are what mutually cancel each other out, as Aristotle himself says, then I can't see how two moving objects that meet on a circular line clash any less than if they meet on a straight line.

SAGREDO. Wait a moment, please. Simplicio, if two knights meet when they are jousting in an open field, or if two naval squadrons or fleets clash and they break up and sink, would you call these encounters contraries?

SIMPLICIO. We can call them contraries, yes.

SAGREDO. How then can there be no contrariety in circular motions? These are motions on the surface of the land or on the sea, which as you know are spherical, so they must be circular motions. There are some circular motions which are not contraries; do you know what they are? When two circles touch on their outer rims and one of them turns, it naturally turns the other in the opposite direction. But if one circle is inside the other and they move in opposite directions, then they must necessarily work against each other.

SALVIATI. Whether you call these contraries or not is just playing with words. What I know is that in reality, it's much simpler and more natural if you can account for everything with a single motion, rather than introducing two which are—let's say opposites, if you don't want to call them contraries. I don't say that introducing a second motion is impossible, or claim to make this the basis of a necessary demonstration; I simply say that a single motion is more probable.

Third argument confirming the same.

Larger orbits take a longer time to complete their revolutions.

The improbability is further increased by a third consideration. It would be a wholly disproportionate disruption of the order which we see in the heavenly bodies whose revolutions are established beyond any possible doubt. This order is such that a larger orbit requires a longer period to complete its revolution, and a smaller orbit a shorter one. Hence, Saturn, which has the largest orbit of all the planets, completes a revolution in thirty years; Jupiter revolves in its smaller orbit in twelve years, and Mars in two. The Moon, with a much smaller orbit, completes a revolution in just a month. The evidence of our senses shows equally clearly that of the four Medicean stars, the one closest to Jupiter makes its revolution in a very short time, about forty-two

Orbital periods of the Medicean planets.

An orbital period of 24 hours attributed to the highest sphere is disproportionate to the periods of the lower spheres.

hours; the next in three and a half days, the third in seven days, and the most distant in sixteen. This harmonious pattern would not be disrupted at all if the revolution in twenty-four hours is [145] assumed to be that of the terrestrial globe on its axis. But if we want to consider the Earth to be at rest, we must pass from the very short orbital period of the Moon to those which are successively longer—to Mars in two years, thence to the larger sphere of Jupiter in twelve years, and beyond that to the still larger sphere of Saturn with its period of thirty years—to pass beyond all these to another sphere which is incomparably larger, and suppose that it completes its revolution in twenty-four hours. And this is to minimize the disruption that would be introduced; if in passing from the sphere of Saturn to that of the fixed stars one were to assume that the latter's size in relation to Saturn is in proportion to its much slower motion, which extends over thousands of years,* then the leap from one to the other would be even more enormous, and we would still have to assume it could complete a revolution in twenty-four hours.

If, on the other hand, we attribute mobility to the Earth, the principle of their orbital periods is perfectly maintained, and we pass from the very slow-moving sphere of Saturn to the fixed stars, which are entirely immobile. This avoids the fourth difficulty which must be admitted if the sphere of the fixed stars is in motion, and that is the huge disparity in the motions of the stars themselves. Some of them would have to be in very rapid motion in vast circles, and others in very slow motion in tiny circles, depending on how near or far they were from the poles. And this presents a further problem, both because the stars whose motion is not in doubt can all be seen to move in very large circles, and also because it doesn't seem a very sound principle to set up bodies to move in circular motion at very great distances from the centre, and then to make them move in very small circles.

Fourth confirming argument: the great disparity in the motions of individual fixed stars, assuming that their sphere is in motion.

What's more—and this is the fifth objection—not only would the size of the orbits, and hence the velocity of motion, of these stars vary enormously from one to another; the same star would vary in its orbit and velocity at different times. The stars which were on the celestial equator two thousand years ago, and therefore moved in very wide circles, are now many

The motions of the fixed stars become faster and slower at different times, if their sphere is in motion.

degrees away from it, so their motion must have become slower and the size of their orbit smaller; and it's quite possible there will come a time when a star that has always been in motion in the past will reach the pole and so will stand still, and then after [146] a period of rest start to move again. But as we know, all the other stars whose motion is not in doubt describe very large circles in their orbits, and never vary from it.

Sixth confirmation.

The improbability is increased still further, on more rigorous examination, by the sixth objection. It is impossible to conceive what ought to be the solidity of that enormous sphere in whose depths so many stars are so firmly held in place, as they are harmoniously carried around with such disparate motions without the slightest change in their relative positions. But if, as seems much more reasonable to suppose, the sphere is not solid but fluid, so that every star moves around in it independently, what law can there be to determine their motions, and to what end, so that they appear from the Earth as if they belonged to a single sphere? It seems to me that such an effect is much more easily and naturally explained by saying that the stars are fixed rather than mobile, just as it's easier to keep in place all the paving stones in a public square than the hordes of children running about on them.

Seventh confirmation.

The seventh and last consideration is that, if we attribute the diurnal revolution to the highest heaven, we must assume that it has such strength and power that it can carry with it the innumerable multitude of the fixed stars, all of them vast bodies and much larger than the Earth, and also all the planetary spheres, even though their natural motion is in the opposite direction. Moreover, we must concede that the element of fire and the greater part of the air are also similarly swept along, so that the little globe of the Earth alone is able to remain resistant and impervious to this power. I find this very difficult to believe, and I can't see how the Earth, a body suspended and balanced on its centre, indifferent to motion or rest, and placed in the midst of a liquid medium, does not also succumb to being swept around. None of these difficulties arise, however, if we posit the motion of the Earth, a small and insignificant body in comparison with the universe, and therefore unable to exert any force on it.

It is not clear how the Earth, being suspended and balanced in a fluid medium, can resist the force of the diurnal motion.

SAGREDO. My head is swimming with a confusion of ideas aroused by the arguments so far. If I'm to apply my mind to what needs to be said next, I must see if I can put them in better order and follow through the thread of the argument, if indeed there is any. It may help me to explain myself more easily if I proceed by asking a series of questions. So first let me ask Simplicio whether he believes that the same simple movable body can have several different natural motions, or whether it has just one that is proper and natural to it.

[147] SIMPLICIO. A simple movable body has one, and only one, motion that belongs to it naturally; it shares any other motion incidentally and by participation. Take someone walking up and down on a ship: walking is their own proper motion, and they share the motion of the ship carrying them towards port by participation; they couldn't reach the port by means of walking, but only if the motion of the ship carries them there.

A simple movable body has only one natural motion; it shares others by participation.

SAGREDO. Tell me then: when a movable body receives a motion by participation, this being different from its own natural motion, does that motion necessarily have to reside in some proper subject of its own, or can it exist in nature without any other support?

SIMPLICIO. Aristotle provides an answer to all these questions. He says that as a movable body has one proper motion, so every motion is proper to one movable body, and therefore it is impossible to conceive of any motion which is not inherent in its proper subject.

Motion cannot exist without a movable subject.

SAGREDO. My third question, then, is this: do you believe that the Moon and the other planets and celestial bodies have their own proper motions, and if so, what are they?

SIMPLICIO. Yes, they do: the motions which they follow as they run through the zodiac—the Moon in a month, the Sun in a year, Mars in two years, and the sphere of the fixed stars in many thousands of years. These are their natural and proper motions.

SAGREDO. So what is the motion I can see when I observe the fixed stars and all the planets moving together from east to west and returning to the east in twenty-four hours? In what way does this belong to them?

SIMPLICIO. It belongs to them by participation.

SAGREDO. In that case it doesn't reside in them; and if it doesn't reside in them, and it can't exist without some subject body to reside in, it must be the proper and natural motion of some other sphere.

SIMPLICIO. That's why astronomers and philosophers have discovered another higher sphere, without stars, to which the diurnal rotation belongs, which they have called the Primum Mobile.* This sphere carries all the lower spheres along with it, sharing its motion with them by participation.

[148] SAGREDO. But why resort to such far-fetched and laboured conditions, bringing in other vast and unknown spheres, other participatory or forced motions, if everything proceeds and responds in perfect harmony simply by allowing each sphere its own one simple motion, without mixing contrary motions but just allowing them all to move in the same direction, as they must if they all depend upon a single principle? Why reject this solution?

SIMPLICIO. The problem is finding this simple, straightforward explanation.

SAGREDO. I think it's staring us in the face. Make the Earth the Primum Mobile—that is, make the Earth turn on its axis in twenty-four hours, in the same direction as all the other spheres. Then, without its motion being shared by participation with any other planet or stars, they will all have their risings, settings, and in a word all their other appearances.

SIMPLICIO. The important thing is making the Earth move without causing a thousand other difficulties.

SALVIATI. All the difficulties will be resolved as you put them forward. So far we have considered only the first and most general grounds which make it not entirely improbable that the daily rotation belongs to the Earth rather than to the rest of the universe. I don't propose these as absolute rules but simply as reasons which have some plausibility. I entirely accept that a single experiment or conclusive proof to the contrary would be enough to knock down this and any number of other probable arguments. So let's not stop here, but proceed to listen to Simplicio's response, and hear what more probable or firmly-established reasons he offers in favour of the contrary view.

A single experiment or conclusive proof prevails over all probable arguments.

SIMPLICIO. I shall start with some general remarks on these arguments as a whole, and then go on to deal with some specific issues. Your whole reasoning seems to be based on the view that it is simpler and easier to produce the same effects by moving the Earth alone rather than the rest of the universe apart from the Earth, given that they could equally well be caused by either. My reply to this is that I would take the same view if I were considering my own strength, which is not just finite but insignificant; but in relation to the power of the Mover, which is infinite, it's no less easy to move the universe than the Earth or for that matter a straw. In fact, if power is infinite, wouldn't we expect a large part of it to be exercised, rather than only a tiny part? So your general argument seems to me to be invalid.

If power is infinite, it seems proper to exercise a large part of it rather than only a little.

[149] SALVIATI. If I'd ever said that the universe does not move because of a lack of power in the Mover, I would have been wrong and I would accept your correction. I agree that it is as easy for an infinite power to move a hundred thousand things as to move just one. But what I said wasn't in relation to the Mover but only to the movable bodies, and not just because of their resistance—which is certainly less in the case of the Earth than for the universe as a whole—but because of all the other points we have been considering. As for saying that it's better to exercise a large part of an infinite power than a small part, I reply that no part of the infinite is larger than any other part, given that they are both finite. You can't say that a hundred thousand is a larger part of infinity than two, even though it's fifty thousand times more. And if the power it would take to move the universe is finite, even though it's incomparably greater than it would take just to move the Earth, it still wouldn't be a larger part of infinity, and the amount of power left unused would still be infinite. So whether it takes a bit more or a bit less power to produce a particular effect is neither here nor there. In any case, the effect of this power isn't limited to just bringing about the diurnal motion. There are many other motions in the universe that we know about, and there may well be many more that we know nothing of.

No part of the infinite is greater than any other part, even if they are unequal.

Having regard, then, to the movable bodies, and given that it is undoubtedly a shorter and more expeditious procedure to move the Earth than the whole universe; and with an eye, also, to the many other short cuts and simplifications that follow from this; we can cite a very true maxim of Aristotle, *Frustra fit per plura quod potest fieri per pauciora*—it is vain to do by many means what can be done by fewer—which teaches us that it is more probable to conclude that the diurnal motion belongs to the Earth alone, than to the rest of the universe apart from the Earth.

SIMPLICIO. You've left out a crucial phrase in your quotation of Aristotle, which makes all the difference in the present context: he adds *aequo bene*. So we must consider whether both assumptions can satisfy us equally well with one assumption in every respect.

SALVIATI. We shall see whether both positions satisfy us equally well when we look in detail at the appearances which must be saved. So far we've been speaking hypothetically, presupposing that both positions are equally effective in saving [150] the appearances. In fact, I suspect that the phrase you say I've left out is rather one which you have superfluously introduced. 'Equally well' establishes a comparison, which necessarily requires at least two terms, since you can't compare something with itself; you can't say, for instance, that rest is equally good as rest. Now when we say 'it is vain to do by many means what can be done by fewer' we understand that this means doing the same thing, not two different things; and since you can't say that the same thing is equally well done as itself, the phrase 'equally well' is superfluous, as there is only one term of comparison.

In the axiom 'Frustra fit per plura', etc., the addition 'aequo bene' is superfluous.

SAGREDO. Please, if we want to avoid a repetition of what happened yesterday, let's get back to the subject, and let Simplicio start to put forward the objections which seem to him to contradict this new ordering of the universe.

SIMPLICIO. It's not a new ordering; in fact it's very ancient, and Aristotle refutes it, his arguments being as follows.* 'First, whether the Earth moves around its axis, if it is in the centre, or else in a circle, if it is not in the centre, such motion must be violent, since this is not its natural motion. For if this was its own motion every part of it would be affected; but every part moves in a straight line towards the centre. Since then this motion is violent and not natural, it cannot be eternal; but the ordering of the universe is eternal; therefore, etc. Second, all other bodies that move in circular

Aristotle's arguments for the Earth being at rest.

motion appear to lag behind and to move with more than one motion, with the exception of the Primum Mobile. Hence the Earth also would necessarily move with two motions, and if this were the case, there would necessarily be changes in the fixed stars. But we see no such changes; on the contrary, the same stars always rise and set in the same places, without any variation. Third, the natural motion of the whole and of the parts is naturally towards the centre of the universe, and for this reason the Earth also is at rest there.' He then goes on to consider the question of whether the natural motion of the parts is towards the centre of the universe or the centre of the Earth, and concludes that their own inclination is towards the centre of the universe, and coincidentally towards the centre of the Earth. We discussed this question at length yesterday.

[151] Finally he confirms this with a fourth argument, based on experiments with heavy objects which, when they are dropped from a height, fall vertically to the Earth's surface. In the same way, projectiles thrown vertically up into the air fall back down perpendicularly by the same line, even if they have been thrown upwards to a great height. This proves conclusively that their motion is towards the centre of the Earth, which waits motionless to receive them.

He refers, lastly, to other reasons which astronomers have given to confirm the same conclusion, namely that the Earth is at the centre of the universe and is at rest. He reproduces just one of these, which is that all the appearances we observe in the motions of the stars agree with the Earth being at the centre, and there would be no such correspondence if it were not in the centre. As for the other reasons given by Ptolemy and other astronomers, I can give you these now, if you wish, or else after you have said whatever you think necessary in response to Aristotle's arguments.

Two kinds of arguments on the question of whether the Earth is in motion or at rest.

SALVIATI. The arguments which are put forward on this matter are of two kinds: those which refer to terrestrial phenomena without reference to the stars, and those drawn from the observed appearances of the heavens. Aristotle mostly draws his arguments from the things around us, leaving the others to astronomers; so if you agree I suggest we start with those based on experiments on Earth, and then move on to the other kind. Since Ptolemy, Tycho, and other astronomers and

Arguments put forward by Ptolemy, Tycho and others, in addition to those of Aristotle.

philosophers have not only confirmed and reinforced Aristotle's arguments but also advanced others of their own, we can consider them all together, so as not to have to give the same or similar responses more than once. So, Simplicio, perhaps you would like to rehearse these arguments yourself, or else let me save you the trouble, whichever you prefer.

SIMPLICIO. It's better if you put them forward; you've studied them more thoroughly and you will have them more readily to hand, and more of them as well.

First argument, based on heavy objects falling downwards.

SALVIATI. The reason which everyone puts forward as the strongest is the one based on the fact that heavy objects fall in a straight line perpendicular to the surface of the Earth. This is considered to be an unassailable argument for the Earth's being at rest, on the grounds that, if it were affected by the diurnal rotation, then a stone dropped from the top of a tower would land hundreds of *braccia* away from its base, because in the time [152] the stone was falling the tower would have been carried by the Earth's rotation a corresponding distance towards the east.

Confirmed by the example of an object falling from the top of a ship's mast.

They confirm this with another experiment, which is to drop a lead ball from the top of the mast of a stationary ship and note the place where it lands, which is close to the foot of the mast; but if the same ball is dropped from the same place when the ship is in motion, it will land as far away from the mast as the ship has travelled in the time that it was falling, for the simple reason that its natural motion when it is released is in a straight line towards the centre of the Earth.

Second argument, based on projectiles fired upwards to a great height.

This argument is further strengthened by an experiment with a projectile fired a very great distance up into the air, such as a ball fired from a cannon pointing straight upwards. The time it takes for the ball to rise and fall is such that, at our latitude, both we and the cannon would be carried by the Earth many miles towards the east, so that the ball could never land close to the cannon but would fall as far to the west as the Earth had run on ahead while it was in the air.

Third argument, based on guns fired towards the east and towards the west.

To this they add a third and very effective experiment, which is as follows. If a ball is fired from a cannon pointing towards the east, and then another ball with the same charge and at the same elevation is fired pointing towards the west, the one towards the west ought to travel much further than the one towards the east. This is because while the ball was travelling west the cannon, carried by the Earth, would be travelling east, and so the distance from the cannon at which the ball falls back to Earth would be the sum of both distances, the distance travelled by the ball itself towards the west and that travelled by the cannon, carried by the Earth, towards the east. Conversely, for the ball fired towards the east you would have to subtract the distance travelled by the cannon following after it. So, for example, if the distance covered by the ball itself was five miles, and the Earth at that latitude travelled three miles in the time that the ball was in flight, then the ball fired towards the west would land eight miles away from the cannon—its own five miles towards the west and the cannon's three miles towards the east. But the shot fired towards the east would land only two miles away from the cannon, this being what remains from the five miles covered by the shot when the three miles in the same direction covered by the cannon are subtracted from it. And yet experiment shows that the shots would be equal; therefore the cannon is at rest, and so too is the Earth.

[153] Moreover, shots fired towards the south and north confirm the stability of the Earth no less than those we have been considering. No shot would ever hit the target it was aimed at, but would always veer towards the west, because the target would have moved eastwards, carried along by the Earth, while the ball was in the air. Nor does this apply only to shots fired along lines of longitude; those fired towards the east or west would be off-target as well, because eastward shots would be too high, and westward shots too low, always assuming they were fired horizontally. This is because in both cases the shots would be fired along the tangent, i.e. in a line parallel to the horizon, whereas if the diurnal motion belongs to the Earth the horizon is continually sinking towards the east and rising in the west—this being why the stars appear to rise in the east and sink in the west. Hence the target in the east would be always sinking below the shot, which would therefore be too high, and the one in the west would be rising and therefore making the shots too low. The result would be that no one could ever shoot accurately in any direction; and since experience shows that the opposite is the case, we must conclude that the Earth is at rest.

The argument confirmed by shots fired towards the north and towards the south.

The same point confirmed by shots towards the east and west.

SIMPLICIO. There can't possibly be any valid response to these arguments.

SALVIATI. Are they new to you, then?

SIMPLICIO. Yes indeed; and I see now how nature has graciously given us all these experiments to help us to recognize the truth. How well one truth agrees with another, and how they all work together so that they are impregnable!

SAGREDO. What a shame there were no cannons in Aristotle's time! He would certainly have used them to defeat ignorance, and wouldn't have hesitated in his pronouncements about the universe.

Those who follow Copernicus do not do so out of ignorance of the arguments to the contrary.

SALVIATI. I'm very glad that these reasons are new to you, because that means you won't share the view of most Aristotelians, who think that if anyone disagrees with Aristotle's teaching it's because they haven't understood or grasped his proofs properly. You will certainly hear other things that are new to you, and what's more you will hear the followers of the new system produce arguments against their own position which are much stronger than those put forward by Aristotle or Ptolemy or any of their other opponents. This will leave you in no doubt that it's not ignorance or lack of experience that has led them to adopt the new theory.

[154]

Lectures given by Christian Wurstisen on the Copernican theory, and their effect.

Those who hold the Copernican theory were all previously opposed to it, but the followers of Aristotle and Ptolemy have never taken the contrary view.

SAGREDO. I must tell you at this point about what happened to me when I first heard about this theory. I was very young and had just finished my course in philosophy, which I then neglected because I devoted myself to other activities. A German from Rostock, whose name I think was Christian Wurstisen,* a follower of the Copernican theory, arrived in these parts and gave two or three lectures on this subject in an academy here. They attracted a large audience, I think more because of the novelty of the subject than for any other reason, but I didn't go because I'd formed a firm impression that this theory could only be sheer madness. When I asked some of those who'd been there about it, they all treated it as a joke except one, who said that it wasn't a ridiculous idea at all. Since this was someone whom I regarded as intelligent and prudent I regretted not having gone, and from then on, whenever I met someone who held the Copernican theory I asked them if they had always taken that view. Without exception, of the large number of people I asked, they all said that they had long been of the opposite opinion, and had adopted the new theory because of the strength of the arguments in its favour. I then questioned them all closely to see how well they were acquainted with the arguments on the other side, and I found that they all had them at their fingertips. So I couldn't truthfully say that they had rushed into this new theory out of ignorance, or from a vain wish to appear as advanced thinkers. In contrast, my curiosity led me to question many of the supporters of Aristotle and Ptolemy as well, and when I asked them how much study they had made of Copernicus's book, I found very few who had even seen it, and I don't think there was a single one who had understood it. And when I tried to find out whether any of the Aristotelians had ever held the contrary opinion, again I didn't find any.

The upshot, then, was that there wasn't a single supporter of the Copernican theory who hadn't originally taken the opposite view and who wasn't extremely well informed of the arguments of Aristotle and Ptolemy, whereas, in contrast, not one of the supporters of Aristotle and Ptolemy had previously followed the Copernican view and abandoned it in favour of Aristotle. In [155] view of this, I came to believe that anyone who abandons a view that they absorbed with their mother's milk and which almost everyone holds, in favour of one which is held by very few, is refuted by all schools of thought and truly seems a tremendous paradox, must indeed have been moved, not to say compelled, by more convincing arguments. As a result, I have become very eager to get to the bottom of this matter, and I count it my great good fortune to have met you both. I expect to hear from you all that has been said, and perhaps all that can be said, on this question, and I'm confident that by listening to your discussion my doubts will be resolved and I will be able to reach a trustworthy conclusion.

SIMPLICIO. As long as your hopes aren't disappointed, and you don't end up even more confused than you were before.

SAGREDO. I'm quite certain that won't happen.

SIMPLICIO. How can you be so sure? Speaking for myself, the further we go in this discussion the more confused I become.

SAGREDO. That shows that the arguments you used to find conclusive, and which gave you confidence that your view was the right one, are beginning to look different to you now, and are allowing you gradually to be led towards the opposite view—perhaps not to adopt it fully, but to incline in that direction. As for me, I've been undecided up to now, and still am, but I have every confidence that I shall be brought to a position of assured certainty. You will agree with me, I'm sure, if you allow me to tell you the grounds for my hope.

SIMPLICIO. I shall be very glad to hear them, and I would be equally pleased if they had the same effect on me.

SAGREDO. Then be so kind as to answer some questions for me, starting with the conclusion we are seeking to understand. This is, is it not, whether we should follow Aristotle and Ptolemy in believing that the Earth alone is fixed at the centre of the universe while all the celestial bodies are in motion, or whether the stellar sphere is at rest, the Sun and not the Earth is at the centre, and the motions which appear to us to be in the Sun and the fixed stars in fact belong to the Earth?

SIMPLICIO. Yes, these are the conclusions we are debating.

SAGREDO. And are not these two conclusions such that one of them must necessarily be true and the other false?

[156] SIMPLICIO. Indeed they are. We are in a dilemma, one side of which must necessarily be true and the other false. Motion and rest are contraries, between which no third alternative is possible; one can't say 'the Earth is not in motion and not at rest' or 'the Sun and stars are not in motion and not at rest'.

SAGREDO. What kinds of things are the Earth, the Sun, and the stars in nature? Are they trifling things or are they important?

SIMPLICIO. They are the most principal, most noble, constitutive bodies in the universe. They are immense, and are hugely important.

SAGREDO. And motion and rest: what kind of events are they in nature?

Motion and rest the principal events in nature.

SIMPLICIO. They are such great and principal events that nature itself is defined by them.

SAGREDO. Being eternally in motion or wholly at rest are, then, two very important conditions in nature, and they indicate fundamental differences, especially when applied to the most principal bodies in the universe; so only the most radically different results can follow from them.

SIMPLICIO. That's undoubtedly the case.

SAGREDO. My next question is this: do you think that any arguments in dialectics, rhetoric, physics, metaphysics, mathematics, or in any other form of discourse, are so strong and demonstrable that they can persuade anyone of false conclusions just as much as of true ones?

SIMPLICIO. Certainly not. On the contrary, I'm quite convinced that a true and necessary conclusion is confirmed by not one but many proofs in nature, and that however much you discuss it and approach it from every possible angle, you will never run into any contradiction; and the more anyone tries to obscure it with sophistry, the more clearly its truth will stand out. But any attempt to make a false proposition appear true has to rely on fallacies, sophisms, non-sequiturs and empty verbiage, full of inconsistencies and contradictions.

Falsehoods cannot be proved like truths.

There can be many conclusive proofs of true propositions, but not of false ones.

SAGREDO. We have established, then, that being eternally in motion or eternally at rest are fundamental conditions in nature, and are so different that the consequences of each must diverge radically, especially when applied to the Sun and the Earth, which are such vast and significant bodies in the universe. We have established, further, that of two contradictory statements one must necessarily be true and the other false, and that the false can only be supported by fallacies, whereas the true can be proved by every kind of conclusive and demonstrative argument. How then can whichever of you has undertaken to uphold the true position fail to convince me? I would have to be dim-witted, warped in judgement, intellectually stubborn and blind to reasoned argument, not to distinguish light from darkness, gems from coal, truth from falsehood.

[157]

SIMPLICIO. I repeat what I've said before, that the best teacher for learning how to recognize sophisms, non-sequiturs and other fallacies is Aristotle. He could never be deceived in such matters.

SAGREDO. You insist on Aristotle, who can't speak for himself; but I say that if Aristotle were here, he would either be persuaded by us or expose the errors in our arguments and convince us with better ones. Didn't you yourself, when Salviati was describing the experiments with cannon shots, express your admiration for them and admit that you found them more conclusive proofs than Aristotle's? And yet, even though he put

Aristotle would either expose the errors in the opposing arguments or change his view.

them forward himself and has undoubtedly examined and scrutinized them carefully, I haven't heard Salviati say that he was persuaded by them, or for that matter by other even more effective arguments that he says he will expound to us. Why do you want to reproach nature, as if in her dotage she could no longer produce great intellects, but only servants of Aristotle who think with his mind and perceive with his senses? But let's hear the other arguments in favour of Aristotle's view, before we come to put them to the test, refining and weighing them in the assayer's balance.

SALVIATI. Before we go any further I must make it clear to Sagredo that I am playing the part of a Copernican in our discussions, and I wouldn't want you to judge my personal view of the arguments I put forward in his favour on the basis of what [158] I say while we're acting out our play. When we put our roles aside you may well find me different from how I appeared on the stage. That said, we can continue.

Argument based on clouds and birds.

Ptolemy and his followers cite another example similar to the experiment with projectiles. They refer to things which remain in the air for a long time, separated from the Earth, such as clouds and birds in flight. Since these are not in contact with the Earth and so can't be said to be carried along by it, it seems impossible that they could keep up with the speed of its motion; in fact they all ought to appear to us to be moving very rapidly westwards. And if we, carried along by the Earth, complete a revolution in twenty-four hours, which at our latitude is at least sixteen thousand miles, how can the birds possibly keep up with us? Yet we see them flying eastwards, westwards or in any other direction with no detectable difference at all. What's more, if we feel the air blowing strongly in our face when we ride on horseback, what kind of wind ought we constantly to feel from the east, since we are so rapidly carried in the opposite direction? And yet we feel no such effect.

Argument based on the wind we feel when we ride on horseback.

Argument based on spinning, which has the effect of extruding and dispersing.

Another very ingenious argument, based on experience, is as follows. Circular motion has the effect of extruding, dispersing, and forcing the parts of a moving body away from its centre, unless the motion is very slow or the parts are very firmly fixed together. So if we were, for example, to spin very fast one of those great treadmills turned by one or two men walking inside them in order to move very heavy weights, such as the massive stones in a press or fully-loaded boats that they drag overland from one body of water to another, then unless the parts of this rapidly turning wheel were very tightly bound together they would all fly off. And any stones or other heavy objects attached to its outer rim, however firmly, would be unable to resist this force which would fling them violently away from the wheel, and consequently away from its centre. Now if the Earth were in motion with so much greater velocity, what weight, what tenacity of lime or mortar, would be able to prevent rocks, buildings, and whole cities from being hurled into the sky by such a rapid spinning motion? How would men and animals, which aren't attached to the Earth in any way, be able to resist this force? And yet we see that on the contrary, not just these but things with much less resistance—small stones, sand, and leaves—rest quietly on the ground, or when they fall back down to earth, do so very slowly.

These, Simplicio, are the very powerful arguments taken [159] from what we might call terrestrial things. There remain those of the other kind, relating to celestial appearances, which it must be said tend more towards proving that the Earth is at the centre, and therefore to deprive it of the annual motion around the centre which Copernicus attributes to it. Since the subject matter of these is rather different, we can consider them after we have examined the validity of those we've put forward so far.

SAGREDO. What do you think, Simplicio? Has Salviati done justice to Ptolemy's and Aristotle's arguments? Do you think any Peripatetic philosopher has such a good grasp of the Copernican proofs?

SIMPLICIO. If it wasn't for the high opinion I have formed from our discussions so far of your learning, Salviati, and your perceptiveness, Sagredo, I would ask you to excuse me now, without hearing another word. I don't see how anyone can argue against such tangible evidence, and I'm inclined to reaffirm my long-held view without any further discussion. Even if it were wrong, the fact that it's supported by such convincing reasons would make it excusable. If these are fallacies, what demonstrations of the truth could possibly be finer?

SAGREDO. I think we should hear Salviati's responses. If they are true, they must also be finer—infinitely so, and the opposing arguments correspondingly ugly, if we are to believe the metaphysical principle that beauty and truth are one and the same, and similarly ugliness and falsehood. So, Salviati, let's not lose any more time.

Truth and beauty are one and the same, as are also falsehood and ugliness.

SALVIATI. The first argument put forward by Simplicio, if I remember rightly, was the following: the Earth cannot move with circular motion because such motion would be forced, and therefore not perpetual. It would be forced because, if it were natural, its parts would also all move naturally in a circle, which is impossible because the natural motion of the parts is downward motion in a straight line. My response to this is to say that I wish Aristotle had expressed himself more clearly when he said 'its parts would also move naturally in a circle', because this circular motion can be understood in two ways. It [160] could mean that every particle separated from the whole would move in a circle around its own centre, each describing its own tiny circles; or it could mean that since the whole globe revolves on its axis in twenty-four hours, then its parts also revolve around the same axis in twenty-four hours. The first would be nonsense, equivalent to saying that every point on the circumference of a circle must itself be a circle, or that because the Earth is a globe then every part of it must also be spherical, in keeping with the axiom that *Eadem est ratio totius et partium*: the same reasoning applies to the whole and to the parts. But if he meant it in the second sense, i.e. that the parts imitate the whole, moving naturally around the centre of the globe in twenty-four hours, I say that this is exactly what they do, and that it's up to you, speaking for Aristotle, to prove that they do not.

Response to Aristotle's first argument.

SIMPLICIO. Aristotle proves this in this same text. He says that the natural motion of the parts is rectilinear motion towards the centre of the universe, and that therefore circular motion cannot be natural to them.

SALVIATI. Don't you see that these same words contain the rebuttal of this response?

SIMPLICIO. How? Where?

SALVIATI. Doesn't he say that circular motion for the Earth would be forced, and therefore not eternal; and that this is absurd, because the ordering of the universe is eternal?

SIMPLICIO. Yes, that's what he says.

SALVIATI. If what is forced cannot be eternal, then conversely what is not eternal cannot be natural. But the downward motion of the Earth cannot possibly be eternal, and therefore cannot be natural, since no motion can be natural that is not also eternal. If, on the other hand, we say that the Earth moves with circular motion, this motion can be eternal for both the whole and the parts, and therefore natural.

What is forced cannot be eternal, and what cannot be eternal cannot be natural.

SIMPLICIO. The natural motion for the parts of the Earth is rectilinear motion. This motion is eternal, and the parts can never fail to move in a straight line, always assuming that any impediments are removed.

SALVIATI. You are in some confusion, Simplicio, so let me try to resolve it for you. Consider this: do you think that if a ship were sailing from Gibraltar towards Palestine, it could sail eternally towards the Palestinian coast, always following the same course?

SIMPLICIO. Certainly not. [161]

SALVIATI. And why not?

SIMPLICIO. Because such a voyage is confined by the Pillars of Hercules at one end and the coast of Palestine at the other, and since this is a fixed distance it must be travelled in a finite period of time. The ship could then turn back and make the same journey in reverse, but this would be interrupted and not continuous motion.

SALVIATI. You are absolutely right. What if the ship sailed through the Magellan strait into the Pacific, then to the East Indies, around the Cape of Good Hope and then through the Magellan strait into the Pacific again: could it then carry on perpetually?

SIMPLICIO. Yes, it could, because this would be a circular voyage turning back on itself; so if it was repeated an infinite number of times it could continue perpetually, without any interruption.

SALVIATI. So a ship making this journey could continue to sail for eternity.

SIMPLICIO. It could, provided the ship was indestructible. If the ship broke up, then its journey would necessarily come to an end.

Two conditions are necessary for motion to be perpetual: unlimited space, and a mobile body that is incorruptible.

SALVIATI. Whereas in the Mediterranean, even if the ship was indestructible it couldn't sail perpetually towards Palestine, since this is a journey of fixed length. It follows then that two conditions are necessary for a movable body to be in motion eternally without interruption: one, that the motion is naturally unlimited and infinite; and the other, that the movable body itself is similarly incorruptible and eternal.

SIMPLICIO. Yes, these conditions are necessary.

Rectilinear motion cannot be eternal, and therefore cannot be natural to the Earth.

SALVIATI. In that case, you yourself have admitted that it is impossible for any mobile body to move eternally in a straight line, since you say that motion in a straight line, whether upwards or downwards, is bounded by the circumference and the centre. So even if the movable body, in this case the Earth, is eternal, rectilinear motion cannot be natural to it because such motion is not eternal but has definite limits. Indeed we saw yesterday that Aristotle himself had to say that the terrestrial globe is eternally at rest.

You are completely mistaken when you say that the parts of the Earth will always move downwards provided any impediments are removed. On the contrary, they have to be impeded, opposed, and forced in order to make them move, because once [162] they have fallen they must be forcibly thrown upwards again before they can fall a second time. Any impediments simply prevent them from reaching the centre of the Earth; if there were a hole deep enough to go beyond the centre, a clod of earth falling down it would only pass the centre by as much as its impetus carried it, before falling back towards the centre and finally coming to rest there. So give up any idea of maintaining that rectilinear motion suits or could naturally suit the Earth or any other movable body, while the universe remains in its perfectly ordered state. If you won't grant that the Earth has a circular motion, then concentrate your efforts on upholding and defending its immobility.

SIMPLICIO. Aristotle's arguments for the Earth's immobility, and even more the additional arguments you brought forward, seem to me to be entirely conclusive so far, and I think it will take a great deal to refute them.

SALVIATI. In that case, let's come to the second argument. This was that the bodies of whose circular motion we are certain have more than one motion, with the exception of the Primum

Mobile, and that therefore if the Earth had a circular motion it would also have two distinct motions. If this were the case, it would produce changes in the rising and setting of the fixed stars; but we see no such change; therefore, etc. The response to this is in the argument itself, and it couldn't be simpler or more to the point. Aristotle himself puts it into words for us, and I'm quite sure, Simplicio, that you have already seen it.

Response to the second argument.

SIMPLICIO. No, I haven't seen it, and I don't see it now.

SALVIATI. Surely you have; it's as clear as can be.

SIMPLICIO. If you don't mind, I'll have a look at the text.

SALVIATI. Of course; we'll have a copy brought straight away.

SIMPLICIO. I always carry a copy in my pocket. Here it is, and I know exactly the relevant place: *On the Heavens*, book 2, chapter 14. Here we are: paragraph 97:* 'All things that move with circular motion, except for the first sphere, are seen to be passed and to move with more than one motion. Therefore the Earth also, whether it moves about the centre or is placed at the centre, must necessarily move with two motions. If this were the case, there would necessarily be mutations and changes in the fixed stars; but no such changes appear, but rather the same stars always rise and set in the same place.' I don't see any flaw in this argument, which seems to me to be entirely conclusive.

SALVIATI. To me, hearing you read it again not only confirms the flaw in the argument but reveals another error besides. Consider this. There are two views, or conclusions, that Aristotle seeks to rebut here. One is the view that the Earth is placed at the centre and turns on its own axis; the other is that it is far removed from the centre and moves around it in circular motion. He uses the same argument to rebut both positions, and I find him to be in error in both cases: in the first because of faulty logic or paralogism, and in the second by drawing a false conclusion. [163]

Aristotle's argument against the motion of the Earth errs in two ways.

Let's start with the first position, which states that the Earth is at the centre and turns on its own axis, and Aristotle's criticism of it. He argues thus: all bodies which move in circular motion seem to lag behind, and move with more than one motion, with the exception of the first sphere or Primum Mobile. Therefore the Earth, if it is placed at the centre and turns on its own axis, must also move with two motions and fall behind. But if this happened, there would have to be some variation in the rising and setting of the fixed stars, but we observe no such thing; therefore the Earth does not move, etc. To expose the faulty logic here, I argue with Aristotle as follows: you say, Aristotle, that if the Earth is placed at the centre it cannot move upon itself, because we would have to attribute two motions to it. If, therefore, it were sufficient to attribute just one motion to it, you wouldn't consider this one motion to be impossible, because there would be no point in arguing for impossibility on the grounds of multiple motions if even a single motion was impossible. Now, you state that of all the movable bodies in the universe there is just one that moves with a single motion, while all the others move with more than one; and you say that this one body is the first sphere, the one by which all the fixed stars and planets appear to be moving together from east to west. Now if the Earth could be this first sphere, which with a single motion of its own makes the stars appear to move from east to west, you would not deny it this role. But those who say that the Earth is placed at the centre and [164] turns on itself attribute just such a motion to it, making the stars all appear to move from east to west; so the Earth becomes that first sphere which you yourself allow has just a single motion. Hence, Aristotle, the conclusion of your argument must be either that the Earth, being placed at the centre, cannot move with even a single motion, or else that not even the first sphere can move with just a single motion. Otherwise your syllogism is shown to be false, because you both allow and deny the same thing.

I come now to the second position, which states that the Earth is far removed from the centre, and that it is in motion around the centre, in other words that it is a planet or wandering star. Your argument against this is conclusive as regards its form, but erroneous in its content. Granted that the Earth moves in this way, and that it has two motions, it doesn't necessarily follow that there must be changes in the rising and setting of the fixed stars, as I shall demonstrate in due course. Aristotle's error is entirely excusable here; in fact it's to his credit that he has identified the most subtle argument that can be brought against Copernicus's position. If the argument is telling and apparently conclusive, you will see how much more subtle and ingenious is the solution, and how it took an intelligence as sharp as Copernicus's to discover it; and from the difficulty of understanding it you can judge how much more difficult it was to discover it. So let's leave the response to this point to one side for now; you will hear it in its proper time and place, after we have repeated Aristotle's own argument and indeed reinforced it greatly to his advantage.

Response to the third argument.

On Aristotle's third argument there is no need for us to respond any further, since we have said enough about it in the course of yesterday and today. He repeats that the natural motion of heavy bodies is in a straight line towards the centre, and he asks whether this is the centre of the Earth or the centre of the universe; he concludes that it is naturally the centre of the universe, which also happens to be the centre of the Earth.

Response to the fourth argument.

So we can move on to the fourth argument, on which we shall have to dwell at some length, since it is based on the experience which provides the foundation for most of the remaining arguments. Aristotle considers it a decisive argument for the immobility of the Earth that when projectiles are thrown straight up in the air, even to a great height, they fall straight back down again to the place from which they were thrown. This, he says, could not happen if the Earth were in motion, because in the time the projectile was in the air and separated from the Earth, the place from which it was thrown would have moved a long way towards the east because of the Earth's rotation, so that the falling projectile would come to Earth a corresponding distance from that place. This is the basis of the argument taken from shots fired into the air from a cannon, and the other argument used by Aristotle and Ptolemy, taken from seeing bodies falling from a great height along a straight line perpendicular to the surface of the Earth. Now, to begin to unravel these knots let me ask Simplicio a question. If someone were to deny Ptolemy's and Aristotle's claim that freely falling bodies fall vertically in a straight line, i.e. directly towards the centre, by what means could this be proved?

[165]

SIMPLICIO. By means of the senses, which assure us that the tower is straight and perpendicular, and that when a stone falls it skims the side of the tower, not deviating a hair's breadth on either side, and lands at the foot of the tower directly below the point from which it was dropped.

SALVIATI. Supposing the Earth's globe did rotate, and hence carried the tower along with it, and yet we still observed the stone to fall skimming the side of the tower, what would the stone's motion have to be?

SIMPLICIO. In that case we would have to speak of its motions in the plural, because its downwards motion would be one and the motion it would need to follow the course of the tower would be another.

SALVIATI. Then it would have a compound of two motions, one measuring the length of the tower and another needed for following its course. This compound motion would mean that the stone was no longer following a simple perpendicular straight line, but a slanting one, and not necessarily straight.

SIMPLICIO. Whether it would be straight or not I don't know, but I can see that it would have to be a slanting line, and not the same as the vertical straight line it would follow if the Earth was at rest.

SALVIATI. That means that you can't be certain, simply from seeing the stone fall skimming the side of the tower, that it follows a vertical straight line, unless you presuppose that the Earth is at rest.

SIMPLICIO. Yes, because if the Earth was in motion the stone's motion would be oblique and not vertical.

Aristotle's and Ptolemy's logical error in taking as known what is in question.

SALVIATI. There you have Aristotle's and Ptolemy's logical error plain to see. You've exposed it yourself: how they take as known what they are trying to prove. [166]

SIMPLICIO. In what way? As far as I can see it's a valid syllogism, not a begging of the question.

SALVIATI. I'll show you in what way. Tell me: in a proof, the conclusion is assumed to be unknown, is it not?

SIMPLICIO. Of course, otherwise there would be no point in trying to prove it.

SALVIATI. And the middle term* of the proof should be known?

SIMPLICIO. It has to be; if it isn't, then we are trying to prove *ignotum per aeque ignotum*: to find an unknown by means of another equally unknown.

SALVIATI. What is the unknown we are trying to prove? Isn't it the immobility of the Earth?

SIMPLICIO. Yes, that's right.

SALVIATI. And the middle term, which needs to be known, is the stone falling vertically in a straight line?

SIMPLICIO. That's the middle term, yes.

SALVIATI. But haven't we just established that we can't be sure that it does fall vertically in a straight line unless we know that the Earth is at rest? So in your syllogism, you base the certainty of the middle term on the uncertainty of the conclusion. That's the nature and extent of the logical error here.

SAGREDO. On behalf of Simplicio I should like, if possible, to defend Aristotle, or at least to get a better idea of the strength of your deduction. Your argument is this: the fact that we observe the stone skimming the side of the tower does not suffice to prove that its motion is perpendicular, this being the middle term of the syllogism, unless we presuppose that the Earth is at rest, which is the conclusion we are seeking to prove. For if the tower was moved along together with the Earth, and the stone skimmed its side, its motion would be oblique, not perpendicular. To this I will reply that if the tower was in motion it would be impossible for the stone to skim its side as it fell; therefore we can infer from the fact that it does skim the side of the tower that the Earth is at rest.

SIMPLICIO. Just so. If the tower was carried along by the Earth's motion and the stone skimmed its side as it fell, the stone would have to have two natural motions—rectilinear motion towards the centre and circular motion around the centre—which is impossible.

SALVIATI. Aristotle's defence, then, is that it is impossible—or at least, that he considers it impossible—for the stone to have a compound motion which is both rectilinear and circular. If he had not excluded the possibility that the stone could move both towards the centre and around it, he would have realized that it could equally well fall skimming the side of the tower regardless of whether the tower was in motion or at rest, and hence he would have understood that the stone skimming the tower doesn't allow us to infer anything about the motion or immobility of the Earth. But this doesn't excuse Aristotle at all. For one thing, if he considered combined rectilinear and circular motion to be impossible he should have said so,

[167]

Aristotle admits that fire moves upwards with its natural motion and in a circle with participatory motion.

especially as it is such a key part of his argument. But in any case, it can't be claimed that such a thing is impossible, or even that Aristotle said it was. On the first point, I shall show shortly that it is not only possible but necessary. As for the second, Aristotle himself allows that fire moves straight upwards with its natural motion, and in a circle with the diurnal motion imparted by the heavens to the whole of the element of fire and to the greater part of the air. If he sees no impossibility in combining rectilinear motion with the circular motion imparted to fire and air up to the sphere of the Moon, there is even less reason why he should consider it impossible to combine the vertical fall of a stone with the circular motion natural to the whole terrestrial globe, of which the stone is a part.

SIMPLICIO. I don't agree. If the element of fire goes around together with the air, then a particle of fire rising from the Earth can easily acquire this circular motion from the air as it passes through it; in fact it's inevitable that it will, since fire is such a light and easily moved body. But it seems inconceivable that a heavy stone or a cannon ball falling through the air could be carried along by the air or by anything else. Then there's the very apt experiment of a stone dropped from the top of a ship's mast, which when the ship is stationary falls to the foot of the mast, but when the ship is travelling falls some distance away from it, corresponding to the distance the ship has moved forward while the stone was falling—and when the ship is moving fast this is a distance of several *braccia*.

Stones falling from the top of a ship's mast and from the summit of a tower are not comparable.

SALVIATI. The example of a ship is quite different from the Earth in the case of the terrestrial globe's daily rotation. The ship's motion is clearly not natural to it, and so it only applies accidentally to all the things that are on the ship; therefore it's [168] not surprising that when a stone which has been held at the top of the mast is dropped it should fall freely, without having to follow the ship's motion. But the daily rotation is assumed to be the proper natural motion of the terrestrial globe, and hence of all its parts; and since it's imparted to it by nature it is ineradicable. So the primary inclination of the stone at the top of the tower is to move around the centre of the body of which it is a part once every twenty-four hours; and it will exercise this natural tendency for all eternity, wherever it happens to be.

To be persuaded of this you simply have to revise your long-standing mental assumption, and say: 'Having thought until now that it is a natural property of the terrestrial globe to remain fixed in relation to its centre, I have never had any difficulty in understanding that any particle of the Earth also remains naturally in the same state of rest. Therefore, if the natural inclination of the terrestrial globe is to turn on its axis every twenty-four hours, it must also be the natural and intrinsic inclination of every part of the globe not to remain at rest, but to follow the same course.' In this way there is no obstacle to concluding that, since the motion imparted by the oars to the ship, and indirectly to everything in it, is not natural but external, it is natural that the stone, once separated from the ship, should return to its natural state and revert to exercising its pure and simple natural tendency.

That part of the air which is lower than the highest mountains follows the motion of the Earth.

A further point is that the air, at least that part of it which is lower than the highest mountains, must be swept and carried along by the unevenness of the Earth's surface—or else that it naturally follows the diurnal motion because it is a mixture of many vapours and exhalations coming from the Earth. In either case this does not apply to the air surrounding the ship driven forward by its oars. Hence it's not possible to argue by analogy from the ship to the tower. The stone falling from the top of the ship's mast enters a medium which does not share the ship's motion, whereas one falling from the top of a tower is surrounded by a medium having the same motion as the whole of the terrestrial globe. So in the latter case it can follow the general motion of the Earth not just unimpeded by the air, but positively aided by the air's motion.

The motion of the air is capable of carrying light objects along with it, but not heavy ones.

SIMPLICIO. I fail to see how the air can impart its motion to a large stone, or to an iron or lead ball weighing, say, two hundred pounds, in the same way as it does to very light things [169] such as feathers or snowflakes. In fact, we can see that such weights don't move as much as an inch when they're exposed to the wind, however strongly it blows; so how much less will they be carried along by the air?

SALVIATI. The experience you describe is very different from the case we are discussing. You postulate the wind blowing on a stone which is at rest, whereas in our case the stone is exposed to the air when both are already moving at the same speed. This means that the air doesn't have to impart a new motion to the stone, but only to maintain—or rather, not to impede—the motion it already has. You are seeking to impel the stone with a motion which is foreign to it and not in its nature; we are just preserving its natural motion. If you wanted to put forward a more relevant example, you would have to propose observing—in our mind's eye if not physically—what would happen if an eagle gliding on the wind were to drop a stone from its talons. I strongly suspect we would find that the stone didn't fall perpendicularly; rather, because it was already flying at the same speed as the wind when it left the eagle's talons, and then falling through a medium that was also moving at the same speed, it would move along a slanting path, following the direction of the wind with the added effect of its own weight.

SIMPLICIO. We'd have to carry out the experiment and judge on the basis of what happened; but for now the result we've described on a ship seems to support my view.

SALVIATI. For now you're right; it may soon appear otherwise. Not to keep you on tenterhooks any longer, Simplicio, tell me: are you quite certain that the example of the ship fits our case so well that we can reasonably suppose that what happens on the ship would also happen in the case of the terrestrial globe?

SIMPLICIO. So far it seems to me that it does. You've pointed out some minor discrepancies, but none that strike me as weighty enough to make me change my mind.

SALVIATI. I hope you won't change your mind. Hold firm to your conviction that the result of the Earth's motion would correspond to that on the ship, as long as you don't have second thoughts if it turns out to undermine your case rather than [170] supporting it. You reason as follows: when the ship is motionless the stone falls to the foot of the mast, and when the ship is in motion it falls some distance away; hence, conversely, when the stone falls to the foot of the mast we can infer that the ship is motionless, and when it falls some distance away we can infer that the ship is in motion. Given that, further, what occurs on the ship must likewise happen in the case of the Earth, we can infer from the fact that the stone falls to the foot of the tower that the terrestrial globe must be at rest. This is your argument, is it not?

SIMPLICIO. Just so, and all the easier to grasp because you have stated it so briefly.

SALVIATI. Now tell me this: if the stone dropped from the top of the mast when the ship is moving at speed were to fall in precisely the same place as it did when the ship was motionless, what value would these falling stones have in determining whether the vessel was in motion or at rest?

SIMPLICIO. Absolutely none. It would be like feeling someone's pulse to tell whether they were awake or asleep, since the pulse beats in the same way in a sleeping person as it does in one who is awake.

SALVIATI. Excellent. Have you ever carried out the experiment on a ship?

SIMPLICIO. No, I haven't, but I'm quite sure that the authors who cite it have observed it carefully. Besides, the cause of the difference between the two cases is so self-evident that it leaves no room for doubt.

SALVIATI. You yourself are a witness to the fact that these authors cite the experiment without having carried it out, since you haven't tried it yourself but report it as a certainty because you take what they say on trust. In the same way, they in turn could have done the same thing, relying on their predecessors without ever getting back to someone who's actually made the experiment. In fact they must have done so, because anyone who tries it will find that it demonstrates the very opposite of what everyone writes. It will show, in fact, that the stone always lands at the same place on the ship, whether the ship is stationary or moving, however fast. And since the same reasoning applies to the Earth as to the ship, the fact that the stone always falls vertically to the foot of the tower proves nothing about whether the Earth is in motion or at rest.

The stone falling from the ship's mast lands in the same place whether the ship is in motion or at rest.

SIMPLICIO. If you based this on anything other than experimental evidence, I think it would be a long time before we resolved our argument, because it strikes me as so remote from all human reason that it leaves no room at all for finding it believable or plausible.

SALVIATI. I don't have any trouble in finding it so. [171]

SIMPLICIO. Do you mean you've not put it to the test a hundred times or even once, and yet you confidently declare it to be certain? I remain as unpersuaded as I was before, and equally certain that the main authors who cite the experiment have carried it out and that it shows what they say it does.

SALVIATI. And I'm certain without having made the experiment that the outcome would be as I've said, because it can't possibly be otherwise. What's more, I tell you that you know yourself that it can only be as I say, even though you claim or pretend to claim not to know; and I'll make you confess as much, because I'm very good at coaxing the truth out of people.

You're very quiet, Sagredo; I thought I saw you move just now, as if you were going to say something.

SAGREDO. Yes, I was, but you've so aroused my curiosity by saying you'll force Simplicio to reveal the knowledge that he's trying to hide from us, that everything else can wait. So please make good your boast.

SALVIATI. I shall, as long as Simplicio is willing to reply to my questioning.

SIMPLICIO. I'll reply to the best of my knowledge, and I'm confident that I won't have any trouble in doing so. I don't claim to know anything about things I believe to be false, since knowledge is concerned with truth and not with falsehood.

SALVIATI. I ask only that you reply according to what you know for certain. So consider this: suppose you had a plane surface, as smooth and polished as a mirror and as hard as steel, and not horizontal but inclined. Suppose, then, that you took a perfectly spherical ball made of some hard, heavy substance such as bronze, and placed it on this surface. If you let it go, what do you think it would do? Do you believe, as I do, that it would stand still?

SIMPLICIO. If the surface was inclined?

SALVIATI. Yes, that's what I've supposed.

SIMPLICIO. I don't believe it would stand still at all; in fact I'm certain it would spontaneously roll down the slope.

SALVIATI. Think carefully about what you say, Simplicio. I'm certain that it would stand still wherever you placed it.

[172] SIMPLICIO. If this is the kind of supposition you make, Salviati, then I'm no longer surprised that you arrive at such false conclusions.

SALVIATI. You're quite sure then, are you, that the ball would move spontaneously in the direction of the downward slope?

SIMPLICIO. What room is there for doubt?

SALVIATI. And you're convinced of this, not because I taught it to you—in fact I tried to persuade you of the opposite—but because you exercised your own natural judgement.

SIMPLICIO. I see now what you're up to. You weren't saying what you really believed, but you just wanted to tempt me and, if you'll forgive the expression, to get me out on a limb.

SALVIATI. Yes, you're right. How long, then, would the ball continue to move, and with what velocity? Bear in mind that I said it was a perfectly round ball and a highly polished surface, to eliminate any external or accidental impediments. I also want you to disregard the air's resistance to being parted, and any other accidental obstacles there might be.

SIMPLICIO. I understand all this perfectly. In answer to your question, I reply that the ball would continue to move indefinitely, as long as the plane remained inclined, and that its speed would continually increase, it being in the nature of moving bodies that *vires acquirant eundo*:* they gain momentum as they travel. The greater the incline, the greater would be the ball's velocity.

SALVIATI. Suppose we wanted to make the ball move upwards on this same surface, do you believe that it would do so?

SIMPLICIO. Not spontaneously, no, but it would if it was pulled or forcibly thrown upwards.

SALVIATI. If it was impelled by some external force, what kind of motion would it have, and how far would it extend?

SIMPLICIO. Its motion would continually weaken and slow down, because it would be contrary to nature; and its duration would be greater or less depending on the degree of force exerted on it, or the degree to which the plane was inclined.

SALVIATI. Thus far, then, you have described what happens to a mobile body on two different planes. On an inclined plane a heavy body spontaneously moves downwards at an ever-increasing speed, and force is needed to prevent it from moving. On a plane inclined upwards, on the other hand, force is needed [173] both to make it move and to prevent it from moving, and the motion imparted to it continually decreases until it's finally extinguished. You say also that in both cases the effect varies depending on the plane's greater or lesser degree of inclination. The greater the inclination the greater will be the velocity, and conversely the same mobile body impelled with the same force will travel further when the upward inclination is less. Tell me now what would happen to this same mobile body on a surface which had no inclination, either upwards or downwards.

SIMPLICIO. I need to think carefully about my reply. If there was no downward slope there could be no natural tendency to move, and if there was no upward slope there could be no resistance to being moved. So there would be an indifference between the propensity and the resistance to motion; therefore it seems to me that it ought naturally to remain at rest. But I'm forgetting that Sagredo explained a little while ago* that this was what would happen.

SALVIATI. I agree, assuming that it was placed down firmly. But what if it was given an impetus in one direction or another: what would happen then?

SIMPLICIO. It would move in that direction.

SALVIATI. Yes, but with what kind of motion—continually accelerating, as on a plane inclined downwards, or ever more slowly, as on an upward incline?

SIMPLICIO. I can see no cause for either acceleration or deceleration, since there is neither an upward nor a downward incline.

SALVIATI. Indeed. But if there is no cause for deceleration, even less will there be any cause for standing still. So how long do you think the body would continue to move?

SIMPLICIO. As long as the level surface continued.

SALVIATI. In that case, if the surface extended indefinitely, would motion on that surface also be boundless, in other words perpetual?

SIMPLICIO. I think it would, provided the mobile body was made of durable material.

SALVIATI. We've already presupposed that, since we said we would exclude all accidental and external impediments, and lack of durability in the body was one such accidental impediment. My next question, then, is this: what do you think is the reason for the ball to move spontaneously down an inclined plane, but only by force on an upward incline?

SIMPLICIO. This is because the inclination of heavy bodies [174] is to move towards the centre of the Earth, but they only move towards the circumference if they are impelled by force. The downward surface brings them closer to the centre, and the upward one takes them away from it.

SALVIATI. That means that every part of a surface which was not inclined either downwards or upwards must be equidistant from the centre. Do such surfaces exist anywhere in the world?

SIMPLICIO. Of course they do. Our terrestrial globe would have such a surface, if it was smooth rather than mountainous and uneven as it is. Smooth, calm water has such a surface.

SALVIATI. So a ship sailing on a calm sea would be such a body as we have envisaged, moving over a surface which has no incline either upwards or downwards. Once given an initial impulse, its motion would be unending and unchanging, provided all external and accidental impediments were removed.

SIMPLICIO. So it would seem.

SALVIATI. What about the stone at the top of the mast? Wouldn't it also be moving on the circumference of a circle around the centre, carried along by the ship, and therefore wouldn't its motion also be ineradicable, assuming external impediments were removed? Moreover, wouldn't this motion have the same speed as the ship's?

SIMPLICIO. So far so good; now what follows?

SALVIATI. You can draw the final consequence for yourself, since you've already established all the premises.

SIMPLICIO. Your final conclusion is that since the motion of the stone is ineradicably imparted to it, the stone won't relinquish this motion but rather will continue moving with the ship, and so will finally fall at the same place as it would when the ship was stationary. My response is that this would be the case if there were no external impediments to disrupt the stone's motion once it is released, but in fact there are two such impediments. One is the resistance of the air, which the moving body would not be strong enough to overcome with its own impetus, once it had lost the impetus it derived from the force of the oars, which it shared with the ship while it was resting on the mast. The other is the new motion of falling downwards, which necessarily impedes the other motion carrying it forwards.

SALVIATI. I grant you the impediment of the air's resistance. If the falling body was of some very light material, such as a feather or a tuft of wool, this impediment would be considerable; but in [175] the case of a heavy stone it's negligible. You said yourself just now that the strongest wind isn't able to move a heavy stone from its place, so how much less will be the effect of the still air which the stone encounters when it's moving no faster than the rest of the ship? Still, as I've said, I grant you the small effect that this impediment may have; and by the same token, I'm sure you will grant to me that if the air was moving at the same speed as the ship and the stone, the effective impediment would be nil.

As for the second impediment, deriving from the additional downward motion, there are two points to be made. The first is that these two motions—circular motion around the centre and rectilinear motion towards the centre—are clearly not contraries or incompatible with each other, nor do they cancel each other out. The moving body has no resistance to such motion. You yourself have acknowledged that a body resists motion away from the centre, and is attracted to motion which brings it closer to the centre; hence it follows that motion which is neither towards nor away from the centre will not provoke either resistance or attraction, and therefore there is no reason for the impulse the body has received to be diminished. The second point is that the motion has not one cause, which might be weakened by the new effect, but two distinct causes: the body's weight, the sole effect of which is to draw it towards the centre, and the impressed force which leads it around the centre; and so there remains no reason for any impediment.

SIMPLICIO. The argument appears very plausible, but in reality it's undermined by an objection which is hard to overcome. Your whole reasoning depends on a presupposition which Peripatetic philosophers will be very reluctant to accept, because it is completely contrary to what Aristotle says. You take it as self-evident and generally understood that when a projectile separates from its projector, it continues to move because of the force impressed on it by that same projector. This impressed

Aristotle's view that a projectile is moved not by impressed force but by the medium.

force is as unacceptable in Peripatetic philosophy as is any other transfer of an accidental property from one subject to another. The Aristotelian view, as I think you know, is that the projectile is carried forward by the medium, in this case the air; so if a stone dropped from the top of a ship's mast were to follow the motion of the ship, this effect would have to be attributed to the air, not to any impressed force. But you assume that the air does not follow the ship's motion, but is at rest. Moreover, whoever drops the stone does not have to throw it or give it any kind of [176] impetus with their arm, but has simply to open their hand and let it go. So since neither the force impressed upon it by the projector nor the benefit of the air can cause the stone to follow the ship's motion, it must be left behind.

SALVIATI. You seem to be saying that, as the person dropping the stone does not throw it, it is not really a projection at all.

SIMPLICIO. Its motion cannot properly be called projection, no.

SALVIATI. In that case, what Aristotle says about motion, a moving body and the motive force of projectiles has nothing to do with our discussion; so why did you bring it up at all?

SIMPLICIO. Because you introduced this idea of impressed force, which doesn't exist and so can have no effect whatever, because *non entium nullae sunt operationes*: entities which do not exist have no effect. The cause of motion, not just of projectiles but of anything else that does not move naturally, must be attributed to the medium, which we have not considered as we should; so everything that has been said so far is beside the point.

SALVIATI. The point will become clear in due course. Tell me: since your whole objection is based on the non-existence of impressed force, would you accept that this force exists if I demonstrated to you that the medium has nothing to do with the continued motion of projectiles once they are separated from their projector, or would you just move to attack it from another angle?

SIMPLICIO. If the effect of the medium were removed, I don't see that there could be any other explanation apart from the property imparted by the mover.

SALVIATI. To remove as far as possible any grounds for endless arguments, it would be useful if you could explain, as fully as you can, the effect of the medium in maintaining the motion of a projectile.

The effect of the medium in maintaining the motion of a projectile.

SIMPLICIO. The thrower, or projector, has the stone in his hand; he moves his arm with velocity and force, and its motion is imparted both to the stone and to the surrounding air. Hence when the stone leaves the thrower's hand, it is in air which is already moving with the same impetus, and is carried along by it. If the air did not affect it in this way, the stone would fall to the thrower's feet.

SALVIATI. Have you really been so credulous as to be [177] persuaded by these vain ideas, when your own senses should enable you to refute them and perceive the true explanation? You said just now that if a heavy stone and a cannon ball were placed on a table, they would not be moved by the wind, however strong it was. Tell me now: if they had been balls of cork or cotton wool, do you think they would have been moved by the wind?

Many experiences and arguments against the cause of the motion of projectiles as posited by Aristotle.

SIMPLICIO. I have no doubt that they would have been blown away, and the lighter the material the more quickly the wind would have blown it away. This explains why we see clouds being borne along by the wind at the same speed as the wind itself.

SALVIATI. And what is the wind?

SIMPLICIO. Wind is defined simply as air in motion.

SALVIATI. So air which is in motion carries light materials much further and more rapidly than heavy materials?

SIMPLICIO. Certainly.

SALVIATI. If you were to throw a stone and then a ball of cotton wool, which of them would travel further or more rapidly?

SIMPLICIO. The stone, by a long way; the cotton wool would just fall at my feet.

SALVIATI. But if it is just the air moved by the thrower's arm which carries the projectile after it has left his hand, and if air which is in motion carries light materials more easily than heavy ones, why doesn't the cotton-wool projectile travel further and more rapidly than the stone one? The stone must retain something, apart from the motion of the air. Take another example: suppose two cords of equal length were suspended from that beam, and a lead ball was attached to the end of one and a ball of cotton wool to the other. If both were moved an equal distance from the perpendicular and then allowed to fall freely, there is no doubt that they would both move back towards the perpendicular and then, carried by their own impetus, move a certain distance beyond the perpendicular, and then back again. But which of the two pendulums do you think would remain in motion longer, before coming to rest at the perpendicular?

SIMPLICIO. The lead ball would move back and forth for a long time, but the cotton wool only two or three times at most.

SALVIATI. So the impetus and motion, whatever causes it, is conserved for longer in heavy materials than in light ones. Now I come to another point, so let me ask you: why does the air not carry away that lemon on the table there?

SIMPLICIO. Because the air itself is not moving. [178]

SALVIATI. In that case the person who does the throwing must impart the motion to the air, which then moves the projectile. But if this kind of force cannot be impressed, since an accidental property cannot be made to pass from one subject to another, how can it go from the thrower's arm to the air? Surely the air is a subject distinct from the arm?

SIMPLICIO. The answer is that the air is in its proper region and so is neither heavy nor light, and therefore can easily receive and conserve any impulse.

SALVIATI. But the pendulums have just shown us that the less weight a moving body has, the less it is able to conserve motion. So how is it that only the air, which when it is in its proper place has no weight, can conserve the motion which is imparted to it? I believe, and I know that for now you share this belief, that as soon as the thrower's arm stops moving, the air around it stops moving as well. So let's go into this small room and stir the air up as much as we can by shaking a towel; then, as soon as we stop, let a small lighted candle be brought in, or let someone drop a piece of gold leaf. You will see that both move quite steadily, showing that the air has immediately returned to a state of rest. I could suggest any number of other experiments, but if one of these does not suffice then our case is desperate indeed.

SAGREDO. Isn't it incredible that when someone shoots an arrow into the wind, the narrow slice of air that was displaced by the bowstring should accompany the arrow in the face of the storm blowing against it! But there's another detail of what Aristotle says that I'd like Simplicio to explain to me. If someone were to shoot two arrows from the same bow, with the string drawn back to the same extent, one pointing forwards in the usual way and the other crosswise, in other words with the length of the arrow laid along the bowstring, which one would travel further? Do please forgive me, and answer my question even though it may seem ridiculous, because as you can see, I'm not very bright and my ability to speculate doesn't rise very high.

SIMPLICIO. I've never seen anyone shoot an arrow crosswise, but I think that if they did it wouldn't travel so much as a twentieth of the distance of one pointing forwards.

[179] SAGREDO. I thought the same thing, which was what prompted me to wonder if there is an inconsistency between what Aristotle says and what we see from experience. Experience suggests that, if I were to put two arrows on this table when a strong wind was blowing, one aligned with the direction of the wind and the other crosswise to it, the wind would quickly blow away the latter and leave the first one where it is. If Aristotle is correct, the same should happen with the two arrows shot from a bow: the one which was shot crosswise would be propelled by a large amount of air set in motion by the bowstring, extending the full length of the arrow, whereas the one which was shot pointing forwards would have only the propulsion of the tiny circle of air corresponding to the arrow's thickness. I can't account for this difference, and I'd very much like to understand it.

SIMPLICIO. The reason seems quite clear to me: the arrow which is fired pointing forwards has to penetrate only a small amount of air, whereas the other has to divide the air along its whole length.

SAGREDO. So when an arrow is shot it has to penetrate the air? But how can it, if the air travels with it, indeed if the air is the medium which carries it along? If this were the case, surely the arrow should move faster than the air—and where does it get this greater velocity from? Do you mean to say that the air gives it a greater velocity than its own? So you see, Simplicio, what happens is exactly the opposite of what Aristotle says. Not only is it wrong to say that the medium confers motion on the projectile; the medium is actually the one thing that impedes it. Once you have grasped this, you won't have any difficulty in understanding that when the air really does move, it will carry the arrow along with it much better crosswise than point first, because in the former position there is a large volume of air pressing on the arrow, but in the latter very little. When they are shot from a bow, on the other hand, since the air is stationary, the arrow fired crosswise encounters a large volume of air and so meets quite strong resistance; but the arrow fired pointing straight ahead easily overcomes the resistance of the very small amount of air which it encounters.

The medium does not confer motion on projectiles, but rather impedes them.

SALVIATI. How many times have I seen propositions in Aristotle—in natural philosophy, that is—which are not just wrong, but wrong in a way which is diametrically opposed to the truth, as in this case! But to pursue our argument, I think Simplicio is convinced now that seeing a stone fall always in the same place does not provide any basis for conjecturing whether the ship is moving or not. If he is not persuaded by what has [180] been said so far, the experiment itself should remove his doubts. The most he might find from this experiment would be that the falling object would be left behind if it was very light and if the air was not following the ship's motion; but assuming that the air was moving at the same speed, no conceivable difference could be found in this or any other experiment, as I shall now explain to you. Now, given that no difference is apparent in this case, what do you claim to deduce from a stone falling from the top of a tower, where the stone's circular movement is not fortuitous or accidental to it but natural and eternal, and where the air duly follows the movement of the tower as the tower does that of the terrestrial globe? Have you anything more to reply on this point, Simplicio?

SIMPLICIO. Only that I have still to see a proof of the motion of the Earth.

SALVIATI. I haven't claimed to have proved it, only to show that the experiment brought forward by its opponents to prove that the Earth is at rest is inconclusive; and I think I can show the same of the others.

SAGREDO. Salviati, before we go on, please allow me to put forward a difficulty which was going round in my mind while you were so patiently examining the experiment with the ship to Simplicio.

SALVIATI. Of course; we're here to discuss, and it's good that we should each express the difficulties which occur to us. That's how we come to knowledge of the truth. So carry on.

SAGREDO. If it's true that the impetus of the ship's motion remains indelibly impressed in the stone after it has been dropped from the mast, and if, moreover, this motion does not impede or retard in any way the stone's natural motion directly downwards, then a marvellous effect in nature must necessarily follow. Suppose the ship is stationary, and the time taken for the stone to fall from the top of the mast is two pulse beats. Suppose, then, that the ship begins to move, and the same stone is dropped from the same place: from what has been said, it will still take two pulse beats to reach the deck. But in that time, the ship will have moved forwards, say, twenty *braccia*. This means that the stone's true motion will have been in a diagonal line, notably longer than the original perpendicular straight line corresponding to the height of the mast; and yet it will have travelled this distance in the same time. Imagine, then, the ship's motion accelerating further, so that the stone has to follow a diagonal line even longer than before. In short, the more the ship's speed increases, the longer the diagonal lines the falling stone will have to travel, and yet it will always do so in the same space of two pulse beats. Similarly, if a long-range cannon were fired point-blank, in other words horizontally, from the top of a tower, then regardless of the amount of gunpowder that was used, so that the ball fell to the ground a thousand, four thousand, six thousand, or ten thousand *braccia* away, in each case the shot would take exactly the same length of time. And it would always be equal to the time it would take for the ball to fall from the mouth of the cannon to the ground, if without any other impulse it were simply dropped vertically from the top of the tower. It seems extraordinary that in the same short interval as it takes to fall vertically from a height of, say, a hundred *braccia*, a ball fired from a gun can travel four hundred, a thousand, four thousand, or ten thousand *braccia*, so that any ball fired horizontally from a cannon remains in the air for the same length of time.

A marvellous phenomenon in the motion of projectiles.

[181]

SALVIATI. That's a very striking and original thought, and if the effect is true—as I don't doubt that it is—it is indeed marvellous. In fact, I'm certain that if it were not for the resistance of the air, another cannon ball which was dropped from the same height at the same moment as the shot was fired would hit the ground at the same time as the one fired from the cannon, even if one ball had travelled ten thousand *braccia* and the other only a hundred—always assuming that the Earth's surface was level, as it would be if you fired the shot over a lake. The effect of air resistance would be to retard the very rapid motion of the ball shot from the cannon.

But let us now turn to resolving the other arguments, since I think I'm right in saying that Simplicio now accepts that this first objection based on falling objects has no validity.

SIMPLICIO. I can't say that all my doubts have been resolved, though I dare say the fault for that is mine, as I don't grasp things as quickly or easily as Sagredo. If it's true, as you say, that the motion which is shared by the stone while it is at the top of a ship's mast remained indelibly conserved in it even when it is no longer in contact with the ship, then the same thing should happen if someone galloping on horseback should [182] drop a ball to the ground. The ball ought to continue moving in the same direction, and follow the horse's movement without lagging behind. But I don't think we see such a thing happen, unless the rider throws the ball forwards; otherwise I think the ball will remain at the point where it hits the ground.

SALVIATI. I think you're much mistaken; I'm sure that experience will show you that on the contrary, when the ball hits the ground it will still run along together with the horse. It would only be left behind if it were hindered by the roughness and unevenness of the road. The reason for this seems quite clear: if you stood still and threw the ball along the ground, surely it would still move forward after it had left your hand, and it would travel further if the ground's surface were smoother? On ice, for instance, would it not go a very long way?

SIMPLICIO. Yes, that's not in doubt, assuming I gave it an impetus with my arm. But in the case of the rider on horseback, it was assumed that the horseman would simply drop the ball.

SALVIATI. That was my assumption as well. But when you throw a ball, the only motion it has once it has left your hand is the motion produced by your arm, which is conserved in the ball and continues to propel it forwards. Now, what difference does it make whether this impetus which is imparted to the ball comes more from your arm or from the horse? When you are on horseback doesn't your hand, and therefore the ball, travel at the same speed as the horse itself? Of course it does. So, if you simply open your hand, the ball is launched with the impetus derived from the horse, which is imparted to you, to your arm, to your hand, and finally to the ball. In fact, I will go further and say that if the rider on the horse throws the ball behind him, it sometimes happens that when the ball hits the ground it will still follow the horse's movement, even though it was thrown in the opposite direction. Sometimes it will just fall to the ground and stay there; and it will only travel in the opposite direction to the horse if the motion it receives from the thrower's arm has a greater velocity than the motion of the horse. And it's quite mistaken to say, as some do, that a horseman can throw a spear in the direction the horse is travelling, and then draw level with it and catch it again; because the only way you can throw a projectile and catch it again is by throwing it straight up [183] in air, just as you would if you were standing still. However fast the horse was running, as long as its motion was uniform and the projectile was not something very light, it would always come back down to the thrower's hand, however high it was thrown.

Various curious problems concerning the motion of projectiles.

SAGREDO. This explanation reminds me of some curious problems concerning this matter of projectiles. The first of these will strike Simplicio as very strange, and it's this: it can happen that when someone moving rapidly in any way simply drops a ball, the ball will not just follow his motion when it hits the ground but will actually run ahead of it. This is connected to another curious fact, that an object thrown horizontally can gain an additional velocity which is much greater than that which it received from the thrower. I have often wondered at this when I have watched people playing with hoops,* which move through the air at a given speed when they leave the thrower's hand, but their speed then increases significantly when they hit the ground. Then, if as they roll they strike an obstacle which makes them bounce up into the air, the hoops move very slowly when they are in the air and then revert to moving fast when they touch the ground again. And I have noticed the strangest thing of all, that not only do they always move more rapidly on the ground than in the air, but when they roll along the ground more than once their motion is sometimes more rapid over the second stretch of ground than over the first. I wonder what Simplicio makes of this?

SIMPLICIO. The first thing I would say is that I have never observed such a thing; second, that I am reluctant to believe it; and third, that if you were able to convince me of this and teach it to me with demonstrative proofs, I would say that you have demonic powers.

SAGREDO. In that case I would be like Socrates' demon,* not one from hell. But you keep talking about us teaching you, to which I reply that if someone doesn't know the truth for themselves, no one else can teach it to them. I could teach you all kinds of things which are neither true nor false, but truths—things which are necessarily true and cannot possibly be otherwise—are things which anyone of average understanding either knows for themselves or else will never be able to know. I'm sure Salviati takes the same view. So in the case of these problems, I say that you know the explanations for them, though you may not have noticed them.

SIMPLICIO. Well, let's not argue about that now: allow me [184] to say that I don't know or understand these matters we are talking about, and see if you can make me understand them.

SAGREDO. This first problem depends on another: how is it that when a hoop is set in motion with a string, it travels much further, and therefore with greater force, than if it is just started off by hand?

SIMPLICIO. Aristotle has a number of problems concerning this matter of projectiles.

SALVIATI. Indeed he does, especially the one explaining how it is that round hoops run more smoothly than square ones.*

SAGREDO. Simplicio, is that a problem that you would be bold enough to explain on your own, without needing anyone to teach you?

SIMPLICIO. Of course, and I think this joke has gone on long enough.

SAGREDO. In that case you know the answer to the other question. So tell me: you know that when a moving object is impeded it comes to a stop?

SIMPLICIO. Yes, provided the impediment is strong enough.

SAGREDO. And you know that moving along the ground is a greater impediment to an object than moving through the air, since the earth is hard and uneven, and the air is soft and unresisting?

SIMPLICIO. Yes, and for this reason I know that the hoop will move more rapidly through the air than along the ground, which is exactly the opposite of what you thought.

SAGREDO. Just a moment. You know that a mobile body turning on its axis has parts moving in all directions—that some parts are rising, others falling, some moving forwards, others backwards?

SIMPLICIO. Yes, I know that because it was taught to me by Aristotle.*

SAGREDO. Do please tell me how he demonstrated it.

SIMPLICIO. By means of the senses.

SAGREDO. You mean that Aristotle made you able to see something which you wouldn't have seen on your own? Did he lend you his eyes? What you mean is that Aristotle told you, made you notice it, reminded you of it, not that he taught it to you.

So when a hoop is not rolling along but spinning on its axis, not parallel to the horizon but vertical to it, some parts of it are rising and the opposite parts are falling; the upper parts are moving in one direction, and the lower parts in the opposite [185] direction. Now imagine such a hoop suspended in the air and spinning rapidly. Still spinning, it is dropped vertically onto the ground. Do you think that when it touches the ground it will continue to spin on itself without rolling along, as it did before?

SIMPLICIO. By no means.

SAGREDO. What will it do, then?

SIMPLICIO. It will run rapidly along the ground.

SAGREDO. In what direction?

SIMPLICIO. In the direction that its spinning carries it.

SAGREDO. But its different parts are spinning in different directions. The upper parts are moving in the opposite direction to the lower parts, so we have to say which it will obey. Neither the ascending nor the descending parts will give way to the other; and the whole hoop can't move downwards, because it's impeded by the earth, or upwards, because of its weight.

SIMPLICIO. The hoop will roll along the ground in the direction towards which its upper parts are moving.

SAGREDO. Why should it not follow the contrary parts, the ones which are in contact with the ground?

SIMPLICIO. Because they are impeded by the roughness of their contact with the ground, in other words by the unevenness of the ground itself; whereas the upper parts, being in contact with the air which is soft and yielding, meet little or no impediment, and so the hoop will move in their direction.

SAGREDO. So the fact that the lower parts are, as it were, attached to the ground, makes them stand still, and only the upper parts press forwards.

SALVIATI. This would mean that if the hoop fell onto ice or some other highly polished surface, it wouldn't move forward so well, but it might well continue spinning on its axis, without gaining any forward motion.

SAGREDO. That could well be the case; but certainly it wouldn't roll as freely as it would if it fell onto a moderately rough surface. But I wonder if Simplicio can tell me why, if the hoop is dropped when it is spinning rapidly on its axis, it doesn't move forwards in the air as it does when it reaches the ground?

SIMPLICIO. Because if it has air both above and beneath it, neither the upper nor the lower part has anything to attach itself to, and as it has no reason to move either backwards or forwards it falls straight downwards.

SAGREDO. So the simple fact of spinning, without any other impetus, is enough to propel the hoop quite rapidly as soon as it hits the ground. Now for the next point. When a player winds a cord around the hoop, then attaches it to his arm and pulls it back, what effect does the cord have on the hoop? [186]

SIMPLICIO. It makes the hoop spin round, to unwind itself from the cord.

SAGREDO. So when the hoop hits the ground it is already spinning because of the cord. Does it not therefore have in itself a reason for moving more rapidly on the ground than in the air?

SIMPLICIO. Yes, certainly: when it was in the air its only impulse was what it received from the thrower's arm; it had a spinning motion as well, but as we have said, this has no propulsive effect when it is in the air. But when it hits the ground, the spinning motion is added to the motion derived from the thrower's arm, and so its speed is redoubled. It's quite clear to me now why the hoop's speed decreases when it bounces up in the air, because then it no longer derives any propulsion from spinning; and when it hits the ground again this propulsive effect returns, and so it once more travels more rapidly than in the air. The only thing I still don't understand is how this second phase of moving along the ground can be more rapid than the first, because if this were the case it would constantly accelerate and its motion would be infinite.

SAGREDO. I didn't say that this second phase of motion was necessarily more rapid than the first, only that this can sometimes happen.

SIMPLICIO. This is the point which I don't follow, and I'd very much like to understand it.

SAGREDO. This too is something you know already. So tell me this: if you were simply to drop the hoop from your hand, without it spinning at all, what would it do when it hit the ground?

SIMPLICIO. Nothing; it would just remain where it was.

SAGREDO. Think again: could it not happen that it would acquire motion when it hit the ground?

SIMPLICIO. Only if we dropped it onto a stone with a sloping surface, as children do when they play with counters;* then it might acquire a spinning motion by hitting the sloping surface obliquely, and this motion could carry it forward along the ground. Apart from this, I can't think of anything else which would prevent it from just standing still where it landed.

[187] SAGREDO. This shows that it is possible for the hoop to acquire a new spinning motion. Is there any reason, then, why a hoop should not bounce up into the air and land obliquely on a stone sloping in the direction in which it's travelling, and so

acquire a new spinning motion in addition to what it originally derived from the cord? And in this case, wouldn't its motion be redoubled, making it move more rapidly than when it first hit the ground?

SIMPLICIO. I understand now how this can easily happen. This has made me think of what would happen if the hoop were made to spin in the opposite direction. When it hit the ground the spinning motion would have the opposite effect, and would slow down the motion derived from the thrower.

SAGREDO. It would slow it down, and sometimes it would cancel it out altogether, if the spinning motion was fast enough. This is the explanation for the effect which expert tennis players can produce to their advantage, when they trick their opponent by cutting the ball (this is the expression they use), hitting it obliquely with the racket so as to give it a spinning motion which is contrary to the motion with which they project it. The result is that when the ball hits the ground, instead of bouncing towards the opposing player as it would if it were not spinning, giving him time to return it, it stops dead or at least bounces much less than it normally would, so that the opponent has no time to return it. The same thing happens with players at bowls,* where the aim is to come as close as possible to a defined goal. If they are playing on a stony road full of obstacles which would deflect the balls all over the place instead of letting them move towards the goal, the players avoid all these obstacles by deliberately throwing the ball like a plate instead of rolling it along the ground. But if the ball is held in the usual way, with the player's hand underneath it, the player's fingers will give it a spin which, added to the propulsive effect of the throw, will carry it a long way forward when it lands near the goal. To prevent this happening and to make it stop where it lands, they skilfully hold the ball with their hand uppermost; then when they throw it, the ball acquires a reverse spin, and when it lands near the goal it stays there or rolls only a little way.

But to return to our original problem, from which all these others arose, I say it is quite possible for someone who is moving rapidly to drop a ball from their hand in such a way that, when it lands, the ball not only follows the thrower's motion but [188] outruns it, moving with more speed than the thrower himself.

To illustrate this, imagine a carriage with a sloping board fixed to the outside, with its lower edge towards the horses and its upper edge towards the back wheels. Now, if someone in the carriage drops a ball on the sloping board when the carriage is moving at full speed, the ball will acquire a spinning motion as it rolls down the board, and this added to the motion imparted to it by the carriage will propel the ball along the ground much faster than the carriage itself. If you were then to fix another board sloping the other way, it would be possible to counteract the carriage's motion so that, when the ball rolled down the board, it stood still when it hit the ground, or sometimes even ran back in the opposite direction. But we've spent too long on this digression, and if Simplicio is now satisfied that we have answered the first objection against the motion of the Earth, based on the motion of falling objects, we can pass on to the others.

SALVIATI. Our digressions so far are not really so remote from the subject in hand as to be considered totally unrelated to it; and in any case, our discussions are prompted by the ideas which occur to all three of us, not just to one person's fancy. Besides, we are conversing for our own enjoyment, and we're not obliged to be as rigorous as someone speaking in a professional capacity and giving a methodical exposition of a subject, perhaps even with the intention of publishing it. I wouldn't want our narrative poem* to follow the principle of unity so strictly that it left us no room for self-contained episodes, which we should be able to introduce on the slightest pretext. It should be as if we were here to tell stories, and I should be entitled to tell whatever story was prompted in my mind by listening to yours.

SAGREDO. That suits me very well. And since we've allowed ourselves this freedom, perhaps I could ask you, Salviati, another question before we move on. Have you ever had occasion to speculate about the line followed by an object falling from the top of a tower? If you have, I'd very much like to hear your thoughts on the matter.

SALVIATI. Yes, I have given it some thought. I have no doubt that, provided one was certain about the nature of the motion with which a heavy object falls towards the centre of the terrestrial [189] globe, this motion combined with the circular motion of the Earth's daily revolution would allow us to define exactly the kind of line followed by the centre of gravity of the falling body. It would be a composite of these two motions.

SAGREDO. I think we can believe with absolute certainty that simple movement towards the centre as the result of gravity follows a straight line, just as it would if the Earth were at rest.

SALVIATI. Indeed, not only can we believe this, but experience makes it quite certain.

SAGREDO. How can we be assured of it by experience, if we only ever see the motion composed of the circular and downward motions combined?

SALVIATI. Actually, Sagredo, we only see the simple downward motion; the other, circular motion which is common to the Earth, the tower and ourselves, is imperceptible to us, as if it didn't exist. The only motion which is apparent to us is the one that we don't share, which is that of the falling stone; and our senses tell us that this follows a straight line, since it remains parallel to the side of the tower, which is built upright and perpendicular to the surface of the Earth.

SAGREDO. Yes, you're right. How foolish of me not to have thought of such a simple fact. But since this is so obvious, what else do you think we need to know to understand the nature of this downward motion?

SALVIATI. It's not enough to know that it is in a straight line; we also need to know whether it is uniform or not—that is, whether it maintains a constant speed or whether it accelerates or decelerates.

SAGREDO. Surely it's clear that it continually accelerates?

SALVIATI. That's still not enough; we need to know at what rate it accelerates. This is a problem* which I don't believe any philosopher or mathematician has yet been able to solve, even though philosophers—especially Aristotelians—have written whole vast volumes on the subject of motion.

SIMPLICIO. Philosophers generally concern themselves with universals. They establish definitions and find their most common applications, and leave the minutiae and trivial details—which are more curiosities than anything else—to mathematicians. So Aristotle confined himself to providing an excellent definition of motion in general, and to identifying the principal attributes of local motion: that it can be natural or forced, simple or [190] composite, uniform or accelerating. As regards accelerating motion, he was content simply to explain the cause of acceleration, leaving it to mechanics or other inferior artisans to work out the rate of acceleration and other such incidental details.

SAGREDO. Yes of course, my dear Simplicio. But you, Salviati, have you ever descended from the Peripatetic throne to amuse yourself with this question of the rate of acceleration of falling objects?

SALVIATI. I haven't needed to think about it, because our mutual friend the Academician has shown me a treatise of his on motion,* in which he demonstrates the answer to this and many other questions. But it would be too much of a digression to interrupt our present discussion, which is itself a digression, in order to explain this; it would be like having a play within a play.

SAGREDO. I am content to excuse you from telling this story now, on condition that this is one of the subjects which we set aside to cover in a separate discussion, as I would very much like to understand it. In the meantime, let's return to the line described by an object falling from the top of a tower to its base.

SALVIATI. If the rectilinear motion towards the centre of the Earth were uniform, and the circular motion towards the east were also uniform, then the composite of these two motions would be a spiral, as described by Archimedes in his book on spirals.* These are generated when a point moves uniformly along a straight line which is being uniformly rotated about a fixed point at one of its extremities. But since the rectilinear motion of the falling weight is continually accelerated, the line resulting from the composition of the two motions must diverge at an ever-increasing rate from the circumference of the circle which the stone's centre of gravity would have described if it had remained on the tower. This divergence must initially be very small, in fact tiny or indeed minute, since a falling body starting from a state of rest—that is, passing from a state where it has no downward motion to one where it is moving directly downwards—must pass through all the infinite degrees of slowness between a state of rest and any given speed, as we have already discussed and established at length.*

[191] Given, then, that a falling object accelerates in this way, and given that its motion would terminate at the centre of the Earth, the line of its compound motion must be such as to travel away from the top of the tower (or rather, from the circumference of the circle traced by the top of the tower due to the rotation of the Earth) at an ever-increasing rate. This distance becomes lesser and lesser *ad infinitum* by how much the falling body is found to be less and less removed from the point where it was first placed. Moreover, the line of this compound motion must terminate at the centre of the Earth.* Making these two assumptions, I draw the circle BI with A as its centre and the radius AB to represent the terrestrial globe. Then I extend the radius AB to C to represent the tower, its height being BC. As the tower is carried by the Earth along the circumference BI, its top describes the arc CD. Now I divide the line CA at its midpoint E, and I draw a semicircle CIA with E as its centre and EC as its radius. I think it highly probable that this semicircle would be the line followed by a stone falling from the top of the tower C, with a motion composed of the Earth's circular motion and the stone's own rectilinear motion. For if CF, FG, GH, and HL are marked on the circumference CD, and straight lines are drawn to the centre A from the points F, G, H, and L, the parts of these lines intercepted between the circumferences CD and BI will always represent the tower CB, as the Earth's rotation carries it towards DI; and the point where each line is intersected by the semicircular arc CI will be the point which the falling stone has reached at a given moment. These points become more distant from the top of the tower in an ever-increasing proportion, which is why the stone's vertical motion

The line described by a naturally falling object, assuming the Earth's rotation on its axis, would probably be the circumference of a circle.

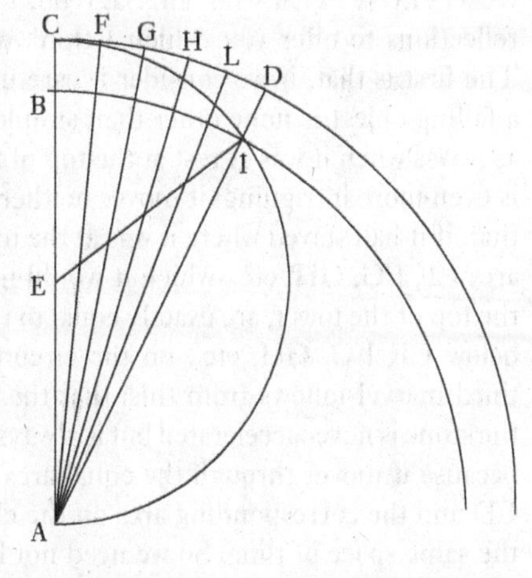

down the side of the tower appears to be constantly accelerating. It can be seen, too, from the infinite acuteness of the angle of contact between the two circles DC and CI, that the distance of the falling object from the circumference CFD—that is, from the top of the tower—is initially very small, which is the same as saying that its motion is initially very slow, and more and more so *ad infinitum* the closer it is to point C, i.e. to a state [192] of rest. Finally, this also shows that its motion would ultimately end at the centre of the Earth, A.

SAGREDO. I understand all this perfectly, and I don't see how the centre of gravity of a falling object could follow any other line than this.

SALVIATI. That's not all, Sagredo; I have three more little reflections to offer you, which I think will not displease you. The first is that, if we consider it carefully, the real motion of a falling object is none other than simple circular motion, just as it was when it was at rest at the top of the tower. The second is even more intriguing: it moves neither more nor less rapidly than if it had stayed where it was at the top of the tower, for the arcs CF, FG, GH, etc., which it would have moved through at the top of the tower, are exactly equal to the corresponding arcs below CF, FG, GH, etc., on the circumference CI. And the third marvel follows from this: that the true actual motion of the stone is never accelerated but is always equable and uniform, because it moves through the equal arcs on the circumference CD and the corresponding arcs on the circumference CI all in the same space of time. So we need not look for new causes of acceleration or any other motions, because a moving object always moves in the same way, i.e. in a circle, and at the same constant speed, whether it stays at the top of the tower or whether it falls. Now, what do you think of this fancy of mine?*

An object falling from the top of a tower moves on the circumference of a circle.

It moves neither more nor less than if it had remained at the top of the tower.

It moves with equal, not accelerating motion.

SAGREDO. I can't find words to express how marvellous I find it, and I can think of nothing at present to suggest that the facts are other than you have described them. If only all philosophers' proofs were half as convincing as this! But to satisfy my curiosity, I would very much like to hear the proof that these arcs are all equal.

SALVIATI. The proof is very easy. Draw a line from I to E. Since the radius of the circle CD, namely the line CA, is twice the length of CE which is the radius of the circle CI, the circumference of CD must be twice the circumference of CI, and every arc of the larger circle will be double the length of a similar arc of the smaller circle. Now the angle CEI, made at the centre E of the smaller circle and subtending the arc CI, is double the angle [193] CAD, made at the centre A of the larger circle and subtending the arc CD. Therefore the arc CD is half the arc of the larger circle similar to the arc CI, and hence the arcs CD and CI are equal. The same proof could be applied to every other part.

However, I don't want to state categorically for now that the motion of falling objects happens in exactly this way; but I will say that if the line followed by a falling object is not exactly as I have described it, it is very close to it.

Rectilinear motion seems to be entirely excluded in nature.

SAGREDO. I have been reflecting on another remarkable thing. If all these considerations are valid, then there is no place for motion in a straight line and nature has no use for it. It doesn't serve even that purpose which you allowed for it at the outset,* of restoring to their proper place any parts of an integral body which had become separated from the whole and were therefore wrongly disposed, since this too is brought about by circular motion.

SALVIATI. This would be a necessary consequence if we had conclusively proved that the terrestrial globe has a circular motion; but I don't claim to have proved this. All we have done so far, and will continue to do, is examine the strength of the arguments which philosophers have put forward to prove the immobility of the Earth. The first of these, based on the motion of falling objects, is open to the objections we have been discussing. I don't know how much weight these carry with Simplicio; so I think it would be good to hear whether he has anything to say in reply to them, before we go on to test the other arguments.

SIMPLICIO. As regards this first argument, I must confess that I've heard a number of subtle points which hadn't occurred to me; and as they are new to me I'm not able to respond to them here and now. But in any case, I don't think this argument based on falling objects is one of the strongest in favour of the immobility of the Earth. I'll be interested to see what we shall make of the question of shots fired from a cannon, especially when they are in the opposite direction to that of the diurnal motion.

SAGREDO. I wish I could understand the flight of birds with no more difficulty than shots from a cannon and all the other examples we've mentioned above! But how birds can fly freely in all directions and, even more, can stay in the air for hours at a time, is something that baffles me completely. I just can't understand how they don't lose the motion of the Earth as they [194] fly back and forth, or how they can keep pace with its velocity, which must be many times greater than the speed of their flight.

SALVIATI. Your doubts about this are well founded. In fact I wonder whether Copernicus himself was able to solve them to his entire satisfaction, or whether that was the reason why he said nothing about them. It's true that he dealt very briefly with the other objections to his theory, I think because, with his great genius, his mind was on greater and higher things—rather as lions pay little attention to the importunate barking of small dogs. I suggest, then, that we leave the example of birds until last, and in the meantime try to resolve Simplicio's other doubts, following our usual method of showing him that he already has the answers to hand although he is not aware of them. So let us begin with shots fired from a cannon. If two shots were fired from the same cannon, with the same ball and the same amount of powder, one eastwards and the other westwards, on what grounds does he believe that (if it is the Earth which rotates every twenty-four hours) the shot fired to the west would travel much further than the one fired towards the east?

The reason why a cannon ball fired in a westerly direction should appear to travel further than one fired towards the east.

SIMPLICIO. I believe that this would happen because, once a ball has been shot towards the east, it is followed by the cannon which, carried along by the Earth, moves rapidly in the same direction; therefore the ball falls to Earth only a short distance from the cannon. But when a shot is fired towards the west, by the time the ball falls to Earth the cannon has moved some considerable distance towards the east. Therefore the length of the shot—that is, the distance between the ball and the cannon—in the second case will appear to exceed the first by the distance which the Earth, and therefore the cannon, has travelled during the time the two balls were in the air.

SALVIATI. I would very much like to devise an experiment corresponding to the motion of these projectiles, as we did with the ship when we were discussing falling objects. I'm just trying to think how it could be done.

SAGREDO. I think a suitable proof could be set up by taking an open carriage and mounting a crossbow on it at half elevation, so as to maximize the length of the shot. Then, as the horses were running, you would fire one shot in the direction in which they were travelling, and then another in the opposite direction. You would have to make a careful note of the carriage's position at the moment when the bolt landed in each case, and from this you would be able to see how much one shot proved to be longer than the other.

An experiment with a moving carriage to illustrate the difference between the shots.

SIMPLICIO. I think this experiment would serve very well, and I have no doubt that the length of the shot—that is, the distance between the arrow and the position of the carriage at the moment when the arrow landed—would be notably less in the case of the shot fired in the direction the carriage was travelling than with the shot fired in the opposite direction. Suppose, for example, that the length of the shot itself was three hundred *braccia*, and that the carriage travelled a hundred *braccia* in the time that the arrow was in the air. With the shot fired in the direction the carriage was travelling, the carriage would have covered one hundred of the three hundred *braccia*, and therefore when the arrow landed its distance from the carriage would be only two hundred *braccia*. But with the shot fired in the opposite direction, the carriage would be moving away from the arrow, and therefore when the arrow had flown its three hundred *braccia* the carriage would have travelled one hundred *braccia* in the other direction, and so the distance between them would be four hundred *braccia*.

[195]

SALVIATI. Would there be any way of making the two shots equal?

SIMPLICIO. Only by making the carriage stand still.

SALVIATI. Of course, but I meant with the carriage travelling at full speed.

SIMPLICIO. You would have to increase the tension of the bow when you were shooting in the direction of travel, and reduce it when you were shooting the other way.

SALVIATI. So then, there is another way of doing it. How much would you have to increase the tension of the bow in one case, and reduce it in the other?

SIMPLICIO. In our example, where we supposed that the bow shot the arrow three hundred *braccia*, we would have to increase the tension so as to shoot four hundred *braccia* for the shot in the direction of travel, and reduce it so as to shoot only two hundred *braccia* for the other. This would mean that both shots would be three hundred *braccia* in relation to the carriage, because the hundred *braccia* travelled by the carriage would be subtracted from the four-hundred-*braccia* shot and added to the one which was only two hundred, making three hundred for both.

SALVIATI. And what effect does the greater or lesser tension of the bow have on the arrow?

SIMPLICIO. When the tension of the bow is increased it shoots the arrow with a greater velocity, and when its tension is reduced it shoots with less. So the same arrow travels a greater distance in one case than the other, in proportion to the greater or lesser velocity with which it is released from the bow in each case.

[196] SALVIATI. So, for the arrow to travel an equal distance from the moving carriage in both directions, in the first case in your example it would need to be shot with, say, four degrees of velocity, and in the second case with only two. But if it was shot with the bow at the same tension, it would have three degrees every time.

SIMPLICIO. That's correct; and that's why, if the arrows are shot with the bow at the same tension from a moving carriage, the shots can never be equal.

SALVIATI. I forgot to ask at what velocity the carriage is assumed to be travelling in this particular experiment.

SIMPLICIO. The velocity of the carriage must be assumed to be one degree, in comparison with the velocity imparted by the bow, which is three.

SALVIATI. Yes indeed; this makes the figures add up correctly. But surely, when the carriage is moving, everything in the carriage is also moving at the same velocity?

SIMPLICIO. Undoubtedly.

SALVIATI. And this applies to the bow, the arrow, and the bowstring from which the arrow is shot.

SIMPLICIO. Yes.

SALVIATI. So when the bolt is shot in the direction in which the carriage is travelling, the bow imparts its three degrees of velocity to a bolt which has one degree already, because this is the speed at which the carriage is carrying it in the same direction; and therefore the bolt is released from the bowstring with four degrees of velocity. When, on the other hand, the bolt is shot the other way, the same bow imparts the same three degrees of velocity to a bolt which is moving with one degree in the opposite direction, and therefore it leaves the bowstring with only two degrees of velocity. But you yourself have already said that for the two shots to be equal, the bolt would need to be shot with four degrees of velocity in one case and two degrees in the other; so without any need to change the tension of the bow, it is the carriage itself which makes the necessary adjustment. The experiment will serve to confirm this to those who are unwilling or unable to be convinced by reason. If you now apply this reasoning to the shots fired from a cannon, you will find that, whether the Earth is in motion or at rest, shots fired with the same force will always be equal, regardless of the direction in which they are fired. The error which Aristotle, Ptolemy, Tycho, you, and everyone else have made, is based on the fixed and deep-seated impression that the Earth is at rest, which you are incapable of shedding even when you want to speculate about what would happen if the Earth were in motion. In the [197] same way in our earlier discussion, it didn't occur to you that when the stone is at the top of the tower, the question of whether or not it is in motion depends on whether or not the terrestrial globe is in motion. Because you have it fixed in your mind that the Earth is at rest, you always talk about the falling stone as if it were starting from a state of rest; whereas you ought to say: 'If the Earth is at rest, the stone starts from a state of rest and falls vertically; but if the Earth is in motion, the stone is also in motion with the same velocity, and it starts not from a state of rest but from one of motion equal to that of the Earth. In this case, its downward motion combines with the Earth's motion to form an oblique motion.'

Solution of the argument based on shots fired eastwards and westwards from a cannon.

SIMPLICIO. But for heaven's sake, if it moves obliquely how is that I see it falling vertically in a straight line? This is simply denying the clear evidence of our senses; and if we can't believe our senses, what other basis is there for our philosophy?

SALVIATI. As far as the Earth, the tower, and we ourselves are concerned, since we all share the same daily motion, together with the stone, the daily motion is as if it did not exist: it remains insensible, imperceptible, and has no effect whatever. The only motion which is observable to us is the one which we do not share, namely the motion of the stone falling down the side of the tower. You are not the first to experience great reluctance in accepting this ineffectiveness of motion among things which share it in common.

A striking case cited by Sagredo to show the ineffectiveness of motion among things which share it in common.

SAGREDO. This reminds me of a fanciful idea which came to me one day as I was sailing to Aleppo, where I was going as our country's ambassador. It might help to explain this ineffectiveness of motion in common, which makes it seem as if it did not exist to those who participate in it; and with Simplicio's permission, I would like to tell him about the fancy which occurred to me then.

SIMPLICIO. Please do; I'm not just willing, but eager to hear this novelty.

SAGREDO. If the nib of a pen which was on the ship all the time it was sailing from Venice to Alexandretta* had been able to leave a visible mark of its whole journey, what kind of trace, or record, or line would it have left?

SIMPLICIO. It would have left a line stretching from Venice to its destination; not a perfectly straight line—or rather, not a perfect arc of a circle—but fluctuating here and there depending on the [198] fluctuating motion of the vessel. But these deviations from a straight line by a few *braccia* to the right or left, up or down, over a distance of many hundreds of miles, would have been insignificant over the whole length of the line and would have been almost imperceptible. So it would not be seriously misleading to say that the line would have been part of a perfect arc.

SAGREDO. So taking away the motion of the waves, if the ship had travelled the whole way over a perfectly calm sea, the absolutely true, real motion of the pen nib would have been a perfect arc of a circle. What if I had had the pen in my hand the whole time, and had occasionally moved it an inch or two this way or that: what difference would that have made to its principal very long line?

SIMPLICIO. Less than if a line a thousand *braccia* long had occasionally deviated from absolute straightness by the breadth of a flea's eye.

SAGREDO. Suppose, then, that an artist had begun drawing with this pen on a piece of paper when the ship left port and continued all the way to Alexandretta. He could have used the pen's motion to draw the whole story of the journey, with lots of figures perfectly outlined and sketched in a thousand different ways, with different countries, buildings, animals, and all kinds of things; and yet the real, true, and essential movement of the pen nib would have been nothing more than one very long, simple line. As far as the artist was concerned, his drawing would have been exactly the same if the ship had never moved; and yet the only trace of the pen's very long journey would be those lines drawn on a sheet of paper. The reason for this is that the ship's motion from Venice to Alexandretta was common to the paper, the pen, and everything else on the ship; whereas the small movements back and forth, to left and right, which the artist's fingers transmitted to the pen and not to the paper, were peculiar to the pen, but left their marks on the paper because the paper was fixed in relation to them. In the same way, since the earth is moving, the motion of the falling stone actually extends over several hundred or even several thousand *braccia*, and if it had traced the course which it followed in motionless air or on some other surface, it would have left a very long oblique line. But that part of its motion which is common to [199] the stone, the tower, and to us remains imperceptible to us, as if it did not exist. The only part of its motion which we can observe is the part in which neither we nor the tower participate, namely the stone's motion as it falls down the height of the tower.

SALVIATI. A very subtle illustration of this point, which many people find very hard to understand. Now, if Simplicio has nothing more to say in response, we can move on to the other experiments, which will be a good deal easier to explain in the light of what we have said so far.

SIMPLICIO. No, I've nothing to add. I was absorbed by this example of drawing, and by the thought that all those lines drawn in all directions, this way and that, up and down, back and forth, with a thousand and one twists and turns, are in reality nothing more than fragments of a single line all drawn in the same direction. The pen may move a little to the right or left, or move more rapidly one moment and more slowly the next, but these are only minimal deviations from a single straight line. It occurs to me that the same is true when we write a letter, and I was thinking of those expert calligraphers who show off their dexterity by embellishing a letter in a single stroke with innumerable flourishes without lifting their pen from the paper. If they were on a rapidly moving ship, they would produce one of their decorations from the motion of the pen, which is essentially a single line all drawn in the same direction, simply by means of minimal deviations or deflections from a perfectly straight line. I'm most grateful to Sagredo for suggesting this idea to me. So let us continue, and I shall be all the more attentive in the hope of hearing more such ideas.

Ironic statement of foolish subtleties taken from an encyclopedia.

SAGREDO. If you should be curious to hear other such witticisms, which not everyone can appreciate, there are plenty of others, especially in this matter of navigation. What do you think of the fine idea which occurred to me on this same voyage, when I realized that the crow's nest on the ship's mast had travelled further than its foot, without the mast bending or breaking? Clearly the top of the mast is further from the centre of the Earth than its foot, and therefore it travelled through an arc of a larger circle than the foot.

SIMPLICIO. So when a man walks, his head travels further than his feet?

[200] SAGREDO. You've understood perfectly, and have worked out the principle for yourself. But we mustn't interrupt Salviati.

SALVIATI. I'm glad to see Simplicio exercising his mental skills—if indeed this idea is his own, and he hasn't learnt it from a certain handbook of assertions* which contains various others no less acute and amusing. But we should continue our discussion of the shot fired from a cannon set up perpendicular to the horizon, in other words towards the zenith, and the fact that the ball falls straight back down onto the cannon again. If, the argument goes, the Earth has carried the cannon several miles towards the east during the long interval in which the ball was separated from it, the ball should fall down to earth an

Argument against the diurnal motion of the Earth based on a shot fired vertically from a cannon.

equal distance to the west of the cannon; but since this does not happen, the cannon must have waited for the ball without being moved. The answer to this objection is the same as with the stone falling from a tower: the misunderstanding and the fallacy in the argument consist in presupposing the truth of what it set out to prove. The objector is always firmly convinced that the cannon ball starts from a state of rest when it is fired from the cannon; but it can only start from a state of rest if we presuppose that the terrestrial globe is at rest, which is precisely the conclusion which is being tested. So I reply that those who maintain the mobility of the Earth point out that the cannon and the ball inside it both share in the motion of the Earth; indeed, like the Earth, they naturally have the same motion. Therefore the ball does not start from a state of rest, but shares in the circular motion around the centre, which is neither removed nor impeded when it is fired from the cannon. So the ball follows the Earth's general motion towards the east, and hence remains continually above the cannon both during its rise and as it comes down. You will see the same effect if you make the experiment on a ship with a ball thrown straight up in the air from a catapult: it will come back down in the same place, whether the ship is moving or not.

Response to this argument, showing the misunderstanding on which it rests.

SAGREDO. This answers the objection entirely. But since I see that Simplicio appreciates cunning arguments to trap the unwary, let me put this to him. Supposing for a moment that the Earth is at rest, and the cannon is aimed vertically straight up into the sky, does he have any difficulty in accepting that such a shot is truly vertical, and that the ball will follow the same straight line both when it is shot upwards and when it comes down, always assuming that any external or accidental impediments are eliminated?

Another reply to the same argument.

SIMPLICIO. Yes, as I understand it, that is exactly what must happen.

[201]

SAGREDO. If instead of being aimed vertically, the cannon were tilted in a given direction, what motion would the ball follow then? Would it still go vertically straight up in the air and follow the same line when it comes down, as with the other shot?

SIMPLICIO. No, it wouldn't. It would follow a straight line in the direction in which the barrel of the gun was pointing, except in so far as its own weight made it decline towards the Earth.

Projectiles continue their motion in a straight line following the direction of the motion which they shared with the projector before they were separated from it.

SAGREDO. Therefore the direction of the gun's barrel determines the motion of the ball, which does not deviate from this line, or would not if it were not pulled down by its own weight. So then, if the barrel is aimed vertically and the ball is shot straight up in the air, it will come down following the same straight line, because the motion due to the heaviness of the ball pulls it down in the same vertical line. When the ball leaves the cannon it continues to travel in the same direction as that portion of its journey which it made inside the barrel, does it not?

SIMPLICIO. So it seems to me, yes.

SAGREDO. Now imagine the barrel aimed vertically, and the Earth rotating in its daily motion on its axis and carrying the cannon with it. What will be the motion of the ball inside the barrel when it is fired?

SIMPLICIO. It will move vertically in a straight line, if the barrel is aimed vertically.

Given the rotation of the Earth, a ball shot from a cannon aimed vertically will not follow a vertical line, but an inclined one.

SAGREDO. Think carefully, because I don't think it would be vertical at all. It would be vertical if the Earth were at rest, because then the ball would have no other motion apart from that imparted to it by the firing of the cannon. But if the Earth rotates, then the ball in the cannon also has this diurnal motion. This means that when it is fired, it travels from the breech of the cannon to its muzzle with two motions, the combined effect of which is that the ball's centre of gravity travels in an inclined line. To make it clearer, let AC be the upright barrel of the cannon, and B the ball inside it. Clearly if the cannon is fired when it is motionless, the ball will emerge from the muzzle A, its centre having travelled the length of the barrel along the vertical line BA, and it will continue in this direction after it has left the cannon, travelling straight upwards. But if the Earth is [202] rotating, and consequently carrying the cannon with it, then when the cannon is fired, it will be carried by the Earth's motion to the new position DE in the time that the ball travels the length of the barrel. So the ball will emerge from the muzzle at D,

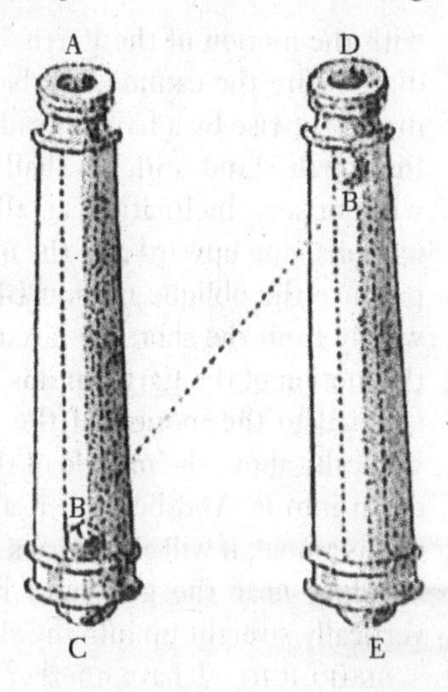

its centre having travelled along the line BD, which is no longer a vertical line but one inclined towards the east. Since we have established that the ball must continue its motion in the air in the same direction as when it was in the barrel, it will continue to follow the inclination of the line BD, not vertically but inclined towards the east. As the cannon is also moving in the same direction, the ball will be able to follow the motion of the Earth and of the cannon. And so you see, Simplicio, how the shot which apparently had to be vertical is in fact not so at all.

SIMPLICIO. I'm not entirely convinced by this argument; what about you, Salviati?

SALVIATI. I'm partly convinced, but I have a difficulty which I'm struggling to put into words. It seems to me to follow from what has been said that, if the cannon is vertical and the Earth is in motion, then not only does the ball not come down some distance to the west of the cannon, as Aristotle and Tycho say it would, but neither does it come down directly above the cannon, as you would have it. Rather, it will fall well to the east of it. According to your explanation it would have two motions that would combine to impel it in that direction, namely the motion which it has in common with the Earth, which carries both the cannon and the ball from CA towards ED, and the motion imparted to it by the shot, which impels it along the inclined line BD. Since both of these motions carry it towards the east, the combined motion should be greater than the motion of the Earth.

SAGREDO. No, you're mistaken. The motion which carries the ball towards the east comes entirely from the Earth, and firing the shot has no part in it; the motion which impels it upwards comes entirely from the shot, and has nothing to do with the motion of the Earth. The proof of this is that, if you did not fire the cannon, the ball would never come out of the muzzle or rise by a hair's breadth; and if you fired it and made the Earth stand still, the ball would shoot straight upwards without any inclination at all. The ball therefore has two motions, one upward and the other circular, which combine to produce the oblique motion BD. The upward motion derives wholly from the shot; the circular motion derives wholly from [203] the motion of the Earth, and is equal to it. Because this motion is equal to the motion of the Earth, the ball always remains vertically above the muzzle of the gun and eventually falls back down into it. And because it always remains above where the gun is aimed, it will also always appear to be overhead to anyone standing near the gun, and hence we perceive it as rising vertically straight up into the sky.

SIMPLICIO. I have another difficulty. The ball travels the length of the gun at such a speed that it seems impossible for the gun to move in that brief moment from CA to ED. How can it impart such a degree of inclination to the oblique line CD that the ball is able to keep up with the motion of the Earth when it is in the air?

SAGREDO. You make several mistakes here. First of all, I think the inclination of the line CD is much greater than you imagine, because I have no doubt that the speed of the Earth's motion—even at our latitude, never mind at the equator—far exceeds that of the cannon ball when it travels the length of the barrel. So the distance CE cannot fail to be greater than the length of the gun barrel, and hence the inclination of the oblique line must be more than forty-five degrees. But in any case, it makes no difference whether the Earth's velocity is greater or less than that of the cannon-shot, because if the Earth moves more slowly and therefore the inclination of the diagonal is less, a smaller inclination will be all that is needed for the ball to maintain its position above the cannon. In fact, if you think carefully about it, you will see that when the Earth's motion transfers the cannon from CA to ED, it bestows exactly the degree of inclination on the diagonal CD that is needed to compensate for the shot. But you also make a second mistake, by attributing the ability of the ball to keep pace with the Earth's motion to the impetus of the shot. This is the same error which Salviati seemed to fall into just now; because the motion by which the ball keeps pace with the Earth is the perennial, perpetual motion which it indelibly and inseparably shares with every other terrestrial object, and which it naturally possesses and will possess in perpetuity.

SALVIATI. We'd better give in, Simplicio; it is indeed just as he says. But this discussion has helped me to resolve a puzzle about wildfowling, which is how marksmen with an arquebus can hit and kill birds in flight. I had always imagined that in order to hit a bird they aimed a certain distance in front of it, allowing for how fast the bird was flying and how far away it was. Then, when they fired the gun and the bullet followed their aim, it reached the same point at the same time as the bird in flight, and so hit it. But when I asked one of them if this was indeed what they did, he said not, and that their technique was much easier and more certain than that. He told me that they use exactly the same method as they would to shoot a bird which is sitting still. They take aim at the bird in flight, and follow it with their gun as it moves, keeping it in their sights until they shoot; and this enables them to hit it as they would a bird sitting still. So the movement of the gun, slow as it is, as they turn it to keep it trained on the bird, must be imparted to the bullet and combined with the impetus imparted by the shot. Thus the bullet receives the upward motion from the shot, and the declining motion following the bird's flight from the barrel of the gun, just as we said just now with the cannon ball. The cannon ball receives its upward motion from the shot, and its inclination towards the east from the motion of the Earth; the two combined make it follow the motion of the Earth while appearing to an observer to be simply moving straight upwards, and to follow the same line when it comes down. In the same way with the huntsman's bullet, keeping the gun continuously trained on the target means that the shot hits the mark; and to keep the gun trained on the target, if the target is stationary then the gun must be held still; if the target is moving, the gun must move to keep it in its sights.

How marksmen are able to kill birds in flight.

[204]

Solution to the objection based on cannon shots fired towards north and south.

The response to the objection based on cannon shots fired at a target to the north or south follows the same principle. It was objected that if the Earth was in motion, such shots would always veer towards the west, because in the time that the ball was in the air between leaving the gun and reaching its target, the target would move eastwards, leaving the ball to the west of it. My response is to ask whether a cannon which is trained on a target and then left in place would continue to be focused on the same target, regardless of whether the Earth moves or not. The answer must be that the aim of the cannon does not change at all, for if the target is stationary, so is the cannon. If the target moves because it is carried along by the Earth, the cannon also moves in the same way. And if it remains trained on the target the shot will always hit the mark, as will be clear from what has been said above.

[205] SAGREDO. Allow me to interrupt you for a moment, Salviati, and to put forward a thought which has occurred to me about these marksmen and birds in flight. I don't doubt that their technique is as you describe it, and that the result is that they succeed in hitting the bird. But this doesn't seem to me to be quite the same as the example of shots fired from a cannon, which have to hit their target whether cannon and target are both in motion or whether they are both at rest. The differences seem to me to be as follows. When the cannon is fired, both it and the target are moving at an equal speed, because they are both carried along by the motion of the terrestrial globe. It's true that sometimes the cannon is placed nearer to the pole than the target, so that its movement is somewhat slower because it is turning in a smaller circle; but the short distance between the cannon and the target makes this difference imperceptible. When a marksman is shooting at a bird, on the other hand, the motion of the arquebus as it follows the bird's flight is far slower than the bird itself; so I don't think it is possible that the limited motion which the turning of the gun barrel imparts to the bullet inside the gun can be multiplied when the bullet is in the air—certainly not to the extent of equalling the speed of the bird's flight, and hence staying aimed at the bird. Rather, the bird must fly ahead of the bullet and leave it behind. What's more, there is no suggestion that the air

through which the bullet passes is moving with the speed of the bird; whereas in the case of the cannon, the gun, the target and the intervening air all share equally in the common diurnal motion. So I think there must be several reasons why the marksman succeeds in hitting the bird. He must not just follow the bird's flight with his gun barrel but anticipate it a little, aiming slightly ahead of it. Then, I believe they shoot not just with one bullet but with a good number of small pellets, which spread out in the air and cover quite a wide area. And there is also the great speed with which the pellets are fired from the gun and fly towards the bird.

SALVIATI. See how Sagredo's wit flies ahead of mine and leaves me behind; I might well have noticed these differences, but only after long and careful thought. But to return now to our main topic, we still have to consider what happens with cannon shots fired horizontally towards the east and towards the west. It is argued that if the Earth moved, the former would always pass above their target and the latter would be below it, because the diurnal motion of the Earth means that the east is always sinking below the tangent parallel to the horizon, and the west is always rising. This is why the stars appear to rise in the east and to set in the west. Therefore a shot aimed along this tangent in an easterly direction would be too high, because the target sinks while the ball is travelling, and one aimed in a westerly direction would be too low, because the target rises as the ball moves along the tangent. The answer to this objection is the same as to the others: just as the motion of the Earth means that the target to the east is continually falling in relation to a fixed tangent, so the cannon is also continually declining for the same reason. Hence it remains trained on the same target, and the shot hits the mark. But this may be an opportune moment for me to point out that the followers of Copernicus have perhaps been too generous in what they concede to their opponents. They accept as established fact some experiments which their adversaries have never actually carried out, such as the example of objects falling from the mast of a moving ship, and many others. I'm sure that this claim to prove that cannon shots fired to the east are too high, and those to the west too low, is one such. Since I don't believe they have ever made the

Response to the argument based on shots fired horizontally towards the east and west.

[206]

Solution of the objection based on shots fired towards the east and west.

The followers of Copernicus are too ready to admit the truth of some dubious propositions.

experiment, I would like them to tell me what difference they would expect to find between the same shots, assuming the Earth first to be at rest and then to be in motion. Perhaps Simplicio would respond on their behalf.

SIMPLICIO. I can't claim to give such a well-founded reply as perhaps someone who understands these things better than me. But I think they would say that the result would be as has already been described: that if the Earth moved, shots fired to the east would always be too high, etc., provided the ball moved along the tangent, as seems probable.

SALVIATI. And if I said that this is in fact what happens, how would you set about challenging my claim?

SIMPLICIO. We would have to make the experiment to resolve the matter.

SALVIATI. But do you think we could find a gunner so expert that he could undertake to hit the target with every shot at a distance of, say, five hundred *braccia*?

SIMPLICIO. No, I don't; I doubt whether anyone, however expert, could promise to be accurate to within less than a *braccio* over this distance.

[207] SALVIATI. So if shooting is so unreliable, how can we assure ourselves of the point which is in doubt?

SIMPLICIO. We could assure ourselves of it in two ways: first, by firing a large number of shots; and second, because the velocity of the Earth's motion is such that the shot's deviation from the target would, in my view, be very large.

SALVIATI. Very large, meaning much more than a *braccio*, since we grant that this much and even more would be the normal margin of error even if the Earth is at rest.

SIMPLICIO. I have no doubt that the variation would be significantly more than this.

Calculation of how far cannon shots would have to deviate from their target, given the motion of the Earth.

SALVIATI. I suggest we make a very rough calculation, just for our own amusement. If, as I hope, the figures add up, it will also serve to warn us on other occasions not to be too ready to take figures on trust, or to assent straight away to whatever we are invited to imagine. To give every possible advantage to the followers of Aristotle and Tycho, let's assume that we are at the equator, and that we shoot horizontally towards the west from a long-range cannon at a target five hundred *braccia* distant.

Let's start by estimating—approximately, as I have said—the time that elapses from when the ball leaves the gun to when it reaches the target. We know that this time is very short, certainly no longer than it would take someone to walk two paces. This is less than a second, assuming that a person walks three miles in an hour: three miles is nine thousand *braccia*, and an hour contains three thousand six hundred seconds; so in a second the walker covers two and a half paces. Therefore a second is longer than the time that the cannon ball is in the air. Since the diurnal revolution of the Earth is twenty-four hours, the western horizon rises by fifteen degrees in an hour, which is fifteen minutes of arc in one minute of time, or fifteen seconds of arc in one second of time. If we say that the time taken by the shot is one second, then in this time the western horizon, and also the target, rises by fifteen seconds of arc. This is fifteen seconds of a circle whose radius is five hundred *braccia*, this being what we assumed as the distance from the cannon to the target. Now let us look in a table of arcs and chords to find what the chord of fifteen seconds is for a radius of five hundred *braccia*. Here it is in Copernicus's book,* where we can see that the chord of one minute of arc is less [208] than 30 parts of a radius of 100,000. Therefore the chord of one second of arc will be less than half of one such part, i.e. less than 1 part of 200,000, and hence the chord of fifteen seconds will be less than 15 parts of 200,000. But 15 parts of 200,000 is still more than four-hundredths of five hundred; so the distance by which the target will have risen while the cannon ball is in the air will be less than four-hundredths, i.e. a twenty-fifth, of a *braccio*, in other words about an inch. Now since such a variation—that is, of shooting an inch lower than one would if the Earth were at rest—in fact occurs in all cannon shots, how, Simplicio, are you going to convince me experimentally that it does not occur? Is it not clear that in order to refute me, you would first have to find a way of shooting at a target with such a degree of accuracy that shots never missed by as much as a hair's breadth? As long as shots can vary by a *braccio*, as in fact they do, I will always be able to say that this variation includes the inch which is due to the motion of the Earth.

A very subtle demonstration that, assuming the motion of the Earth, there is no reason for a cannon shot to vary any more than if the Earth is at rest.

SAGREDO. Forgive me, Salviati, but I think you're being too generous. I would say to the Aristotelians that even if every shot hit the very centre of the target, that would still not invalidate the motion of the Earth. After all, gunners have always trained to adjust their sights so as to hit the target, and are skilled in pointing the cannon at the mark so as to hit it despite the motion of the Earth. If the Earth were to stand still, their shots would not hit the mark, but those towards the west would be too high and those towards the east too low. Let Simplicio try to convince me otherwise.

SALVIATI. That's a subtlety worthy of Sagredo. But the point is that since this variation depending on whether the Earth is in motion or at rest must necessarily be extremely small, it is bound to remain submerged among the much larger variations which occur all the time because of any number of accidental factors. So let us concede this to Simplicio, simply as a warning of how cautious we must be in accepting as true many experiments which those who cite them have never carried out, although that doesn't stop them boldly bringing them out when they serve to bolster their case. I say let us grant this to Simplicio, because the plain truth is that the outcome of these [209] cannon shots would be exactly the same whether the terrestrial globe is in motion or at rest. The same will be true of all the other experiments which have been or could be cited; they all appear at first sight to have some semblance of truth, in so far as the old idea that the Earth is at rest continues to perpetuate our misunderstandings.

The need to be very cautious in accepting as true experiments which those who cite them have never carried out.

Experiments and arguments against the motion of the Earth appear conclusive in so far as they perpetuate misunderstandings.

SAGREDO. For my part, I'm fully convinced by the arguments so far. It's quite clear to me now that, once our imagination has grasped the fact that this general participation in the daily rotation is common to all terrestrial things, and that this is their natural state—just as in the old world view their natural state was assumed to be a state of rest around the centre—then one has no difficulty in seeing through the false reasoning and misunderstandings which made the objections raised appear convincing. I just have one remaining doubt, as I mentioned earlier, about the flight of birds. As living creatures, they have the faculty of moving at will in any number of ways and of staying airborne, separated from the Earth, for long spells at a time. They fly about in a completely random way, and I can't understand how in all this confusion of motion they don't lose their sense of the primary common motion, or how having lost it they find it again and compensate for it in their flight. And how do they keep up with the towers and trees which rush so precipitously towards the east? I say precipitously, because for the great circle of the terrestrial globe this is almost a thousand miles per hour, and I don't think a swallow's flight is even fifty miles per hour.

SALVIATI. The birds would certainly struggle to keep up with the motion of the trees if they had to rely on their own wings; and if they no longer shared in the universal rotation, they would be so far left behind, and their westward course would be so furious, that it would look as if they were flying much faster than an arrow—if indeed you could see them at all. In fact, I don't think you would be able to see them, any more than we can see a cannon ball in flight when it is propelled by the force of the gun's charge. But the truth is that the birds' own motion—their flight, that is—has nothing to do with the universal motion, which neither helps nor hinders them. This motion remains constant for the birds because of the air in which they fly, which naturally follows the spinning of the Earth, carrying with it the clouds, the birds and anything else that is [210] suspended in it. So the birds don't need to worry about following the motion of the Earth; they can simply remain oblivious of it.

SAGREDO. I have no problem at all in understanding how the air carries the clouds along with it. They are easily moved, since they are so light and have no disposition to any contrary motion, and in any case they are of a matter which shares in the conditions and properties of the Earth. But the birds, as living creatures, are capable of motion which may even be contrary to the diurnal motion, and I find it quite hard to see how the air can restore this motion to them once they have interrupted it. What's more, they are solid bodies which have weight, and we've already seen how stones and other heavy bodies resist the impetus of the air, and even if they succumb to it they never acquire the velocity of the wind which drives them.

SALVIATI. Sagredo, let's not underestimate the strength of air in motion, which has the power when it moves rapidly to drive fully-laden ships, to uproot forests and to overturn towers; and yet even these violent actions of the wind can't be said to come anywhere near the velocity of the diurnal revolution.

SIMPLICIO. Ah, so air in motion can maintain the motion of projectiles, as Aristotle said. I thought it was very strange that he should have erred on this point.

SALVIATI. No doubt it would be able to, if it could maintain its own motion. But when the wind drops, ships no longer move and trees are no longer uprooted; so, since the air is no longer in motion when the thrower lets go of the stone and stops moving his arm, the fact remains that it must be something other than the air which keeps the projectile in motion.

SIMPLICIO. What do you mean, when the wind drops the ship no longer moves? On the contrary, we can see that without any wind, and even with its sails furled, a vessel can carry on moving for miles at a time.

SALVIATI. But Simplicio, it contradicts your argument if, when the air which was driving the ship by filling the sails is no longer in motion, the ship continues on its way without any help from the medium.

[211] SIMPLICIO. It could be said that the water is the medium which drives the ship and maintains its motion.

SALVIATI. It could, but it would be the opposite of the truth. The truth is that the water strongly resists being divided by the body of the vessel, and noisily opposes it, largely depriving it of the velocity which the boat would acquire from the wind if the obstacle of the water were not there. I don't think, Simplicio, you can ever have observed how the water foams as it runs past the side of a boat, when the boat is propelled rapidly through still water by oars or the wind. If you had noted this effect you would not have thought of putting forward such a foolish suggestion now. It's becoming clear to me that you have hitherto been one of that flock who, when you want to find out about such effects and to understand natural phenomena, withdraw to your study and look through an index and a table of contents to see if Aristotle said anything about them, instead of going out and observing boats, or catapults, or cannons. And once you've satisfied yourselves that you've established the true meaning of the text, you look no further, and it doesn't occur to you that there is any more to be known on the subject.

SAGREDO. What a happy and enviable position to be in! If all humans naturally desire knowledge,* and if being informed is the same as believing oneself to have understood, then such people enjoy a huge privilege. They are able to persuade themselves that they know everything, and make fools of those who are aware of how little they know and who, recognizing that they know only a tiny part of what there is to be known, wear themselves out with long hours of study and reflection, and labour over experiments and observations.

Great and enviable happiness of those who are convinced that they know everything.

But please let's return to our discussion of birds. You were saying that the air, moving at great speed, could restore to the birds whatever part of the daily rotation they might lose in the course of their erratic flying hither and thither. My reply to this is that air in motion does not seem to be able to impart to a solid heavy body a speed equal to its own; and since the motion of the air is the same as that of the Earth, the air doesn't seem sufficient to restore the motion which the birds might have lost in their flying about.

SALVIATI. Your argument appears very plausible, and the doubt you express is not a trivial one. But if we look beyond appearances, I don't think that in essence it has any more force than the others which we have already considered and disposed of.

SAGREDO. It's clear that if it can't be shown to be necessarily conclusive then it's invalid, because only then can no valid counter-argument be produced against it.

SALVIATI. It seems to me that the greater difficulty you have with this example than with the others derives from the fact that birds are living creatures, and that therefore they can use their strength at will to resist the primary motion which is inherent in all terrestrial things. Hence, we can see birds flying upwards as long as they are alive, although this motion is impossible for them as heavy bodies; and indeed dead birds can only fall to the ground. You conclude, therefore, that the principles applying to all kinds of projectiles which we have discussed above, cannot apply to birds. This is quite true, which is why we don't see other projectiles behaving as birds do. If you [212] were to let go of a dead bird and a live one from the top of a tower, the dead bird would behave in the same way as a stone: it would follow first the general diurnal motion, and then, as a heavy body, a downward motion. But there would be nothing to stop the live bird, while still participating in the daily motion, from using its wings to fly in whatever direction it liked; and this new motion would be apparent to us because it would be specific to the bird and we would have no part in it. And if it chose to fly towards the west, there would be nothing to stop it from using its wings once more to fly back to the top of the tower. This is because, in the last analysis, the effect of the bird's flying towards the west was simply to subtract one degree of velocity from, say, ten degrees of the diurnal motion, of which the bird would still have nine degrees while it was flying. Then, if it settled on the ground, it would have the common ten degrees again; and by flying back towards the east it could add one more degree, and these eleven degrees would carry it back to the tower. So in short, if we examine the effects of birds in flight more closely, we find that the only way in which they differ from the motion of other projectiles in any direction is that whereas other projectiles are moved by an external projecting force, the flight of birds derives from an internal principle.

Solution to the objection based on birds flying contrary to the motion of the Earth.

And now, to set the seal on the invalidity of all the experiments which have been cited so far, I think this is the time and place to show how they can all be very easily put to the test. Shut yourself in with a friend in the largest cabin below deck on some large ship. In the cabin, see that you have some flies, butterflies, and other such flying creatures. Have also a large tank of water, with some fish swimming in it. Suspend from the ceiling a small bucket, so that it lets water fall a drop at a time into another container with a narrow opening placed below it. Then, while the ship is not moving, observe carefully how the flying creatures fly at the same speed towards every part of the cabin; the fish swim equally well in every direction; the drops of water all fall into the container below. If you throw something to your companion, you will not need to throw it with more force in one direction than another, provided the distances are the same. Jump with both feet across the floor, and you will jump the same distance in any direction.

Experiment to show the invalidity of all the examples cited as objections to the motion of the Earth.

[213]

Observe all these effects carefully, even though as long as the ship is stationary there is no reason to expect them to be otherwise than as they are. Then have the ship move, at whatever speed you choose. Provided its motion is uniform and not rocking to and fro, you will not notice the slightest change in any of these effects, and none of them would give you any indication of whether the ship was moving or at rest. If you jump across the floor, you will jump the same distance as before, and however fast the ship is travelling you will not jump any further towards the stern than towards the bow, even though the floor will have moved in the opposite direction to your jump while your feet were off the ground. If you throw something to your companion, you will not have to throw it any harder to reach him whether he is nearer the bow and you nearer the stern, or vice versa. The drops of water will still fall into the container below as they did before, even though the ship travels several feet as the drop is falling. The fish will swim equally easily towards the forward-facing end of the tank as towards the aft-facing end, and will readily come to food wherever it is placed around the edge. Finally, the flies and butterflies will continue to fly equally in all directions, and you will not notice any fewer of them in the forward part of the cabin, as if they were struggling to keep up with the rapid course of the ship, even though they had been separated from it by flying around for a long time. And if you were to produce a little smoke by burning a few grains of incense, you would see it rise in the air and form a small cloud, which would not move any more in one direction than the other.

The reason why none of these effects changes is that the motion of the ship is common to everything in the cabin, including the air, which is why I said that it should be below deck. If you were on deck in the open air, where the air does not follow the course of the ship, you would see more or less notable [214] differences in some of the effects I have mentioned. The smoke, undoubtedly, would be left behind, as would the air itself. The flies and butterflies, too, would encounter resistance from the air, and so would not be able to keep up with the motion of the ship once they had been separated from it for a significant length of time. If they stayed close to the ship, however, they would be able to follow it without difficulty, because the ship, being an irregularly shaped construction, carries some of the air close to it along with it—in the same way as we sometimes see flies and horseflies following the horses, flying now on one side of their bodies and now on the other, when a stage coach is travelling at full speed. With the falling drops of water there would be very little difference, and with jumping and throwing heavy objects it would be quite imperceptible.

SAGREDO. It never occurred to me to put this to the test when I was travelling on a ship, but I'm quite sure that these observations would be exactly as you have said. In fact, I can recall dozens of times when I was in my cabin and I had to ask whether the ship was moving or not; and sometimes if I was lost in thought I imagined it was moving in one direction when in fact it was moving in the other. So I'm entirely convinced of the worthlessness of all the experiments that have been cited as evidence against rather than in favour of the rotation of the Earth. There remains the question raised by the observation that a rapid spinning motion has the effect of extruding and dispersing any material adhering to the device which is spinning. It seemed to many, including Ptolemy,* that if the Earth rotated at such a great speed, stones and living creatures would be hurled out towards the stars, and no cement would be strong enough to attach buildings to their foundations and stop them being swept away.

Foolishness of those who believe that the Earth began to move only when Pythagoras began to say that it moved.

SALVIATI. Before I come to the solution of this objection, I can't resist recalling how often I've had occasion to laugh at the reaction of almost everyone when they first hear it suggested that the Earth which they believed to be fixed and immobile is actually in motion. Not only had they never doubted that the Earth was at rest, but they were also firmly convinced that everyone else shared their belief that it had been created immobile and had remained so throughout all the past centuries. With this idea firmly fixed in their mind, they are then astonished [215] when they hear that someone says that it moves, as if that person had originally considered it motionless and had then foolishly imagined that it began to move only when Pythagoras or whoever else it was first suggested that it moved. It doesn't surprise me that the common masses, with their shallow understanding, should make such a stupid mistake—as if those who accept the motion of the Earth imagined that it was motionless from its creation up to the time of Pythagoras, and that it only became mobile when Pythagoras described it as such. But for thinkers like Aristotle and Ptolemy to have fallen into such a childish error seems to me to show extraordinary and quite inexcusable simple-mindedness.

SAGREDO. Do you mean to say, Salviati, that you think that when Ptolemy argued for the immobility of the Earth, he thought he had to argue against those who conceded that it had been immobile up to the time of Pythagoras, and claimed that it had only begun to move when Pythagoras attributed motion to it?

SALVIATI. No other interpretation is possible, if we consider the argument he uses to refute their claim. He refutes it by saying that buildings would be destroyed, and that stones, animals, and humans themselves would be hurled into the sky. But since this destruction and being thrust into the sky could only happen to buildings and animals which were on the Earth in the first place, and since humans could only settle on the Earth and construct buildings when the Earth was at rest, it's clear that Ptolemy must be arguing against those who conceded that Earth had once been at rest. Only then could animals and stones and builders have established themselves there, and constructed palaces and cities; and then it must suddenly have started to move, causing the ruin and destruction of buildings, animals, etc. But if Ptolemy had set out to argue against those who said that the Earth had been spinning ever since it was created, he would have done so by saying that if the Earth had always been in motion it would never have been possible for animals, humans, or stones to exist there, much less to construct buildings, found cities, etc.

When Aristotle and Ptolemy argue against the mobility of the Earth, they appear to argue against those who believed that it had long been at rest, and had only begun to move at the time of Pythagoras.

SIMPLICIO. I'm not sure that I see this inconsistency which you attribute to Aristotle and Ptolemy here.

SALVIATI. Ptolemy argues either against those who consider that the Earth has always been in motion, or against those who say it was once at rest and then started to move. If he was arguing against the former, he should have said: 'The Earth has not always been in motion, because if it had there would never have been humans, animals, or buildings on the Earth, as the [216] spinning of the terrestrial globe would have made it impossible for them to remain there.' But what he actually says is: 'The Earth does not move, because the animals, humans, and buildings which are on the Earth would collapse.' This presupposes that the condition of the Earth was once such that animals and humans were able to live there and construct buildings, and consequently that it was once at rest, this being the condition in which animals could live there and houses and buildings could be constructed. Now do you see what I meant?

SIMPLICIO. Yes, though I'm not entirely convinced by it. But this has little to do with the merits of the case, and an inadvertent slip on Ptolemy's part isn't enough to establish the motion of the Earth if in fact it's at rest. Now let's set joking aside and come to the heart of the objection, which seems to me to be insoluble.

SALVIATI. And I, Simplicio, far from wanting to solve it, intend to pull the knot even tighter. I will provide sense evidence to show that when heavy bodies are spun rapidly around a fixed centre they acquire an impetus to move away from that centre, even if their natural propensity would be to move towards it. Suppose we take a small bucket with some water in it, and tie one end of a rope to it. Then, holding firmly on to the other end of the rope, make it go rapidly round in a circle, the rope and your arm being the radius and your shoulder joint its centre. You will find that, whether the circle is horizontal, or vertical, or at any inclination you please, the water will not spill out of the container; rather, the person who swings it will feel a constant tension on the rope, pulling it away from their shoulder. If you make a hole in the bottom of the bucket, you will see water spurting out from it equally whether it is towards the sky, laterally, or towards the ground. If you were to replace the water with small stones and spin it in the same way, you would feel the same force pulling on the rope. Or again, you can see boys throwing stones a great distance by whirling round a slotted stick with a stone inserted at one end. All these arguments support the truth of the conclusion that a spinning motion, provided it is fast enough, confers an impetus on a moving body towards the circumference. This explains why, if the Earth revolves on its axis, the motion of its surface, which—especially near the equator—is incomparably faster

Rapid spinning has the effect of extruding and dispersing.

[217]

than those we have cited, ought to extrude everything outwards towards the sky.

SIMPLICIO. The objection seems to me very well established, and I think it will be hard for anyone to remove it and untie the knot.

SALVIATI. Untying the knot depends on some facts which you know and assent to no less than I do; but since you don't recall them, you can't see how it can be untied. So I shan't teach you these things, because you know them already, but simply remind you of them, so that you can resolve the objection for yourself.

SIMPLICIO. I've noted your method of reasoning several times, and it has reminded me of what Plato says, that *nostrum scire sit quoddam reminisci*—our knowledge is a kind of remembering.* So please explain your thinking, and remove this doubt for me.

Our knowledge is a kind of remembering, according to Plato.

SALVIATI. I will show you my view of what Plato says by what I say and by what I do. I've already done so by what I've done several times in the course of our discussions, and I'll follow the same method with the matter we are dealing with here. This may serve as an example to help you understand my view of how we gain knowledge, provided we have time for another day's discussion and we don't try Sagredo's patience with this digression.

SAGREDO. On the contrary, I shall be delighted to hear it. I remember that when I studied logic, I was never entirely convinced by this much-vaunted powerful method of Aristotle's.

SALVIATI. Then let us proceed. First, I would like Simplicio to tell me what kind of motion the boy imparts to the stone slotted in the cleft of his stick when he moves to throw it.

SIMPLICIO. The motion of the stone while it is in the stick is circular; that is, it travels along the arc of a circle whose fixed centre is the thrower's shoulder joint and whose radius is the stick and the thrower's arm.

SALVIATI. When the stone escapes from the stick, what is its motion then? Does it continue with its previous circular motion, or does it follow a different course?

SIMPLICIO. It clearly doesn't continue its circular motion, since this wouldn't allow it to move away from the thrower's shoulder, whereas we see that it travels a long way.

[218] SALVIATI. What motion does it follow, then?

SIMPLICIO. Let me think about it for a moment; I haven't tried to imagine it before.

SAGREDO. A word in your ear, Salviati: here is the *quoddam reminisci*, something remembered, in action, to be sure. You're giving it a lot of thought, Simplicio!

The motion imparted by the projector is solely in a straight line.

SIMPLICIO. It seems to me that the motion which the stone receives when it leaves the sling can only be in a straight line—or rather, the extra impetus imparted to it must be in a straight line. I was bothered by seeing it describing an arc; but since this arc always bends downwards and not in any other direction, I understand that this must come from the weight of the stone, which naturally pulls it downwards. So I say that the impressed impetus is definitely in a straight line.

SALVIATI. Yes, but what straight line? An infinite number of straight lines in all directions can be drawn from the end of the stick and the point at which the stone is separated from it.

SIMPLICIO. It moves in the direction of the motion which the stone had when it was moving with the stick.

SALVIATI. But you have already said that the motion of the stone when it was in the stick was circular; and circular motion cannot also be in a straight line, since no part of a circular line is straight.

SIMPLICIO. I don't mean that it is projected along the whole of the circle, but only of the last point at which its circular motion ceased. I know what I mean but I can't express it properly.

SALVIATI. It's clear to me too that you understand the fact, but you don't have the proper terms to express it. I can teach you these—teach you the words, that is, but not the truths themselves, which are facts. So to show you that you know the fact but lack only the terms to express it, tell me: when you shoot a bullet with a gun, in what direction does it acquire the impetus to travel?

SIMPLICIO. It acquires the impetus to travel along the straight line which continues the direction of the gun's barrel; it deviates from this neither to left or right, nor up or down.

SALVIATI. Which is as much as to say that it forms no angle whatsoever with the straight line of motion through the gun's barrel.

SIMPLICIO. Yes, that was what I meant.

SALVIATI. If, then, the line followed by the projectile must extend without forming an angle with the circular line which it described when it was with the projector, and if it must pass from circular motion to motion in a straight line, what must this straight line be? [219]

SIMPLICIO. It can only be the line touching the circle at the point of separation. It seems to me that any other line, if it were extended, would intersect the circumference, and therefore would form an angle with it.

SALVIATI. You have argued very well, and shown yourself to have the makings of a geometer. So here are the words which you can commit to memory to express your understanding of the fact: that a projectile acquires an impetus to move along the tangent to the arc described by the motion of the thrower at the point at which it separates from what projected it.

SIMPLICIO. I understand this entirely, and it is exactly what I meant.

SALVIATI. On a straight line which touches a circle, what is the closest point to the centre of the circle?

SIMPLICIO. It must clearly be the point of contact, since this is a point on the circumference of the circle, and all the other points are outside it; and points on the circumference are all equally distant from the centre.

SALVIATI. Therefore when an object leaves the point of contact and moves along a tangential straight line, it will move continually away from the point of contact and also from the centre of the circle.

SIMPLICIO. Certainly.

SALVIATI. Now, if you have kept in mind the propositions which you have explained to me, bring them together and tell me what can be gathered from them.

SIMPLICIO. I don't think my memory is so defective that I can't recall them. From what has been said we can gather that when a projectile which is moved rapidly in a circle separates from the thrower, it retains the impetus to continue its motion along the straight line which touches the circle described by the motion of the thrower at the point of separation. This motion takes the projectile continually further away from the centre of the circle followed by the motion which projected it.

A projectile moves along the tangent to its previous circle of motion at the point of separation.

SALVIATI. You now know the reason why heavy objects adhering to the surface of a rapidly turning wheel are extruded and thrown out from its circumference, at an ever-increasing distance from the centre.

[220] SIMPLICIO. Yes, this is quite clear to me. But this new understanding increases rather than diminishes my incredulity that the Earth can rotate at such a great speed, without extruding stones, animals, etc., towards the skies.

SALVIATI. You will come to know the rest in the same way as you have come to know everything so far; in fact, you know it already, and if you thought about it you would be able to recall it for yourself. But to save time I will help you to recall it. So far, you have worked out for yourself that the circular motion of the projector impresses an impetus upon the projectile to move when it separates from the projector along the straight line tangent to the circle of motion at the point of separation. Its motion continues along this line, taking it continually further away from the projector. Then, you have said that the projectile would continue its motion along this straight line if it were not for its own weight, which gives it an additional inclination downwards, thus making the line of motion into a curve. I think you also knew for yourself that this inclination tends always towards the centre of the Earth, because this is the point to which all heavy objects tend. Now for the next step: can you tell me whether the moving object, continuing its rectilinear motion after its separation from the projector, moves at a constant rate away from the centre (or, if you prefer, from the circumference) of the circle whose motion it had previously shared? Putting the question another way, when a moving object leaves from the point of tangency and continues to move along that tangent, does it move away from the point of contact at the same rate as it does from the circle's circumference?

SIMPLICIO. No, it does not. When the tangent is close to the point of contact its distance from the circumference is very small, and it forms a very narrow angle with the circumference; but as it moves further away, its distance from the circumference grows at an ever-increasing rate. So, for example, on a circle with a diameter of, say, ten *braccia*, a point on the tangent two feet away from the point of contact would be three or four times as far from the circumference as a point only one foot away; and the latter point in turn would be almost four times as far from the circumference as a point half a foot from the point of contact. So within an inch or two of the point of contact, the separation of the tangent from the circumference is barely visible.

SALVIATI. So, then, the initial distance of the projectile from the circumference of its previous circular motion is extremely small?

SIMPLICIO. Yes, almost imperceptible.

SALVIATI. Now the thrower imparts an impetus to the projectile to move straight along the tangent, and the projectile would continue to move along this line if it were not pulled down by its own weight. So my next question is this: how long after the moment of separation is it before the projectile begins to decline downwards? [221]

SIMPLICIO. I think it starts to decline immediately. If there is nothing to hold it up, its weight cannot fail to act on it.

A heavy projectile begins to decline immediately on its separation from the projector.

SALVIATI. So then, if a stone thrown out by a rapidly turning wheel had the same natural propensity to move towards the centre of the wheel as it does towards the centre of the Earth, it would readily return to the wheel—or rather, it would never separate from it. This is because, at the initial moment of separation, its distance from the wheel is so tiny, because of the infinite acuteness of the angle of contact, that the slightest inclination to draw it towards the centre of the wheel would be enough to keep it on the circumference.

SIMPLICIO. I don't doubt that, granted such an impossibility, namely that the inclination of these heavy objects is to go towards the centre of the wheel, they would not be extruded or flung out.

SALVIATI. I don't assume an impossibility, nor do I need to, because I don't deny that the stones are flung out by the wheel. I simply put it as a hypothetical case, so that you could tell me what follows. Imagine now the Earth as a great wheel which is turning so rapidly that it throws out the stones. You've already told me, quite rightly, that the motion of the projectile will be along the straight line which touches the Earth at the point of separation. Now how perceptibly does this tangent move away from the surface of the terrestrial globe?

SIMPLICIO. I doubt whether the distance is as much as an inch in a thousand *braccia*.

SALVIATI. And you said, did you not, that the projectile is drawn by its own weight to decline from the tangent towards the centre of the Earth?

SIMPLICIO. I did indeed, and now I will say what follows from it. I see perfectly that the stone will not separate from the Earth, because its initial separation would be so minimal that its inclination to move towards the centre of the Earth would be a thousand times stronger; and in this case the centre of the Earth is also the centre of the wheel. So it must be conceded that stones, animals, and other heavy objects cannot be extruded by the rotation of the Earth. But now I have a new difficulty with very light objects, whose inclination to fall towards the [222] centre is very weak. Since they do not have the property of being drawn back to the surface, I see no reason why they should not be extruded. And as you know, *ad destruendum sufficit unum*: one objection is enough to destroy an argument.

SALVIATI. We shall deal with this objection as well. But tell me first of all what you mean by light objects: do you mean those materials which are truly so light that they move upwards, or do you mean those which are not absolutely light, but which have so little weight that they do fall to the ground, but only very slowly? Because if you mean those which are absolutely light, I will grant that they are extruded even more than you say.

SIMPLICIO. I mean the latter—things such as feathers, wool, cotton, and the like. The slightest force is enough to lift them into the air, and yet we see them at rest quietly on the ground.

SALVIATI. If a feather has a natural propensity, however weak, to fall towards the surface of the Earth, this will be enough to prevent it from being thrown out. You know this as well as I do. So tell me: if the feather were extruded by the spinning of the Earth, what would be its line of motion?

SIMPLICIO. The tangent at the point of separation.

SALVIATI. And if it were to return to the surface, what would its line of motion be?

SIMPLICIO. The line from the feather to the centre of the Earth.

SALVIATI. So there are two motions to be considered: one from the projection, which starts from the point of contact and continues along the tangent; the other from the inclination downwards, which starts from the projectile and follows the secant towards the centre. For the projection to continue, the impetus along the tangent must be greater than the inclination along the secant, must it not?

SIMPLICIO. So it seems to me.

SALVIATI. What do you think is necessary in the motion from the projection if it is to prevail over the tendency to move downwards, so that the feather is separated and carried away from the Earth?

SIMPLICIO. I don't know.

SALVIATI. Of course you do. The moving object is the same in both cases, namely the feather; so how can the same moving object surpass and prevail over itself?

SIMPLICIO. The only way it can prevail over or be surpassed by itself is by moving either more rapidly or more slowly.

SALVIATI. You see, you did know this for yourself. So then, [223] for the projection of the feather to continue and for its motion along the tangent to prevail over its motion along the secant, what must their velocities be?

SIMPLICIO. The velocity along the tangent must be greater than the velocity along the secant. Oh, what a fool I have been! Isn't this velocity a hundred thousand times greater than that of a falling stone, let alone of a feather? And I was naïve enough to let myself be persuaded that stones couldn't be extruded by the spinning of the earth! So I take back what I said, and I declare that if the Earth were in motion, stones, elephants, towers, and cities would all necessarily fly off into the sky; and since this does not happen, I say that the Earth is not in motion.

SALVIATI. Oh, Simplicio, you are so easily carried away that I shall start worrying more about you than about the feather. Calm down a little and listen. If keeping the stone or the feather in contact with the Earth's surface depended on their downward motion being greater than or equal to the motion along the tangent, you would be quite right. The downward motion along the secant would have to be as rapid as, or more rapid than, the motion eastwards along the tangent. But did you not say just now that a distance of a thousand *braccia* from the point of contact along the tangent would barely give rise to a distance of an inch from the circumference? So it's not enough for the motion along the tangent (that of the diurnal rotation) to be simply greater than the motion along the secant (the downward motion of the feather). It would have to exceed it by so much that the time taken to carry the feather, say, a thousand *braccia* along the tangent, would not suffice for it to fall even an inch along the secant. I say that this could never happen, however fast you make one motion and however slow you make the other.

SIMPLICIO. Why could the motion along the tangent not be so rapid that there was not time for the feather to reach the surface of the Earth?

SALVIATI. Try to frame the question in terms of proportions, and I will reply. So tell me by how much you think the velocity of the former motion would have to exceed the latter.

SIMPLICIO. I would say that if, for example, the former motion was a million times faster than the latter, both the feather and the stone would be extruded.

[224] SALVIATI. In saying this you're mistaken, not because of any fault in logic or physics or metaphysics, but simply in geometry. If you knew just the basic elements of geometry, you would know that if you drew a straight line from the centre of a circle to a tangent, it wouldn't matter if it intersected it in such a way that the part of the tangent between the secant and the point of contact* was a million, two million, or three million times greater than the part of the secant between the tangent and the circumference. The nearer the secant comes to the point of contact, the more this ratio grows, *ad infinitum*. Therefore, however rapid the rotation may be, and however slow the downward motion, there is never any reason to fear that the feather, or any even lighter object, might begin to rise from the surface, because the downward inclination will always be greater than the velocity of projection.

SIMPLICIO. I'm not entirely convinced by this argument.

Geometrical proof that the Earth's rotation cannot possibly cause extrusion.

SALVIATI. I'll give you a proof which is both universally valid and very easy. Let the ratio of BA to C be given, with BA being as much greater than C as you like. From a circle with centre D we want to draw a secant such that the tangent shall have the same ratio to it as BA has to C. From BA and C take the third proportional, AI. As BI is to IA make the diameter FE to EG. From G draw the tangent GH. Now HG is to GE as BA is to C, which is what we were looking for. Since FE is to EG as BI is to IA, by composition FG is to GE as BA is to AI; and since C is the mean proportional between BA and AI, GH is the mean between FG and GE. Therefore as BA is to C, so FG is to GH, i.e. HG to GE, which was what had to be proved.

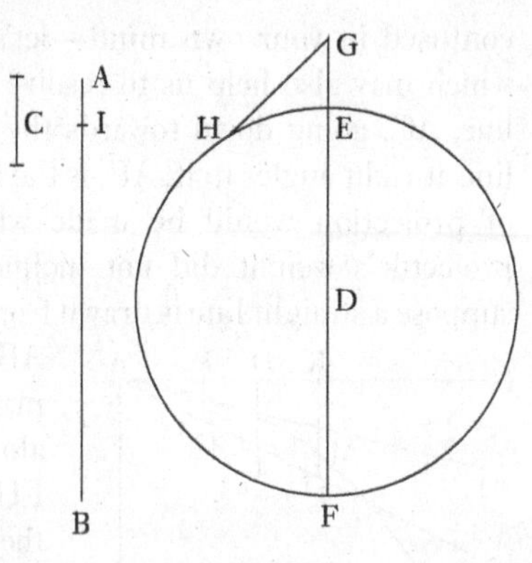

SAGREDO. I understand this proof, but it doesn't entirely remove my doubts. There's still some confusion in my mind, like a thick fog that prevents me from seeing the necessity of the conclusion as clearly as one expects in a mathematical proof. What confuses me is this: it's true that the space between the tangent and the circumference decreases *ad infinitum* as they approach the point of contact. But it's also true that the inclination of the moving body to descend always diminishes the nearer it comes to the immediate goal of its descent, namely a state of rest. This is clear from what you told us when you showed how a falling body starting from a state of rest must pass through every degree of slowness between rest and any given degree of velocity, and that these become less and less *ad infinitum*. What's more, this velocity and tendency to motion can also infinitely decrease for another reason, which is that the body's weight can infinitely decrease. This means that there are two causes, both potentially infinite, which can diminish the body's tendency to fall, and hence to favour its projection, namely its lightness and its closeness to a state of rest. There is only one cause to oppose these two and make it easier for the body to be projected, but I don't see how this single cause can resist the combined force of the other two, given that all three causes are potentially infinite. [225]

SALVIATI. Your doubt does you credit, Sagredo. To elucidate it and help us to grasp it more clearly—since you say it's rather confused in your own mind—let's set it down in a diagram, which may also help us to resolve it. We draw here a vertical line, AC, going down towards the centre; and this horizontal line at right angles to it, AB, is the line along which the motion of projection would be made with uniform motion if the projectile's weight did not incline it downwards. Now let's suppose a straight line is drawn from A at any chosen angle with AB, and call it AE. We then mark a number of equal spaces along AB, calling them AF, FH, HK, and from them drop the perpendiculars FG, HI, KL as far as AE. Since, as we've said before, a falling body starting from a state of rest gains an ever-increasing degree of velocity in successive intervals of time, we can take the intervals AF, FH, HK as representing equal intervals of time, and the perpendiculars FG, HI, KL the degrees of velocity gained in these intervals. So the line KL represents the degree of velocity [226] gained in the whole of the period AK, compared to the degree HI in the time period AH and the degree FG in the time period AF. It's clear that these degrees, KL, HI, FG, are in the same proportions as the time intervals KA, HA, FA. We could drop more perpendiculars from any other points on the line AF, and they would always show smaller and smaller degrees, *ad infinitum*, as they approached point A, which represents the initial instant of time and the initial state of rest. This withdrawal towards point A represents the first inclination to downward motion, which diminishes *ad infinitum* as the body moves closer to its initial state of rest, a moving closer which itself can extend *ad infinitum*.

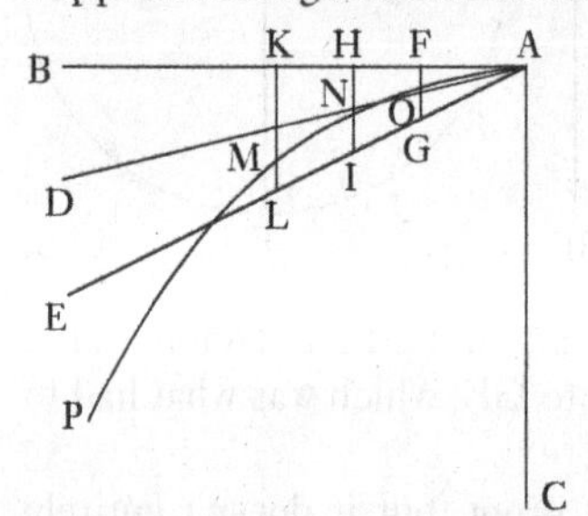

Now we can find the other decrease in the body's velocity, which is also potentially infinite, caused by the decrease in the body's weight. We can represent this by drawing more lines from point A, forming angles smaller than BAE, such as this line AD. This line intersects the parallel lines KL, HI, FG at points M, N, O, so that FO, HN, KM represent degrees of velocity gained in the time intervals AF, AH, AK; these are less than the degrees FG, HI, KL gained in the same time intervals, but by a heavier body, this being a lighter one. It's clear that by moving the line EA closer to AB, reducing the angle EAB—which can be done *ad infinitum*, since weight can be reduced *ad infinitum*—the velocity of the falling body, and hence its effect in impeding its projection, is similarly reduced *ad infinitum*. Thus it appears that when these two factors which work against projection, combined, are reduced *ad infinitum*, they are not able to prevent it.

The whole argument can be summed up in a few words as follows: the degrees of velocity LK, IH, GF are reduced by narrowing the angle EAB; the same degrees are also reduced by moving the parallels KL, HI, FG closer to the angle A; and both these reductions can be extended *ad infinitum*. Therefore the velocity of the body's downward motion can indeed be reduced to such an extent—since it can be reduced *ad infinitum* twice over—that it no longer suffices to bring the body back to the circumference of the wheel, and hence to impede and prevent its projection.

For projection not to happen, then, the spaces through which the projectile must fall to return to the wheel must be made so small and narrow that its downward motion suffices to bring it back, however reduced this motion might be—indeed even if it is diminished *ad infinitum*. So these spaces must be reduced not just *ad infinitum*, but to such a degree of infinity that it exceeds the double infinity by which the falling body's downward [227] motion is reduced. But how can a value be reduced by a greater amount than one which is doubly reduced *ad infinitum*? Now you can see, Simplicio, how far one can get in natural philosophy without reference to geometry!

When degrees of velocity are diminished *ad infinitum*, as a result of both the decrease in the body's weight and its proximity to the earliest point of its motion, i.e. to a state of rest, these degrees are always determinate, and are in proportion to the parallel lines contained between two straight lines meeting at an angle such as BAE or BAD, or any angle which can be infinitely more acute, provided always that it's rectilinear. The spaces through which a body must move to return to the surface of the wheel, however, are proportional to a different kind of diminution, included between lines which contain an angle infinitely narrower and more acute than any rectilinear angle whatever. We can represent it thus: take any point C on the vertical line AC, and with this point as the centre describe the arc AMP of radius CA. This arc will intersect the parallel lines which determine degrees of velocity, however short they may be and however acute the rectilinear angle within which they are contained. The parts of these parallel lines remaining between the arc and the tangent AB represent the size of the spaces through which the body must pass to return to the wheel. They constantly diminish in relation to the parallel lines of which they are part, and proportionately more so the closer they come to the point of contact.

The parallels included between the straight lines decrease in the same ratio as they withdraw towards the angle. So, for example, if the line AH is divided in half at point F, the parallel HI will be twice the length of FG; and if FA in turn is sub-divided in half, the parallel dropped from the dividing point will be half the length of FG, and so on *ad infinitum*, with each successive parallel being half the length of the one before it. The same is not true, however, of the lines intercepted between the tangent and the circumference of the circle. Here, if we make the same sub-division of FA and assume, for the sake of argument, that the parallel dropped from H is twice the length of the one from F, then the parallel from F will be more than twice the length of the one after that. Continuing towards the meeting point A we will find that each successive line contains the ones that follow it three, four, ten, a hundred, a thousand, a hundred thousand, and a hundred million times, *ad infinitum*. Therefore the shortness of these lines far exceeds what is needed to make the projectile return to the circumference of the wheel, or rather never to depart from it, however light the projectile may be.

[228] SAGREDO. I'm entirely convinced by your whole exposition and by the conclusiveness of the argument; nonetheless I think that someone who wanted to test it further could put forward an objection. They could argue that, while the first cause which diminishes the body's downward motion *ad infinitum*—the one deriving from its proximity to the very beginning of its descent—clearly increases in a constant ratio, just as parallel lines always keep the same ratio to each other, etc., the same is not true of the second cause. It's not equally clear that the reduction in velocity due to the reduction in the body's weight

follows the same proportion. How can we be sure that it doesn't follow the ratio of the lines intercepted between the tangent and the circumference, or indeed some even greater proportion?

SALVIATI. I was making the assumption that the velocity of freely falling bodies is proportional to their weight out of consideration for Simplicio and Aristotle, since Aristotle in several places affirms this as a self-evident truth. You question this for the sake of the contrary point of view, saying that a body's velocity could increase in a greater ratio than that of its weight, even infinitely greater, in which case the whole earlier argument would fall to the ground. In its defence, my reply is that velocity increases by a much smaller proportion than weight, which means that the argument is not just defended but strengthened. I'll prove this by reference to an experiment which will show that a body even thirty or forty times heavier than another, such as, for instance, a ball of lead compared to a ball of cork, will fall not much more than twice as fast. Now if projection would be prevented when the falling body's velocity decreased in proportion to its weight, how much less will it occur if the loss of velocity is much smaller than the decrease in weight?

But supposing for the sake of argument that velocity did decrease by a greater ratio than the decrease in weight—even if it decreased in the same ratio as the parallel lines between the tangent and the circumference—I see nothing to persuade me that even the lightest material would be projected. In fact I positively affirm that it would not, always assuming that by the lightest material we mean one which has very little weight so [229] that it falls very slowly, not something which has no weight at all and so naturally moves upwards. My grounds for saying this are as follows. If weight decreases in the same proportion as the parallel lines between the tangent and the circumference, the final and absolute term of this diminution is total weightlessness, just as the final term of the diminution of the parallel lines is contact with the circumference, which is an indivisible point. But the weight of a body never does diminish to its final term, since this would be a body which had no weight. The distance by which a projectile has to return to the circumference, on the other hand, does diminish to its ultimate smallness, since when the body rests on the circumference at its point of contact there is no distance between them. Therefore, however small the impulse to downward motion may be, it is always more than enough to bring the body back to the circumference from which it is separated by the minimum possible distance, i.e. no distance at all.

SAGREDO. This is certainly a very subtle argument, but no less convincing for that; and it must be admitted that trying to deal with questions of natural philosophy without reference to geometry is to attempt the impossible.

SALVIATI. I don't suppose Simplicio will agree—although I don't believe he is one of those Peripatetics who discourage their pupils from studying mathematics on the grounds that it would corrupt their powers of reason and make them less apt for contemplation.

SIMPLICIO. I wouldn't do such an injustice to Plato, but I do agree with Aristotle that Plato became too absorbed and fascinated by this geometry of his. In the end, Salviati, these mathematical subtleties are all very well in the abstract, but they don't correspond to physical material reality. To give an example similar to our present discussion, mathematicians may use their principles to prove that *sphaera tangit planum in puncto*—a sphere touches a plane surface in a single point—but, when it comes to the material world, the reality is quite different.

SALVIATI. Do you not believe, then, that the tangent touches the surface of the terrestrial globe in a single point?

SIMPLICIO. Not in a single point, no. I think that a straight line would remain in contact with the surface of water, never mind of the Earth, for many tens if not hundreds of *braccia* before it separated from it.

[230] *Truth is sometimes strengthened by being contradicted.*

SALVIATI. Don't you see that if I grant you this point it only undermines your argument? If we've established that a projectile wouldn't separate from the surface of the Earth if the tangent had only a single point of contact with the surface, because of the extreme acuteness of the angle of contact—if indeed it can be called an angle at all—then how much less cause will it have to separate if this angle is completely closed and the tangent and the surface remain joined? Isn't it clear that the projection in this case would be along the surface of the Earth itself, in other words that there would be no projection at all? This shows the strength of the truth, that when you try to knock it down your own attacks just support and strengthen it.

Now that I've disabused you of this error, let me also correct your other error, of saying that a material sphere doesn't touch a plane surface in a single point. I'd like to think that even a few hours' conversation with those who have some knowledge of geometry might enable you to appear slightly more knowledgeable among those who know nothing of it at all. To show you the enormity of the error of those who assert that a sphere of, say, bronze, does not touch a plane surface of, say, steel, in a single point, tell me what you would think of someone who insisted that a sphere was not really a sphere.

SIMPLICIO. I would think they were entirely devoid of reason.

SALVIATI. This is exactly the case of someone who says that a material sphere does not touch a plane surface which is also material in a single point, because this is tantamount to saying that the sphere is not a sphere. To show the truth of this, tell me what in your view constitutes the essence of a sphere, in other words what it is that distinguishes a sphere from all other solid bodies.

Even a material sphere touches a material plane surface in a single point.

SIMPLICIO. I take it to be the essence of a sphere that all the straight lines from its centre to its circumference are equal.

Definition of a sphere.

SALVIATI. So that if such lines were not equal, such a solid body could not be a sphere.

SIMPLICIO. No.

SALVIATI. Now tell me whether you think that, of many lines that can be drawn between two points, there can be more than one that is straight.

SIMPLICIO. Certainly not.

[231]

SALVIATI. And you also accept that this one straight line must necessarily be shorter than all the other lines between these two points.

SIMPLICIO. I do, and I have a clear proof of it, provided by a great Peripatetic philosopher. If I remember rightly, he puts it forward as a reproach to Archimedes, who takes it as given when he could have proved it.

SALVIATI. He must be a great mathematician indeed, if he could find a proof for something that Archimedes was quite unable to prove. If you can recall the proof I'd very much like to hear it, because I remember quite clearly that in his books on the sphere and the cylinder Archimedes includes this proposition among the postulates, and I'm quite certain that he considered it to be incapable of demonstration.

SIMPLICIO. I think I can recall it, because it's very short and simple.

SALVIATI. All the more shame on Archimedes and credit to this philosopher.

A Peripatetic philosopher's proof that a straight line is the shortest of all.

SIMPLICIO. I'll draw a figure for it. Between points A and B, draw a straight line AB and a curved line ACB. The proof that the straight line is shorter is as follows. On the curved line take a point C, and draw two more straight lines, AC and CB. Euclid proves that these two lines are longer than the single line AB. But the curve ACB is longer than the two straight lines AC and CB; therefore, *a fortiori*, the curve ACB is much longer than the straight line AB, which was what had to be proved.

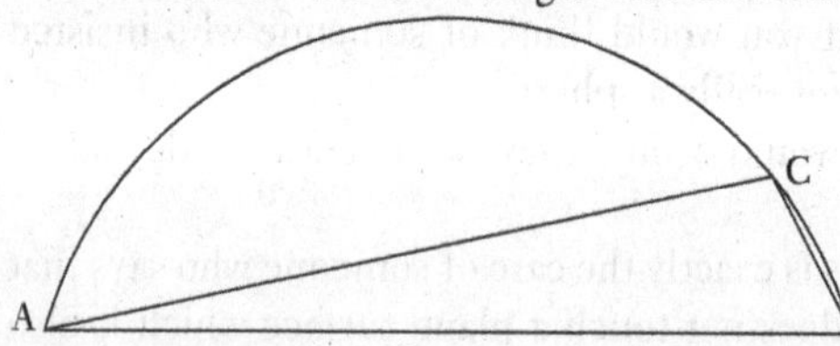

The Peripatetic philosopher's paralogism in proving ignotum per ignotius.

SALVIATI. If you searched through all the paralogisms in the world, I don't think you could find a better example of the most glaring of all fallacies—that of proving *ignotum per ignotius*, finding an unknown by means of a greater unknown.

SIMPLICIO. In what way?

SALVIATI. In what way? The unknown that you're trying to prove is that the curve ACB is longer than the straight line AB. The middle term, which you take as known, is that the curve ACB is greater than the two lines AC and CB, which are known to be greater than AB. But if it's unknown whether the curve is greater than the straight line AB, isn't it even more unknown whether it is greater than the two straight lines AC and CB, which are known to be greater than AB? And yet this is what you take as known.

SIMPLICIO. I still don't quite see where the fallacy lies.

[232] SALVIATI. Since the two straight lines are greater than AB, as we know from Euclid, then the curve, provided it's greater

than the two straight lines ACB, must be much greater than the single line AB, must it not?

SIMPLICIO. Yes, that's right.

SALVIATI. That the curve ACB is greater than the straight line AB is the conclusion; this is better known than the middle term, which is that the same curve is greater than the two straight lines AC and CB. Now when the middle term is less known than the conclusion, this is attempting to prove *ignotum per ignotius*. Now, to come back to what we were saying, what matters is that you accept that a straight line is the shortest of all possible lines between two points. And as regards the main point we were discussing, you say that a material sphere doesn't touch a plane surface in a single point. In that case, what kind of contact does it have?

SIMPLICIO. It must be in contact with part of its surface area.

SALVIATI. What about the sphere's contact with another sphere which is equal to it? Will that also be with a similar part of its surface?

SIMPLICIO. I don't see any reason why it should be otherwise.

Proof that a sphere touches a plane surface in a single point.

SALVIATI. This means that two spheres which are touching each other must be in contact over the same part of their surfaces, since if they both adapt to being in contact with the same plane surface they must also be adapted to each other. Now imagine two spheres in contact with each other, with centres A and B, and join their centres with the straight line AB passing through their contact at point C. Then take another point in this contact, D, and join the two straight lines AD and DB, forming the triangle ADB. The two sides of the triangle ADB must be equal to the single side ACB, since in both cases they contain two radii, which according to the definition of a sphere must all

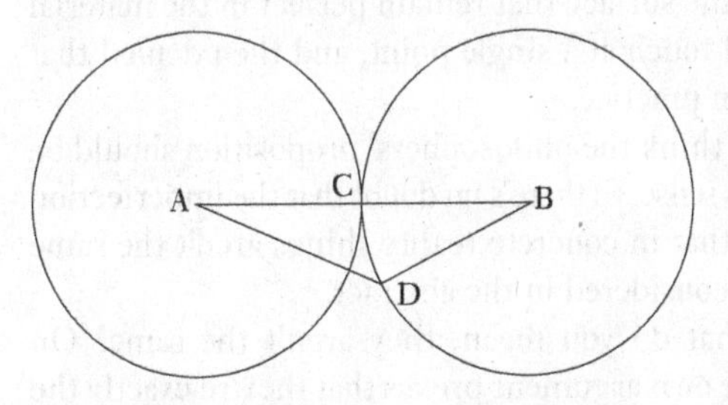

be equal. Therefore the straight line AB drawn between the two centres A and B is not the shortest of all possible lines, since the two lines AD and DB are equal to it; but what you have conceded shows that this is absurd.

SIMPLICIO. This proof is conclusive for spheres in the abstract, but not for actual material spheres.

SALVIATI. In that case, show me the fallacy in my argument, since it is not valid for material spheres but only for abstract, immaterial ones.

[233] SIMPLICIO. Material spheres are subject to many accidents to which immaterial spheres are not. Isn't it possible that, when a metal sphere is placed on a plane, its own weight will press down so that the plane yields a little, or else that the sphere is dented slightly at the contact? Besides, the plane surface is unlikely to be perfectly smooth, if only because its material is porous; and I dare say it's equally unlikely that a sphere will be so perfect that all the lines from its centre to its surface are exactly equal.

Why a sphere in the abstract touches a plane surface in a single point, but the same is not true of a material sphere in reality.

SALVIATI. Yes, I readily concede all these things, but they are all beside the point. In your attempt to show me that a material sphere doesn't touch a material plane surface at a single point, you are using a sphere that isn't a sphere and a plane surface that isn't a plane, because you say that either such things don't exist in the real world, or if they exist they are imperfect in any practical application. So it would have been better if you'd accepted my proof conditionally, and said that if we could have a sphere and a plane surface that remain perfect in the material world they would touch at a single point, and then denied that they could exist in practice.

SIMPLICIO. I think the philosophers' proposition should be understood in this sense, as there's no doubt that the imperfection of matter means that in concrete reality things aren't the same as when they are considered in the abstract.

SALVIATI. What do you mean, they aren't the same? On the contrary, your own argument proves that they're exactly the same.

SIMPLICIO. How's that?

SALVIATI. Aren't you saying that because of the imperfection of matter a body which should be perfectly spherical, and a plane surface which should be perfectly flat, are not as we imagine them to be in the abstract?

SIMPLICIO. Yes, that's right.

SALVIATI. That means that whenever, in concrete reality, you put a material sphere on a material plane, you are putting an imperfect sphere on an imperfect plane; and you say that they don't touch at a single point. To this I reply that in the abstract too, an immaterial sphere which is imperfect can touch an immaterial plane which is imperfect, not at a single point but over a part of its surface, so that what happens in concrete reality happens in the same way in the abstract.

Things are exactly the same in the abstract as in concrete reality.

It would be strange indeed if calculations and measurements worked out with abstract numbers didn't then correspond to gold and silver coins and actual concrete goods. You know how, in calculating the weight of sugar, silk, or wool, accountants have to make a deduction for the chests, bales, and other packaging? In the same way, geometricians who want to see the concrete effects of their abstract proofs must make allowances for the obstacles inherent in the material; and if they are able to do this, then I assure you that the results will tally exactly with the calculations. So errors come about not because of the difference between abstract and concrete, or between geometry and physics, but simply because the person doing the sums doesn't know how to calculate correctly. In the same way, if you had a perfect sphere and a perfect plane surface, even though they were material, you can be quite sure that they would touch at just a single point; and if such things were and are impossible, then it was completely irrelevant to say that *sphaera aenea non tangit in puncto*, that a bronze sphere does not touch at a single point.

[234]

I'll go further, Simplicio. Granted that a perfect material sphere and a perfect material plane can't exist materially, do you believe that there can be two material bodies whose surfaces are partly curved, however irregularly?

SIMPLICIO. I don't think there is any lack of such bodies.

SALVIATI. If there are, they too will touch at just a single point, because touching at a single point isn't the exclusive preserve of a perfect sphere and a perfect plane. In fact, if you think more carefully about it, you'll see that it's much more difficult to find two bodies which are in contact over part of their

To touch at a single point is not the exclusive property of perfect spheres, but of all curved figures.

surface than at just a single point. For two bodies to fit perfectly together, either they must both be perfectly flat, or if one is convex the other must be concave, with a concavity corresponding exactly to the convexity of the other. It's much harder to find bodies with these qualities, since they are so strictly determined, than irregular bodies whose random shapes are potentially infinite.

It is more difficult to find figures which touch over part of their surface area than ones which touch at only a single point.

SIMPLICIO. Do you think, then, that two stones or pieces of iron taken at random and brought together, will in the majority of cases touch at just a single point?

[235] SALVIATI. If they are just casually brought together, I think not. This is because there will usually be a small amount of dirt on them which will yield on contact, and also because we don't usually take care not to bang them together, and even the smallest impact makes one surface yield slightly to the other so that they mirror each other's shape, at least in some minimal part of their surface. But if the two surfaces were well polished and they were both placed on a table so that neither was pressing against the other, then I don't doubt that they could be brought into contact at a single point.

SAGREDO. With your permission, I must put forward a problem which occurred to me when I heard Simplicio say it was impossible to find a solid material body in the form of a perfect sphere, and Salviati appeared to agree since he didn't contradict him. My question is whether the same difficulty would arise in forming a solid of some other shape; or to put it more clearly, whether it is more difficult to shape a block of marble into a perfect sphere than into a perfect pyramid, or a perfect figure of a horse or a locust.

SALVIATI. I'll answer first, and I'll start by apologizing if I appeared to agree with Simplicio. It was only because, before I went on to saying something else, I wanted to express the same thought as you, or something very like it. In answer to your first question, I reply that of all the shapes that can be given to a solid body a sphere is the easiest, since it's the simplest of all solid shapes. It has the same status among solid figures as the circle has among plane figures: as the easiest of all figures to describe, the circle is the only one which mathematicians have seen fit to include among the postulates underlying the description of all other figures. As an indication

The figure of a sphere is easier to form than any other.

The circle is the only figure to be included among the postulates.

of how easy it is to form a sphere, if you make a circular hole in a flat metal plate and turn any roughly rounded solid body randomly inside it, the body will be reduced to a spherical shape of its own accord, as near perfect as it's possible to be, without being worked in any other way, provided it's not smaller than the sphere which will fit inside the circle. Even more to the point, spheres of different sizes can be formed in this same circular hole. And as for forming a horse or—to use your own example—a locust, I leave you to judge how difficult this is, since you know how very few sculptors there are in the world who are capable of doing it. I think Simplicio will agree with me on this point.

[236]

Spheres of different sizes can be formed with a single instrument.

SIMPLICIO. I don't think I disagree with you at all. My own view is that none of the shapes we've discussed can be perfectly formed; but to come as close to perfection as possible, I consider it incomparably easier to reduce a solid body to the figure of a sphere than to the form of a horse or a locust.

SAGREDO. What do you think is the reason for this greater difficulty?

SIMPLICIO. In the same way as it's easy to form a sphere because of its absolute simplicity and uniformity, so the other shapes are difficult to produce because they are so irregular.

Irregular forms are difficult to produce.

SAGREDO. If it's irregularity that causes the difficulty, then the shape of a stone broken haphazardly with a hammer must be among those which are difficult to produce, since it's very likely to be even more irregular than the shape of a horse.

SIMPLICIO. Yes, that must be so.

SAGREDO. But tell me: does the stone have this shape, whatever it is, perfectly or not?

SIMPLICIO. Whatever shape it has, it has to perfection, and no other shape matches it so closely.

SAGREDO. If, then, there are so many irregular shapes, all perfectly realized, even though they are hard to produce, on what grounds can we say that the simplest shape, which is therefore the easiest of all, is impossible to achieve?

SALVIATI. Forgive me, gentlemen, but I think we're getting sidetracked into splitting hairs. We're wasting our time on frivolous arguments of no consequence when we should be pursuing our discussion of weighty and serious matters. Please let's remember that seeking to understand the constitution of the universe is among the greatest and noblest problems to be found in nature, and all the more so if it leads us to solve that other great problem, the cause of the ebb and flow of the tides—something that all the great men who have ever lived have tried to discover, and perhaps none have yet succeeded in finding it. So if nothing more remains to be said to deal with the objection based on the spinning of the Earth, which was the last of the arguments for concluding that it is fixed on its own centre, we can go on to examine the factors for and against the Earth's annual motion.

The constitution of the universe is among the noblest problems.

[237]

SAGREDO. Salviati, please don't judge our mental capacity by the same standard as yours. You are used to always devoting yourself to contemplating the most elevated matters, and so something which seems to us a worthy subject for our intellect strikes you as trifling and frivolous. So, just occasionally, please indulge us by stooping to our level and making some concession to our curiosity.

As for answering the last objection, based on the casting off of things caused by the Earth's diurnal rotation, you have given far more than was needed to convince me; and yet the things you have said over and above what was necessary have so fascinated me that my imagination, far from being wearied, has been absorbed by their novelty, and has given me the greatest pleasure I could wish for. So if you have any more speculations to add please let us have them, as for my part I will be very glad to hear them.

SALVIATI. My greatest delight has always been in the things I have discovered, but second only to this is the pleasure of discussing them with friends who understand them and show their appreciation of them. As you are one such friend, I'll give rein to my ambition—which is always gratified by proving myself more perceptive than others who have a reputation for clear-sightedness—and add for good measure the crowning touch to our earlier discussion, by pointing out another fallacy in the argument put forward by the followers of Ptolemy and Aristotle.

SAGREDO. I can't wait to hear it.

SALVIATI. So far we haven't challenged Ptolemy in taking as an undisputed fact that, since the extrusion of the stone is produced by the velocity of the wheel's rotation, the cause of the extrusion will increase in proportion to the velocity of the wheel's spinning. From this it was inferred that, since the velocity of the Earth's spinning is immeasurably greater than that of any rotating device we can produce artificially, the force with which stones, animals, etc., are extruded must be exceptionally great. But I must point out now that there is a major fallacy in this argument, if it involves comparing different velocities [238] indiscriminately and absolutely. It's true that if I compare the same wheel, or two equal wheels, rotating at different velocities, then the one which rotates more rapidly will extrude the stone with greater force, and the cause of the projection will increase in proportion to the increase in the wheel's velocity. But increasing the velocity of the same wheel, so that it completes a larger number of revolutions in the same time, is not the same as increasing the wheel's diameter to make it larger, so that the large wheel completes a revolution in the same time as a smaller one. In this case the only reason the larger wheel's velocity is greater is that its circumference is larger, and it's quite mistaken to suppose that the force of extrusion of the larger wheel will increase in the same proportion as the velocity of its circumference to that of the smaller wheel. A very simple experiment will provide a rough and ready demonstration that this is false. We could throw a stone further with a stick a *braccio* long than with one six *braccia* long, even though the end of the longer stick, and hence of the stone lodged there, was moving more than twice as fast as the end of the shorter stick—as would be the case if the shorter stick completed three revolutions in the time it took for the longer stick to complete one.

The cause of projection does not increase in proportion to its velocity when it is increased by enlarging the wheel.

SAGREDO. I see that the outcome must indeed be as you say, Salviati, but it's still not clear to me why equal velocities should not produce an equal effect in extruding projectiles, or why a smaller wheel is so much more effective than a larger one. Please explain to me how this comes about.

SIMPLICIO. You don't seem to be your usual self on this occasion, Sagredo. You generally grasp everything instantly, but this time you've overlooked a fallacy which I've spotted in the experiment with sticks, and that is that we throw in different ways depending on whether we're using a long stick or a short one. To make the stone fly away from the stick you mustn't maintain its motion uniformly; when it's at its fastest, you have to hold back your arm and restrain the velocity of the stick, so that the stone flies out impetuously. But you can't do that with [239] a longer stick, because its length and flexibility mean that it doesn't entirely obey the restraint of your arm, but travels with the stone a bit further, holding it back slightly and maintaining its contact with it. It wouldn't be the same if it collided with some hard obstacle; then it would let the stone fly off. So I think that if both sticks collided with something which made them stop abruptly, then the stone would fly off in the same way from both of them, even if they were both moving at the same speed.

SAGREDO. If you'll allow me, Salviati, I shall respond to Simplicio, since he has addressed his remarks to me. It seems to me that there is both good and bad in what he says; good, because almost all of what he says is true; bad, because it's entirely beside the point. It's perfectly true that if stones are being carried forward at speed, and the means by which they're being carried collides with an immovable obstacle, the impetus will make the stones rush forwards. We see the same effect every day when a boat which is moving at a good speed runs aground or collides with an obstruction of some kind, and all the people in the boat are caught unawares and suddenly stumble and fall, in the direction in which the boat was travelling. If the terrestrial globe were to encounter an obstacle capable of resisting its spinning motion and bringing it to a standstill, then I agree that not just animals, buildings, and cities, but mountains, lakes, and seas would all be shaken out of their place, if indeed the globe itself didn't fall apart. But none of this is to our purpose; we're talking about the effects that might follow from the motion of the Earth as it turns steadily and calmly on its axis, albeit at a very great speed.

If the daily spinning motion belongs to the Earth, and if it were brought to a standstill by some sudden obstacle or obstruction, then buildings, mountains, and perhaps the whole globe would disintegrate.

In the same way, what you say about the sticks is partly true; but Salviati didn't introduce them as an exact parallel to the subject of our discussion, but simply as a rough example to provoke us to think more carefully about whether an increase in speed, however it arises, increases the cause of projection in the same ratio. If a wheel with a diameter of, say, ten *braccia* is in motion so that a point on its circumference moves a hundred *braccia* in a minute, and this gives it the impetus to extrude a stone, would that impetus be a hundred thousand times greater in a wheel with a diameter of a million *braccia*? Salviati says that it wouldn't, and I'm inclined to agree with him; but as I don't know why this should be, I've asked him to explain the cause to me, and I'm eagerly awaiting his answer.

SALVIATI. And I'm ready to satisfy you to the best of my [240] ability. You may think that what I say to begin with has nothing to do with our discussion, but I think we'll find as we proceed that it's not unrelated. So I'd like Sagredo to tell me what he has observed about the nature of any moving body's resistance to being moved.

SAGREDO. For the moment, the only internal resistance to being moved that I can see in a movable body is its natural inclination and tendency to move in the opposite direction; so, for instance, a heavy body has a tendency to move downwards and a resistance to being moved upwards. I say 'internal resistance' because I assume that's what you meant, rather than the very many external causes of resistance which are quite fortuitous.

SALVIATI. Yes, that's what I meant, and your perceptiveness has defeated the stratagem behind my question. But if I was less than frank in posing my question, I wonder whether Sagredo may not have omitted something in his answer, and whether there may not be another natural and intrinsic quality in a movable body which makes it resistant to motion, quite apart from its natural tendency to move in the opposite direction. Let me ask you again, then: do you not think that the inclination, for instance, of heavy bodies to move downwards is equal to their resistance to being driven upwards?

The inclination of heavy bodies to downward motion is equal to their resistance to upward motion.

SAGREDO. Yes, I think it's exactly so. That's why I see two equal weights in a balance remain motionless in equilibrium, because the heaviness of one resists being raised by the heaviness of the other, which seeks to raise it as it presses downwards.

SALVIATI. Very good. This means that in order for one to raise the other, we must either add some weight to the one pressing down or take some away from the other. But if it's only heaviness that constitutes this resistance to upward motion, how is it that in a steelyard balance—one with unequal arms—the downward pressure of a hundred-pound weight is sometimes not enough to raise a four-pound weight on the other arm, while the downward motion of the four-pound weight can raise the hundred-pound one? We can see that this is the effect of the counterweight on the heavy body we want to weigh. If resistance to motion depends only on heaviness, how can the counterweight, weighing only four pounds, resist the weight of a bale of wool or silk which may weigh eight hundred or a thousand pounds, and even prevail over the bale with its momentum and lift it up? So we must admit, Sagredo, that we're dealing with some other force and resistance here than simply that of heaviness.

[241] SAGREDO. This must indeed be the case; so tell me what this second force is.

SALVIATI. It's whatever was not to be found in the balance which had arms of equal length. Think what it is that's new in the steelyard, and this must necessarily be the cause which produces the new effect.

SAGREDO. I think your prompting has brought something to my mind. In both instruments we're dealing with weight and motion. In the balance, the two motions are equal, and therefore one weight must surpass the other in heaviness in order to move it. In the steelyard, the lighter weight will move the heavier one only if the latter moves just a little, being suspended at a shorter distance, while the former moves a great deal, being suspended at a greater distance. We must conclude, then, that the lesser weight overcomes the resistance of the greater weight by moving more while the greater weight moves less.

SALVIATI. Which is tantamount to saying that the velocity of the less heavy body offsets the weight of the heavier and slower body.

Greater velocity exactly counteracts greater heaviness.

SAGREDO. Do you think, though, that velocity exactly compensates for heaviness, so that the moment and force of a body weighing, say, four pounds, is equal to that of a body weighing a hundred pounds, as long as the former has a hundred degrees of speed while the latter has only four?

SALVIATI. Certainly, and I could show it to you by many experiments; but for the moment let the example of the steelyard suffice. This shows that a small counterweight can hold up and balance a heavy bale provided its distance from the centre, on which the steelyard is suspended and about which it turns, exceeds the distance at which the bale is suspended by the same amount as the absolute weight of the bale exceeds that of the counterweight. The fact that a large bale, with all its weight, can't raise the counterweight which is so much lighter, can only be explained by the disparity in the distance which each weight has to travel. If the bale weighs a hundred times more than the counterweight, and the distance of the counterweight from the centre of the steelyard is a hundred times greater than the distance from the centre to the point from which the bale is suspended, then for every inch the bale falls the counterweight will be raised a hundred inches. And if the counterweight travels a hundred inches in the same time as the bale travels just an inch, this is the same as saying that the velocity of motion of the counterweight is a hundred times greater than that of the bale.

We should note this, then, as a true and generally understood [242] principle: that the resistance which derives from velocity of motion compensates for the resistance that depends on the weight of another moving body. It follows that a body weighing one pound and moving at a hundred degrees of velocity will have the same resistance to being held back as another body weighing a hundred pounds and moving at only one degree of velocity. It also follows that two equal bodies will equally resist being moved if they are made to move at equal velocities; but if one is to move faster than the other, its resistance will be greater, in proportion to the greater velocity being imparted to it.

Having established these points, let's come to the elucidation of our problem. It will be easier to understand if we draw a little diagram. Let there be two unequal wheels around this centre A, the smaller on the circumference BG and the larger on the circumference CEH, the radius ABC being vertical to the horizon. Through points B and C we draw the tangent lines BF and CD, and in the arcs BG and CE take two equal parts, BG and CE. Suppose that the two wheels are rotating around their centres

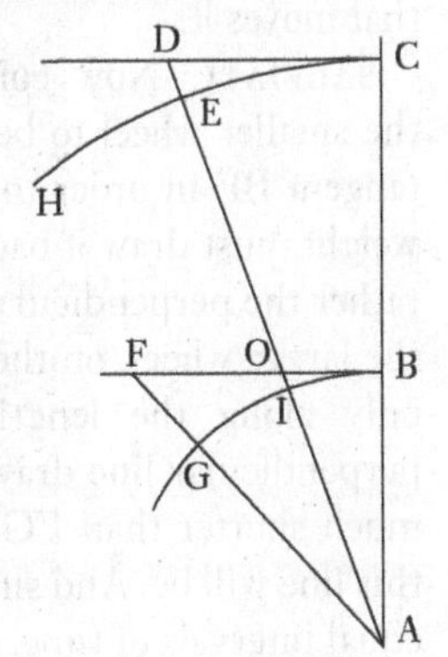

with equal speeds, so that two moving bodies, such as stones, placed at B and C are carried along the circumferences BG and CE at equal speeds; thus in the same time as stone B travels through the arc BG, stone C moves through the arc CE. Now I say that the spinning of the smaller wheel is much more powerful at projecting stone B than the spinning of the larger wheel at projecting stone C. Since we have established that projection occurs along the tangent, if the stones B and C should be separated from their wheels and begin their projected motion from points B and C, they would be thrown out along the tangents BF and CD. Hence the two stones have an equal impetus to travel along the tangents BF and CD, and they would do so unless some other force were to deviate them from it. That's right, isn't it, Sagredo?

SAGREDO. Yes, that's what I think would happen.

SALVIATI. What force, then, do you think there can be to deviate the stones from moving along the tangent, since this is how the impetus derived from spinning will cast them?

[243] SAGREDO. Either their own heaviness, or some glue that holds them in place and attached to the wheels.

SALVIATI. But if a moving body has an impetus to a given motion, surely there must be a greater or lesser degree of force to deviate it from this motion, depending on the greater or lesser extent of the deviation—on whether, that is, the deviation involves travelling through a greater or lesser space in the same time?

SAGREDO. Yes, because we've already established that the faster a body must be made to move the greater must be the force that moves it.

SALVIATI. Now consider what it takes for the stone on the smaller wheel to be deviated from its projection along the tangent BF in order to remain attached to the wheel. Its own weight must draw it back along the length of the secant FG, or rather the perpendicular line drawn from G to the line BF. On the larger wheel, on the other hand, it needs to be drawn back only along the length of the secant DE, or rather the perpendicular line drawn from E to the tangent DC, which is much shorter than FG; and the larger the wheel, the shorter this line will be. And since the drawing back must be effected in equal intervals of time, i.e. the time taken to travel through the equal arcs BG and CE, the motion of stone B, drawing it back along the line FG, must be more rapid than the motion of the other stone along DE. Therefore it will need a much greater force to keep stone B attached to its small wheel than stone C to its large one. This is as much as to say that a small force will impede projection from a large wheel but would not be enough to prevent it from a small one. It's clear, therefore, that the larger the wheel, the more the cause of projection will be diminished.

SAGREDO. Thanks to your detailed exposition, I think I now understand enough to be able to answer my questions with quite a brief argument. Given that the equal velocity of the two wheels imparts an equal impetus to the two stones to move along their tangents, we can see that the larger wheel's greater circumference, by diverging only a small distance from the tangent, in a way favours and only gently reins in what we might call the stone's desire to break away from the circumference. Hence any small restraint, whether from the stone's own inclination or from a glue of some kind, is enough to keep the stone in contact with the wheel. With the smaller wheel, on the other hand, such a restraint will be insufficient to produce the same effect, since the smaller wheel hardly follows the direction of the tangent at all, but tries too greedily to keep the stone to [244] itself; so if the reins or the glue are no stronger than those which kept the stone in contact with the larger wheel, the halter will break and the stone will veer off along the tangent.

I can see now the error of those who believed that the cause of projection increased in proportion to the increase in the velocity of spinning. What's more, this leads me on to another consideration: if the cause of projection diminishes as the wheel is enlarged, provided the velocity of the wheels remains the same, it may be that for a larger wheel to have the same force of projection as a smaller one, its speed must increase in proportion to its diameter, so that both wheels complete one revolution in the same time. If this were the case, then the spinning of the Earth would be no more effective in projecting stones than any small wheel which turned so slowly that it completed just one revolution in twenty-four hours.

SALVIATI. I don't think we should pursue this any further for now. Unless I'm mistaken, we have said more than enough to show the inadequacy of the argument which at first sight seemed so conclusive, as some of the greatest minds have believed it to be. I'll consider it time and words well spent if I've been able to convince Simplicio, not of the mobility of the Earth, but at least that those who uphold this view are not so self-evidently foolish as the great mass of ordinary philosophers believe.

SIMPLICIO. As regards the objections to the daily revolution of the Earth based on weights falling from the top of a tower, and on projectiles launched perpendicularly or with any given lateral inclination, towards east, west, south, or north, etc., the answers you've given thus far have partly dissuaded me from the ancient scepticism about such an idea. But now my mind is contending with other and more substantial difficulties which I can see no way of resolving, and I doubt whether you will be able to extricate yourself from them either. They are quite recent, so it's possible that you may not have heard of them. These are the reflections of two authors* who have written *ex professo* against Copernicus: the first in a booklet of scientific conclusions, and the other, who is both a great philosopher and a mathematician, in a treatise supporting Aristotle and his doctrine of the immutability of the heavens. In this he shows [245] that comets, and also the new stars that appeared in Cassiopeia in 1572 and in Sagittarius in 1604, were not above the spheres of the planets at all but were actually in the elemental region below the sphere of the Moon. He demonstrates this in refutation of Tycho, Kepler, and many other astronomers, and he does so by means of their own arguments, that is by reference to parallax. If you wish I shall rehearse the arguments advanced by both these authors, as I've read them carefully more than once, and you can judge their force for yourselves and tell me what you think of them.

Other arguments against Copernicus by two modern authors.

SALVIATI. Since our main purpose is to put forward and evaluate all that has been said for and against the Ptolemaic and Copernican systems, we ought not to overlook anything that has been written on the subject.

First objection by the modern author of the booklet of conclusions.

SIMPLICIO. In that case I shall start with the objections contained in the booklet of conclusions, and then go on to the others. First of all, then, this author carefully calculates how many miles per hour are travelled by a point on the Earth's surface at the equator, and then by other points located at other latitudes. Not content with establishing these motions to the nearest hour, he goes on to work them out in minutes and then, not content with minutes, he defines them down to the nearest second. Next, he goes so far as to demonstrate exactly how many miles would be travelled in these time intervals by a cannon ball placed on the inner surface of the sphere of the Moon; and to eliminate any possible subterfuge by his opponent, he assumes that this sphere is as large as Copernicus himself took it to be. Then, on the basis of this ingenious and refined calculation, he shows that a body falling from such a height would take more than six days to reach the centre of the Earth, the point to which all heavy bodies naturally incline.

This author contends that a cannon ball would take more than six days to fall from the sphere of the Moon to the centre of the Earth.

Now if by divine omnipotence or by means of some angel a very large cannon ball could be miraculously taken up to the lunar sphere, placed at a point directly above us and then released, it seems quite incredible to this author and also to me that it would remain in a vertical line above us, while continuing to turn with the Earth about its centre all the days that it was falling. It would have to descend in a spiral in the plane of the great circle at the equator, in spirals around cones at other latitudes, and in a simple straight line at the poles.

[246]

He goes on to confirm the gross implausibility of this supposition by putting forward numerous difficulties in the form of questions which the followers of Copernicus can't possibly resolve. These are, if I remember rightly . . .

SALVIATI. Hold on a moment. Please, Simplicio, don't confuse me with so many new ideas all at once; my memory's not very good, so I have to take things a step at a time. I have in fact made the calculation of how long it would take such a heavy object to fall from the sphere of the Moon to the centre of the Earth, and as far as I remember it wasn't such a long time as that. So it would be good if you could tell us the rule this author followed in working out his computation.

SIMPLICIO. He made an assumption which was very favourable to his opponent, in order to prove his point *a fortiori*. He assumed that the body would have the same speed when it was falling vertically towards the centre of the Earth as it had in its circular motion around the great circle of the lunar sphere. On this basis it would travel at twelve thousand six hundred German miles* per hour, which frankly seems impossible; but to err on the side of caution, and to give every possible advantage to the opposite view, he assumes it is correct, and concludes that the time it would take to fall would definitely be more than six days.

SALVIATI. Is this the whole of his argument? Is that how he proves that the duration of the fall must be more than six days?

SAGREDO. I think he's been too modest. He could have attributed any speed he wanted to such a falling body, and so could have said it would take six months or even six years to fall to Earth, but he settled for six days. But Salviati, please restore my good humour by telling me how you carried out your calculation, since you say you've done it on another occasion. I'm sure you wouldn't have applied your mind to it if it wasn't a problem requiring some ingenious thought.

SALVIATI. What matters is not just whether the question is a great and noble one, Sagredo, but whether we deal nobly with it. Everyone knows that dissecting an animal can reveal infinite wonders of a wise and provident nature, and yet for every animal an anatomist dissects there are a thousand that are just chopped up by the butchers. I'm not sure which of those roles [247] I shall play in trying to answer your question; but I'm encouraged by the performance of Simplicio's author to press ahead and tell you the method I followed, if I can remember it.

Before I go on, though, I have to say that I very much doubt whether Simplicio has given a faithful account of the method followed by this author of his, when he concluded that the cannon ball would take more than six days to fall from the sphere of the Moon to the centre of the Earth. Because if he assumed that the ball's speed in falling would be equal to that of the sphere, which is what Simplicio said he assumed, then he would have shown himself ignorant of the most basic and elementary principles of geometry. In fact, I'm surprised that Simplicio himself doesn't see the gross error in this assumption that he reports.

SIMPLICIO. I may have been confused in my account of it, but I certainly can't see any fallacy in it.

SALVIATI. Perhaps I didn't fully understand the account you gave. Didn't you say that this author makes the speed of the ball's motion as it fell equal to that of its motion when it was revolving in the lunar sphere, and that when it fell at this speed it would reach the centre of the Earth in six days?

SIMPLICIO. That's what I think he wrote, yes.

Gross error in the argument based on a ball falling from the sphere of the Moon.

SALVIATI. Don't you see the error in this? You must be pretending; it's not possible you don't know that the radius of a circle is less than a sixth of its circumference, and therefore that the time it takes for a body to travel along the radius must be less than a sixth of what it would take to travel around the circumference at the same speed. It follows that if the ball fell with the velocity it had when it was revolving in the sphere it would reach the centre in less than four hours, assuming that in the sphere it would complete a revolution in twenty-four hours, which must be the assumption if the ball is to remain always in the same vertical line.

SIMPLICIO. Now I understand the error perfectly. I wouldn't want to attribute it wrongly to this author, and I must have made a mistake in reproducing his argument. To avoid foisting any other mistakes on him I'd like to have the book, and I'd be very grateful if someone could fetch it for me.

[248]

SAGREDO. I'll send a servant straight away; meanwhile, so as not to waste any time, perhaps Salviati will be good enough to give us his calculation.

SIMPLICIO. Thank you. He'll find the book open on my desk, together with the other one that also argues against Copernicus.

SAGREDO. We'll have that one fetched as well, just to be sure. I've despatched a servant; now let Salviati make his calculation.

Computation of the time taken for a cannon ball to fall from the sphere of the Moon to the centre of the Earth.

SALVIATI. Before anything else we must consider how the motion of falling objects is not uniform, but starts from a state of rest and constantly accelerates. This fact is known and recognized by everyone except the modern author just cited,* who says it is constant and makes no reference to acceleration. But this general recognition is useless unless we know the rate at which this increase in velocity occurs, and this is something which has been unknown to all philosophers up to our own

The natural motion of heavy objects accelerates by odd numbers, starting from unity.

time. Our mutual friend the Academician has been the first to discover and demonstrate it, in some of his writings which are not yet published,* but which he has shown in confidence to me and a number of his other friends. He shows that the natural motion of heavy objects accelerates by odd numbers, starting from unity. That is, taking any number of equal time intervals, if the moving body starting from a state of rest travels, say, one *canna** in the first time interval, in the second interval it will travel three *canne*, in the third interval five, in the fourth interval seven, and so on, following the sequence of odd numbers. This is the same as saying that the spaces passed over by a body starting from rest have to each other the ratios of the squares of the times in which such spaces were traversed; or in other words, the spaces passed over are to each other as the squares of the times.

The distances travelled by a falling object are proportional to the squares of the times in which they are travelled.

SAGREDO. I am amazed to hear this. And you say that there is a mathematical proof?

SALVIATI. Purely mathematical; and our friend has discovered and demonstrated not just this, but many other beautiful properties of natural motion, and of the motion of projectiles as well. It's been a great source of pleasure and wonder for me to see and study them all, and to witness the rise of a whole new understanding of a subject on which hundreds of volumes have already been written. Not a single one of our friend's infinite admirable conclusions has been observed or understood by anyone before.

The Academician's whole new science concerning local motion.

[249] SAGREDO. You make me abandon my desire to pursue further the discussions we've already begun, solely so as to hear some of the demonstrations you mention. So either tell me about them now, or at least promise that you will devote a separate session to them for me, and for Simplicio as well, if he would like to hear about the properties and attributes of this basic effect in nature.

SIMPLICIO. I would indeed, although I don't consider it necessary for a natural philosopher to descend to the level of minute details; it suffices to have a general understanding of the definition of motion, the distinction between natural and constrained and between constant and accelerated motion, and so forth. If this had not been enough, I don't believe Aristotle would have omitted to teach us whatever we lacked.

SALVIATI. That may be so, but let's not lose any more time on this now. I promise to devote a separate half day to satisfying you about it; in fact, I remember now that I promised on another occasion to do just that. So to return to what we were saying, we had begun to calculate the time it would take for a falling object to travel from the sphere of the Moon to the centre of the Earth. To proceed not at random but following a method that will yield conclusive results, we shall first try to establish, by means of an experiment that can be repeated several times, how long it takes for a ball of, say, iron to reach the Earth from a height of a hundred *braccia*.

SAGREDO. We shall have to take a ball of a specified weight, and use the same weight when we calculate the time taken for it to fall from the Moon.

SALVIATI. That doesn't matter at all: balls weighing one, ten, a hundred, or a thousand pounds will all cover the same hundred *braccia* in the same time.

SIMPLICIO. That I can't believe, and no more did Aristotle. He writes that the velocity of falling bodies is in proportion to their weight.

Aristotle's error in affirming that falling bodies move in proportion to their weight.

SALVIATI. If you want to maintain that this is true, Simplicio, then you must also believe that, if you let fall two balls of the same material, one weighing a hundred pounds and the other one pound, at the same moment from a height of a hundred *braccia*, the larger ball will reach the Earth before the smaller one has fallen a single *braccio*. Now try to imagine, if you can, seeing the large ball reaching the Earth when the small one is still less than one *braccio* from the top of the tower. [250]

SAGREDO. I don't doubt for a moment that this proposition is false, but I'm not entirely convinced that yours is absolutely true. Still, I believe it, since you affirm it so confidently, and I'm sure you wouldn't do so if you were not assured of it either from a conclusive experiment or by a clear demonstration.

SALVIATI. I have both, and I will explain them to you when we have our separate discussion on the subject of motion. But for now, to avoid any further digressions, let's assume that we want to make our calculation for an iron ball of a hundred pounds, which in repeated experiments falls from a height of a hundred *braccia* in five seconds.* Now since, as I've said, the space travelled by a falling object increases in double proportion to the time elapsed, i.e. by the squares of the times elapsed, and since a minute is twelve times five seconds, if we multiply the hundred *braccia* by the square of 12, which is 144, we get 14,400, and this will be the number of *braccia* that the object will fall in a minute. Following the same rule, since an hour is 60 minutes, if we multiply 14,400 (the number of *braccia* travelled in a minute) by the square of 60, that is by 3600, this will give us 51,840,000, which will be the number of *braccia* travelled in an hour, or 17,280 miles. To find out the distance it would travel in four hours we multiply 17,280 by 16 (this being the square of 4), which is 276,480 miles. This is far more than the distance from the sphere of the Moon, which is 196,000 miles assuming, as the modern author does, that the distance of the sphere is 56 times the radius of the Earth,* and that the radius of the Earth is 3500 of our Italian miles, a mile being 3000 *braccia*. So you can see, Simplicio, that the distance from the sphere of the Moon to the centre of the Earth, which your author said would take more than six days to travel, could in fact be travelled in less than four hours, if the calculation is based on experimental evidence and not just by counting on one's fingers. To be precise, the time it would take would be 3 hours, 22 minutes, and 4 seconds.*

SAGREDO. My dear Salviati, please don't deprive me of the details of this calculation, for it must be truly remarkable.

[251] SALVIATI. Indeed it is. So, having first established by careful experiment that a given object falls from a height of a hundred *braccia* in five seconds, we ask: if 100 *braccia* are passed in 5 seconds, how many seconds will it take to cover 588,000,000 *braccia*, this being the length of 56 times the radius of the Earth? The rule for this calculation is to multiply the third number by the square of the second, which gives 14,700,000,000. Divide this by the first number, i.e. by 100, and the number we are looking for is the square root of the quotient, which is 12,124. 12,124 seconds are 3 hours, 22 minutes, and 4 seconds.

SAGREDO. I can see what you've done, but I have no idea of the reasoning behind it. But perhaps this is not the time to ask for it.

SALVIATI. On the contrary, I'll tell you even though you don't ask me, as it's very easy. Let's designate these three numbers* with the letters A for the first, B for the second, and

C for the third. A and C refer to distances, B refers to time; we are looking for the fourth number, which also refers to time. We know that whatever proportion the space A has to space C, the square of the time B must be the square of the time we are seeking. So following the golden rule, we multiply C by the square of B and divide the product by A. The quotient will be the square of the number we want, which will therefore be the square root of the quotient. Now you can see how easy it is to understand.

100	5	588000000	
A	B	C	25
	1	14700000000	
	22	35956	
	241	10	
	2422	60	12124
	24244		202
			3

SAGREDO. So are all truths, once they have been discovered; the difficulty lies in discovering them. I understand perfectly, and I'm grateful to you. If you have any other curious things to say about this topic, please tell me; because to speak frankly, and with no offence to Simplicio, I always learn some interesting new truth from what you say, whereas from his philosophers I don't know that I've yet learnt anything of any great significance.

SALVIATI. There is still all too much to be said on this matter of local motion, but as we've agreed, we'll save it for a separate session. So now I shall say something about the author cited by Simplicio. He thinks he has given a great advantage to his opponent by conceding that the cannon ball, in falling from the [252] sphere of the Moon, could reach a velocity equal to that which it would have had if it had remained up in the lunar sphere and had participated in the sphere's daily motion. My reply is that in falling from the sphere to the centre, the ball will acquire a velocity more than double that of the sphere's daily motion; and I shall demonstrate this on the basis of factual suppositions, not arbitrary ones. As a heavy object falls and continually gains speed at the rate we have established, at any given point in its fall it will have gained a degree of velocity such that, if it were to continue to move uniformly at that speed without accelerating any further, it would travel twice the distance it had travelled up to that point in the same time as it had already been falling. So, for example, if the cannon ball has taken 3 hours, 22 minutes, and 4 seconds to fall from the sphere of the Moon to the centre, then when it reaches the centre it will have gained such a degree

If a falling object continued to move uniformly for an equal time with the degree of motion acquired, it would travel twice the distance that it had already travelled with accelerated motion.

of velocity that if it were to continue moving constantly at that speed, without accelerating any further, in the next 3 hours, 22 minutes, and 4 seconds it would travel twice that distance, which is to say the full diameter of the lunar sphere. Now since the distance from the sphere of the Moon to the centre is 196,000 miles, and the ball travels that distance in 3 hours, 22 minutes, and 4 seconds, then—bearing in mind what has already been said—if it continued to move at the speed it had acquired when it reached the centre, in the next 3 hours, 22 minutes, and 4 seconds it would travel twice this distance, i.e. 392,000 miles. But the circumference of the sphere of the Moon is 1,232,000 miles; so if the ball had remained up in the lunar sphere and had participated in its daily motion, in 3 hours, 22 minutes, and 4 seconds it would have travelled 172,880 miles, which is well under half of 392,000. This shows that the motion of the surface of the sphere does not, as the modern author claims, have a velocity that the falling ball could not possibly share, etc.

SAGREDO. This argument would make perfect sense and would convince me, if I were assured of the assertion that a falling object would travel twice the distance in the same time if it were to continue to move uniformly at the maximum velocity it had already acquired in its descent. You have already assumed the truth of this proposition on another occasion, but you haven't demonstrated it.

SALVIATI. This is one of the propositions that our friend has demonstrated, as you will see in due course. In the meantime I want to put some conjectures to you, not to teach you anything [253] new, but to dissuade you from a contrary opinion and to show you that what I say is indeed possible. I'm sure you have observed how a lead weight suspended from the ceiling by a fine, long thread, if it is moved from the perpendicular and then let go, will fall and then spontaneously move almost as far from the perpendicular on the other side.

The motion of a weighted pendulum would be perpetual if all impediments were removed.

SAGREDO. Yes, I've observed it very well, and I've seen how—especially if the weight is quite heavy—there is so little disparity between the distance it descends and the distance it rises again, that I've sometimes thought that the rising arc was equal to the falling one, and so have wondered whether its oscillations could continue in perpetuity. I could believe that they would if it were not for the impediment of the air, which resists being parted by the motion of the pendulum. But it's a very small resistance, as can be seen from the large number of oscillations the pendulum makes before it comes to a stop.

SALVIATI. The motion wouldn't be perpetual, Sagredo, even if the impediment of the air were completely removed, because there is another which is much less apparent.

SAGREDO. What is it? I can't think of anything else.

SALVIATI. You will appreciate it when you hear it, but I'll tell it to you later; in the meantime let us continue. I have proposed this observation of a pendulum to you to help you understand that the impetus acquired in the descending arc, where the motion is natural, has in itself the power to impel the weight upwards, with constrained motion, by an equivalent distance in the corresponding rising arc; and this power is intrinsic to it, when all external impediments are removed. I think we can also take it as beyond doubt that, just as the velocity increases in the descending arc until the lowest point is reached at the perpendicular, so from the lowest point the velocity decreases in the rising arc until it reaches its highest point; and it decreases in the same proportions as it first increased, so that the degrees of velocity are equal at points which are equidistant from the lowest point.

Hence, arguing by analogy, it seems to me reasonable to believe that if the terrestrial globe were pierced by a hole passing through its centre, a cannon ball falling down this shaft would gain such an impetus from its speed that it would pass the centre and would be impelled upwards by the same distance as it had fallen. Its velocity would decrease after it had passed the centre at the same rate as it increased while it was falling, and I believe the time taken by this rising motion would be [254] equal to the time of its fall. So as the ball's velocity gradually decreases from its maximum point at the centre to the point where it stops altogether, it moves through the same distance in the same time as it had travelled as its velocity increased from complete motionlessness to its maximum. Hence it seems reasonable to suppose that if the object moved constantly at its maximum speed it would travel both distances in the same time. If we mentally divide these speeds into ascending and descending degrees, as in these numbers here, with the ascending numbers from 1 to 10 corresponding to the time the object

If there were a hole passing right through the terrestrial globe, an object falling down this hole would pass the centre and would then rise a distance equal to its descent on the other side.

is falling and the descending numbers from 10 to 1 corresponding to its rise, we can see that when they are all added together they come to the same total as if one of the two parts had been made up of the highest number throughout, namely 10. Therefore the total distance travelled through the whole range of ascending and descending velocities—which is the whole diameter of the Earth—must be equal to the distance travelled at the maximum speed for half the total of ascending and descending numbers. I know that I've explained this very badly, but I hope I've made myself understood.

SAGREDO. I think I've understood it very well, and I think I can show that I've understood in a few words. You mean that if motion starts from a state of rest and increases in speed by equal amounts, corresponding to consecutive numbers starting from 1—or rather from zero, since this represents a state of rest—up to any number you choose, so that the lowest degree is zero and the highest is, say, 5, then all these degrees of velocity with which the object moves add up to 15. But if the object were in motion with the same number of degrees, each degree being the maximum, i.e. 5, the total of all these velocities would be twice as much, namely 30. So if the object is in motion for the same length of time but at a constant speed of the highest degree, 5, then it will travel twice the distance as it did with accelerated motion starting from zero.

1
2
3
4
5
6
7
8
9
10
10
9
8
7
6
5
4
3
2
1
0
1
2
3
4
5

SALVIATI. Your quick and perceptive understanding has enabled you to explain it all much more clearly than me, and you prompt me to add something further. In accelerated motion the [255] increase is continuous, and so velocity which is constantly increasing cannot be broken down into a fixed number of degrees; it is changing every moment, and so its degrees are infinite. So a better illustration of our meaning would be to draw a triangle, such as this triangle ABC, dividing its side AC into as many equal parts as we choose, AD, DE, EF, FG, and drawing straight lines from D, E, F, and G parallel to the base BC. Let us imagine the sections we have marked on the line AC as equal intervals of time, and the parallel lines drawn from D, E, F, and G as representing degrees of velocity, increasing by equal amounts in equal intervals of time. A represents the state of rest, starting from which the moving object gains the degree of velocity DH in the time AD. In the following time interval its speed will have increased from the degree DH to EI, and so on in each successive time interval, corresponding to the increasing length of the lines FK, GL, etc. But acceleration is made continuously from moment to moment, not in steps from one time interval to the next. So, if the point A represents the minimum degree of velocity, namely the state of rest, and also the first instant of the subsequent time AD, it is clear that before the moving object reaches the degree of velocity DH, which it does in the time interval AD, it has passed through an infinite number of smaller degrees, acquiring them in the infinite number of instants in the time interval DA, corresponding to the infinite number of points on the line DA. So if we want to represent the infinite degrees of velocity which come before the degree DH, we must imagine an infinite number of shorter and shorter lines, parallel to DH, drawn from the infinite number of points on the line DA. This infinite number of lines will ultimately be represented by the surface of the triangle AHD. So we shall understand how a body which starts from the state of rest and continues uniformly accelerating has consumed and made use of infinite increasing degrees of velocity, corresponding to the infinite number of lines that can be drawn, starting from point A, parallel to the lines HD, IE, LG, and BC, for as long as the motion continues.

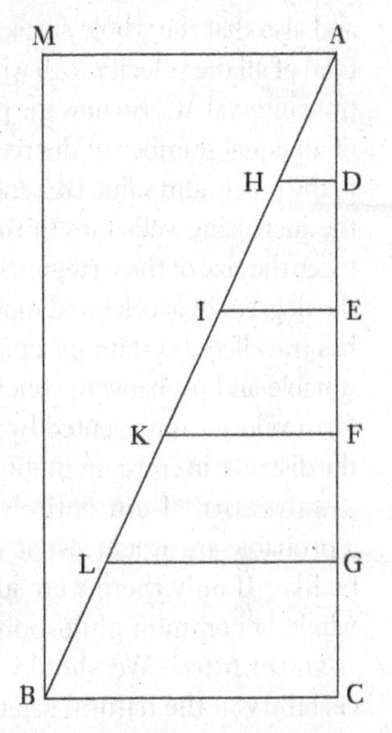

Now let us complete the parallelogram AMBC, and extend to its side BM not just the parallel lines we have marked in the triangle, but the infinite number of lines we have imagined drawn from every point on the side AC. We saw that BC was the longest of the infinite number of lines in the triangle, and that it represented the maximum degree of velocity gained by the object in accelerated motion; [256]

and also that the whole surface of the triangle represented the sum total of all the velocity with which it travelled a given distance in the time interval AC. So now the parallelogram represents the sum total of an equal number of degrees of velocity, but each one now equal to the maximum value BC; and this total velocity is twice the sum of the increasing velocities in the triangle, since the parallelogram is twice the size of the triangle. So if the falling object, passing through the degrees of accelerated motion represented by the triangle ABC, has travelled a certain distance in a given interval of time, it is reasonable and probable to conclude that if it passes through the uniform velocity represented by the parallelogram, it will travel twice the distance in constant motion as it travelled in accelerated motion.

SAGREDO. I am entirely satisfied. If this is what you call a probable argument, what will your necessary demonstrations be like? If only there were just one such conclusive proof in the whole of common philosophy!

Mathematical certainty is not to be sought in the natural sciences.

SIMPLICIO. We should not look for perfect mathematical certainty in the natural sciences.

SAGREDO. Surely motion is a topic in natural science, but I don't recall Aristotle demonstrating even the smallest detail about it. But let's not embark on any further digressions; and Salviati, please keep your promise to explain the reason why the motion of a pendulum ceases, apart from the resistance of the air to being parted.

A longer pendulum oscillates more infrequently than one which is shorter.

SALVIATI. Tell me: of two pendulums of different lengths, doesn't the one whose cord is longer oscillate more infrequently?

SAGREDO. Yes, assuming they move an equal distance from the perpendicular.

The oscillations of a pendulum always have the same frequency, regardless of whether their amplitude is large or small.

SALVIATI. The distance from the perpendicular doesn't make any difference, because the same pendulum will always complete its oscillations in the same time interval, however long or short they are, in other words however much or little it is moved away from the perpendicular. The time intervals may not be exactly the same, but experience will show that the differences are imperceptible, and even if they were substantial this would help rather than hinder our argument. Let us draw a perpendicular [257] AB, with a cord AC fixed at point A, to which we attach a weight, C, and another higher up on the same cord, E: If we draw the cord AC away from the perpendicular and let it go, the weights C and

The cause which impedes the motion of a pendulum and brings it to a stop.

E will move through the arcs CBD and EGF. Weight E, as it hangs from a shorter length of cord and also, as you say, has moved a shorter distance from the perpendicular, will seek to return more quickly and to oscillate with a higher frequency than weight C. Hence it will prevent weight C from moving as far towards its terminal point D as it would if it were unimpeded; and so, by continually exerting this impediment on each oscillation, it will eventually bring it to a stop. Now the cord itself, even without the weights, is made up of many pendulums with a weight of their own, because each part of the cord is such a pendulum, attached closer and closer to point A and therefore with a tendency to oscillate with greater and greater frequency; and consequently it has the effect of exerting a continuous impediment on weight C. This is clear from the fact that if we observe the cord AC, we shall see that it extends in an arc and not in a straight line; and if we were to replace the cord with a chain, this effect would be even more apparent, especially if weight C were moved a long way from the perpendicular AB. For since a chain is made up of many separate links, each of them quite heavy, the arcs AEC and AFD will be seen to have a significant curve. This is because the parts of the chain that are closer to point A seek to oscillate with greater frequency, and so will not let the lower parts move as far as they otherwise would. By continuously detracting from the oscillations of weight C they will eventually bring it to a stop, and would do so even if the air impediment could be removed.

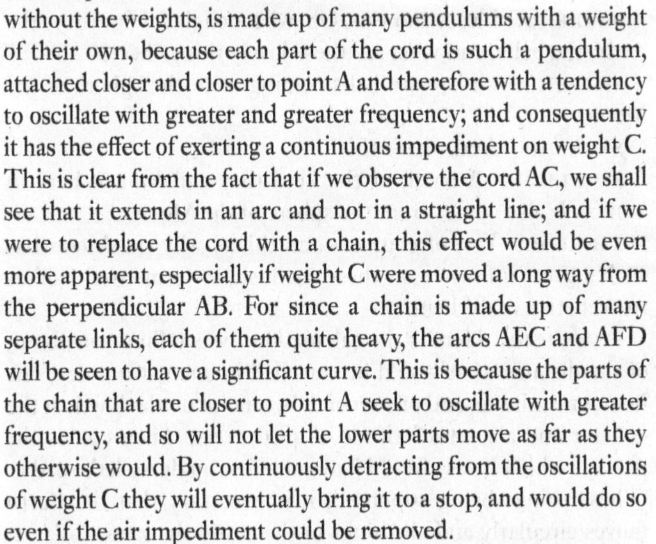

The cord or chain to which the pendulum is attached does not extend in a straight line when the pendulum oscillates, but bends in an arc.

SAGREDO. Here are the books we sent for just now.* Simplicio, take them and see if you can find the passage we were unsure about.

SIMPLICIO. Here it is, where the author begins to argue against the daily motion of the Earth, having first refuted its annual motion: 'The annual motion of the Earth compels the Copernicans to assert its daily rotation, because otherwise the same hemisphere of the Earth would always be turned towards the Sun, and the other side would always be in shadow', meaning that one half of the Earth would never see the Sun.*

[258] SALVIATI. It seems to me from this first comment that your author has failed to understand Copernicus's position correctly. If he had noted how Copernicus makes the terrestrial globe's axis constantly parallel to itself, he would not have said that one half of the Earth would never see the Sun, but rather that a year would be only a single natural day, meaning that the whole Earth would have six months' daylight and six months' night, as in fact happens for the inhabitants of the polar regions. But let's excuse him this oversight, and pass on to what he says next.

SIMPLICIO. He continues: 'That this revolution of the Earth is impossible we demonstrate as follows.' Then he goes on to describe the illustration that follows, which shows many heavy objects falling and light ones rising, birds flying in the air, etc.

SAGREDO. Please show us. Oh, what delightful figures—the birds, the balls, and what are these lovely things here?*

SIMPLICIO. They are balls coming from the sphere of the Moon.

SAGREDO. And what's this?

SIMPLICIO. A snail, of the kind they call *buovoli* here in Venice. It also comes from the sphere of the Moon.

SAGREDO. Yes, of course—and this is why the Moon has such a great influence on these shellfish, or 'armoured fish' as we Venetians call them.

SIMPLICIO. Here is the calculation I was telling you about, of the distance that would be travelled in one natural day, one hour, one minute, and one second, by a point on the Earth at the equator and also at a latitude of 48 degrees. Then comes the passage which I was afraid I had quoted wrongly, so let's read it: 'These things being supposed, it is necessary that the Earth moves circularly and all things in the air do the same . . .', etc. 'So that if we suppose these balls to be equal in weight and size, and they are placed inside the hollow of the Moon's orbit and are allowed to fall freely, then if their downward motion is equal in velocity to their circular motion—although this is not the case since ball A', etc.—'they will fall for at least six days (to concede this much to our adversaries), and in this time they will have circled around the Earth six times', etc.*

SALVIATI. You've reported this author's argument only too faithfully. This just shows, Simplicio, how careful one must be in trying to persuade others to believe what one doesn't believe oneself—because this author can't possibly have failed to realize that he was imagining a circle with a diameter more than twelve times larger than its circumference, and all mathematicians know that the diameter is less than a third of the circumference. This error means giving a value of more than 36 to a measurement that is actually less than 1.

SAGREDO. Perhaps these mathematical proportions, while [259] true in the abstract, don't correspond exactly when applied to real physical circles—although I think when coopers want to find the diameter of the head they have to make for a barrel, which is real and physical enough, they do use the abstract mathematical rule. So perhaps Simplicio would like to defend this author, and tell us whether he thinks that physics can be so very different from mathematics.

SIMPLICIO. I don't think this would be enough to explain such a large discrepancy as this, so all I can say in this case is, *Quandoque bonus* . . . :* sometimes good Homer nods. But assuming that Salviati's calculation is more accurate and that the ball's descent would take not more than three hours, it still seems to me that, if it falls from such a great distance as the orbit of the Moon, it would be extraordinary if it had a natural tendency to keep itself constantly above the point on the Earth's surface that was below it when it started to fall. Would it not rather be left a great distance behind?

SALVIATI. Whether it would be extraordinary, or simply natural and normal, would depend on its state before it started to fall. Your author supposes that the ball, while it remained in the Moon's orbit, would participate in the circular motion every 24 hours which it shares with the Earth and everything else within that orbit. In that case the same force that made it go around before it started to fall would continue to make it do so while it was falling. Then, far from failing to keep up with the Earth's motion and being left behind, it ought rather to run ahead of it, because as it approached the Earth it would be revolving in continually decreasing circles; so if the ball maintained the same speed that it had when it was within the orbit, it would run [260] ahead of the rotation of the Earth. If on the other hand the ball was not affected by the circular motion when it was in the sphere, then there is no reason why it should continue to fall directly above the point on the Earth that was beneath it when it started. Neither Copernicus nor any of his followers would claim that it should.

SIMPLICIO. But the author presses his point, as you can see. He asks what principle produces this circular motion of heavy and light bodies—whether it is an internal principle or an external one.

SALVIATI. Keeping to the question we're discussing, my reply is that the principle that produced the ball's circular motion when it was in the lunar orbit is the same principle as kept it circulating as it was falling. Whether this was an internal or an external principle I leave for the author to decide as he prefers.

SIMPLICIO. He will prove that it cannot be either an internal or an external principle.

SALVIATI. To which I shall reply that when the ball was in the lunar orbit it was not moving. This will exempt me from having to explain how it remains constantly above the same point as it falls, because this doesn't happen.

SIMPLICIO. Very well; but if heavy and light bodies can have no internal or external principle causing them to move in a circle, then neither does the Earth move in a circle, which proves the author's point.

SALVIATI. I didn't say that the Earth has no external or internal principle causing it to move in a circle; I said that I don't know which of the two it is. The fact that I don't know doesn't mean that it doesn't exist. If this author knows what kind of principle causes the circular motion of the other heavenly bodies, whose motion is not in doubt, then I shall say that the principle which moves the Earth is similar to that which causes Mars, Jupiter, and in his view also the sphere of the fixed stars to move. If he can tell me the moving force of one of these bodies in motion, I shall undertake to tell him the force that moves the Earth. In fact I'll do the same if he can identify for me the force that causes individual parts of the Earth to move downwards.

SIMPLICIO. Everyone knows that the cause of this is gravity.

SALVIATI. You're mistaken, Simplicio; you mean that everyone knows that it's called gravity. But I'm not asking you what it's called, but what its true essence is; and of that essence you have no more idea than of the essence which makes the [261] stars turn round. You know only the name that we have given to it, which makes it familiar to us because we see the name a hundred times a day. But we do not really understand the principle or force that makes a stone fall downwards, any more than we do the force that impels it upwards when it separates from its projector, or the force that makes the Moon move around. All we know, as I have said, is the specific name we have given to the first force, which is gravity, and to the second which we call, more generically, impressed force. As for the force that causes the stars and Moon to turn, we call it intelligence, either assisting or informing;* and for any number of other motions we simply attribute their cause to nature.

We have no more knowledge of the force that moves heavy objects downwards than of that which moves the stars in a circle. Of these causes we know only the names that we have given them.

SIMPLICIO. I think the author is asking a much more modest question than the one you refuse to answer. He doesn't ask you to specify and name the principle that makes heavy and light bodies move in a circle, but only whether you think this principle, whatever it is, is internal or external. I may not be able to define gravity, the force that makes Earth move downwards, but I do know that it is an internal principle, because if it's allowed to move freely it will spontaneously move downwards; and I know that the principle that moves it upwards is external, even though I can't define the force impressed on it by the projector.

SALVIATI. We'd be sidetracked into an awful lot of questions if we tried to settle all the problems that keep arising, each one following on from another. You say the principle that makes a heavy projectile move upwards is external, and I dare say you will also call it unnatural and forced; but it may be no less internal and natural than the principle that makes it move downwards. Perhaps it could be called external and forced as long as the body is in contact with the projector; but what is external about the moving force behind an arrow or a cannon ball once it's separated from its projector? We can only conclude that the force that impels it upwards is no less internal than the force that moves it downwards; and in my view the upward motion of heavy bodies deriving from an impetus they have received is no less natural than their downward motion deriving from gravity.

The force that impels heavy projectiles upwards is no less natural than the gravity that makes them move downwards.

SIMPLICIO. That I shall never accept, because the principle of downward motion is internal, natural, and perpetual; upward motion is external, constrained, and finite.

SALVIATI. If you're so reluctant to concede that the principles of upward and downward motion in heavy bodies are equally internal and natural, what would you do if I told you that they could be one and the same?

[262] SIMPLICIO. I leave you to judge.

Contrary principles cannot naturally be found in the same subject.

SALVIATI. No, I want you to be the judge. Tell me, then: do you think internal principles which are contrary to each other can be found in the same natural body?

SIMPLICIO. Absolutely not.

SALVIATI. What do you consider to be the natural internal inclination of earth, lead, gold—of the heaviest materials, in short? That is, what do you think is the motion to which their internal principle draws them?

SIMPLICIO. Motion towards the centre of heavy things, in other words the centre of the Earth and of the universe, to which they would move if they were unimpeded.

SALVIATI. So that if the terrestrial globe was pierced by a hole passing through its centre, and a cannon ball was dropped into the hole, the ball's natural internal principle would carry it to the centre, and its motion would be the spontaneous result of its intrinsic principle; is that right?

SIMPLICIO. Yes, I'm sure that's right.

SALVIATI. When the ball reached the centre, do you think it would pass beyond it or would its motion immediately cease at that point?

SIMPLICIO. I think it would continue in motion for a long way.

Natural motion converts itself into motion which is considered unnatural and constrained.

SALVIATI. Surely once it passed the centre its motion would be upwards, and so would be what you would call unnatural and constrained—and yet it can only derive from the same principle which made it move towards the centre, which you say is internal and natural? Or can you find a new external projector to act on it to drive it upwards?

Now what's true of motion through the centre can also be observed up here in our world. If a heavy body is moving down an inclined plane and the surface is bent upwards, its internal impetus will carry it upwards, without any break in its motion. If a lead ball hanging from a thread is moved away from the perpendicular, its internal inclination will make it spontaneously fall back, and then it will pass through its lowest point and move upwards, without any pause in its motion, and without any new motive force to drive it. And I know you won't deny that the principles which move heavy bodies downwards and light bodies upwards are both equally internal and natural. So consider a wooden ball, falling from a great height and therefore moved by an internal principle, which falls into deep water, and continues to fall and to sink a long way, without any other external force acting on it. Yet it's unnatural for wood to sink in water, but nonetheless this derives from a principle which is not external to it but internal. So you see how a movable body can be impelled by the same internal principle to move in contrary directions. [263]

SIMPLICIO. I believe there are answers to all these arguments, although I can't recall them at present. But be that as it may, the author goes on to ask from what principle this circular motion of heavy and light bodies derives—whether, that is, the principle is internal or external—and then proves that it cannot be either. In his words, *Si ab externo, Deusne illum excitat per continuum miraculum? An vero angelus? An aër? Et hunc quidem multi assignant. Sed contra . . .* [If it is external, is it God who produces it, by a continuous miracle? Or is it an angel, or the air? Many explain it in this way. But against this . . .]

SALVIATI. Don't bother to read the arguments. I'm not one of those who ascribe this principle to the surrounding air. I'm inclined to opt for a miracle or an angel, since something that begins with a divine miracle or the intervention of an angel, like transporting a cannon ball up to the orbit of the Moon, seems likely to do everything else by means of the same principle. As for the air, all I ask is that it doesn't impede the circular motion of the bodies that are assumed to move through it; and for this it suffices that the air moves with the same motion, and completes its revolutions at the same speed, as the terrestrial globe.

SIMPLICIO. He will counter this by asking what it is that moves the air—is it a natural motion or a constrained one? And he will deny that it can be natural, saying that this would be contrary to truth and experience, and indeed to Copernicus himself.

SALVIATI. It certainly isn't contrary to Copernicus, who wrote nothing of the sort; this author is too generous in attributing it to him. What Copernicus does say, rightly in my view, is that the part of the air which is close to the Earth naturally follows its motion, since it consists rather of vapours rising from the Earth, and so shares the Earth's nature; or else that it follows the Earth's motion because it is contiguous with it, in the same way as the Peripatetics say that the upper air and the element of fire follow the motion of the Moon's orbit. So it's up to them to say whether this motion is natural or constrained.

[264] SIMPLICIO. The author will reply that if Copernicus says only the lower part of the air moves, and that the upper part has no such motion, then he can't explain how this motionless air can carry heavy bodies along and make them follow the motion of the Earth.

The tendency of elemental bodies to follow the Earth is effective only in a limited sphere.

SALVIATI. To this Copernicus will say that this natural tendency of elemental bodies to follow the motion of the globe is effective only in a limited sphere, and that no such natural inclination exists beyond it. Besides, I've already said that it's not the air that carries these movable bodies which follow the Earth's motion even when they are separated from it; so all this author's arguments to prove that the air can't produce these effects are invalid.

SIMPLICIO. Well, if it's not an external principle we must conclude that these effects are the product of an internal one; and against such a position *oboriuntur dificillimae, immo inextricabiles, quaestiones secundae*, as follows: *Principium illud internum vel est accidens, vel substantia: si primum, quale nam illud? Nam qualitas loco motiva circum hactenus nulla videtur esse agnita* [there arise very difficult, even insoluble, secondary questions . . . this internal principle is either an accidental property or a substance. If the former, what is it? No quality of local motion around a centre has yet been recognized].

SALVIATI. What does he mean, none has been recognized? What about all this elemental material that moves around together with the Earth? You see, this author is assuming the truth of the point that's at issue.

SIMPLICIO. He says that it can't be seen, and in this I think he's right.

SALVIATI. It can't be seen by us, because we are revolving along with it.

SIMPLICIO. Listen to his other objection: *Quae etiam si esset, quomodo tamen inveniretur in rebus tam contrariis? In igne ut in aqua? In aëre ut in terra? In viventibus ut in anima carentibus?* [If it does exist, how can it be found in such contrary things? In fire as in water? In air as in earth? In living as in inanimate things?]

SALVIATI. Assuming for the moment that water and fire are indeed contraries, as are air and earth—although there's a good deal that could be said about that—the most that can follow from this is that they can't have motions in common which are contrary to one another. So, for instance, upward motion, which naturally belongs to fire, can't also belong to water; and the natural motion of water, as it's by nature the contrary of fire, will be motion *deorsum*, downwards, this being the contrary of the motion of fire. But circular motion isn't contrary to motion either *sursum*, upwards, or *deorsum*; in fact it can be mixed with both of them, as Aristotle himself affirms, so why can it not belong equally to heavy and light bodies? As for living and [265] inanimate beings, the motions that can't be common to them are those which depend on the soul. There's no reason why those which belong to the body, since the body is elemental and therefore shares in the qualities of the elements, shouldn't be common to both a corpse and a living body.

SAGREDO. This author must believe that if a dead cat falls out of a window, it's not possible for a live cat to fall from it too, since it's not proper for a dead body to share in the qualities that belong to a living one.

SALVIATI. We must conclude, then, that this author fails to refute those who say that the principle of circular motion of heavy and light bodies is an internal accident. I wonder how he will get on in proving that it can't be a substance.

SIMPLICIO. He raises many arguments against this, the first of which is as follows: *Si secundum (nempe si dicas, tale principium esse substantiam), illud est aut materia, aut forma, aut compositum: sed repugnant iterum tot diversae rerum naturae, quales sunt aves, limaces, saxa, sagittae, nives, fumi, grandines, pisces, etc., quae tamen omnia, specie et genere differentia, moverentur a natura sua circulariter, ipsa naturis diversissima, etc.* [If the latter—if, that is, you say that this principle is a substance—then it is either matter, or form, or a composite of both. But again, things of so many different natures repel each other: birds, snails, stones, arrows, snow, smoke, hail, fish, etc., are all of different species and kinds, yet they are moved by their nature in circular motion, even though their natures are utterly different, etc.]

SALVIATI. If all these things you've listed have different natures, and things with different natures cannot have a motion in common, we'll have to find many more motions than just straight up and down. To cater for them all we'll need to find one motion for arrows, another for snails, another for stones, another for fish—and we mustn't forget earthworms or topazes or mushrooms, since there's just as much difference between their natures as there is between hail and snow.

SIMPLICIO. I think you're making fun of these arguments.

SALVIATI. Not at all, Simplicio; but we've already answered this point above. If upward or downward motion can belong to the things you've mentioned, then so equally can circular motion. For that matter, still arguing in Peripatetic terms, would you not posit a greater difference between a comet, which is elemental, and a celestial star, than between a fish and a bird? And yet comets and stars both move in circular motion.

Now let's hear the second argument.

[266] SIMPLICIO. *Si terra staret per voluntatem Dei, rotarentne caetera annon? Si hoc, falsum est a natura gyrari: si illud, redeunt priores quaestiones; et sane mirum esset, quod gavia pisciculo, alauda nidulo suo, et corvus limaci petraeque, etiam volens, imminere non posset.* [If the Earth were to stand still by God's will, would everything else still rotate or not? If not, then it is false that they naturally rotate. If so, then the earlier objections arise again: it would be strange indeed if the seagull wanted to hover over the fish, or the lark over its nest, or the crow over the snail and the stone, but they were unable to do so.]

SALVIATI. Speaking for myself, I would reply in general terms and say that if by God's will the Earth were to cease its daily rotation, the birds would do whatever it was God's will for them to do. But if this author were to ask for a more specific response, I would say that they would do exactly the opposite of what they would do in the contrary case—if, by divine will, the Earth were suddenly to start moving at a tremendous speed while they were floating in the air and separated from its surface. Now it's for the author to enlighten us about what would happen in such a case.

SAGREDO. Salviati, I'd be grateful if you would concede on my behalf to this author that, if by God's will the Earth were to come to a standstill, all these other things which were separated from it would continue in their natural circular motion; and let's hear what impossible or difficult consequences would follow. For my part, the only disorders I can see are those which the author himself cites: that if larks wanted to hover above their nests, or crows above a snail or a stone, they would be unable to do so. The consequence would be that crows would have to endure their hunger for snails, and lark chicks would die of hunger and cold because their mother would be unable to feed or cover them. That's the extent of the damage that would result, as far as I can see, in the circumstances that the author describes. Simplicio, can you see if there are any greater disruptions that would ensue?

SIMPLICIO. I can't see any, but it may well be that the author perceived other disruptions in nature besides these, but out of respect preferred not to spell them out.

So I'll go on to the third objection: *Insuper, quî fit ut istae res tam variae tantum moveantur ab occasu in ortum parallelae ad aequatorem? Ut semper moveantur, numquam quiescant?* [Furthermore, how is it that all these diverse things are moved only from west to east and parallel to the equator, and that they are always in motion and never rest?]

SALVIATI. They move from west to east, parallel to the equator, without ceasing, in exactly the same way as you believe the fixed stars to move from east to west, parallel to the equator, without ceasing.

SIMPLICIO. *Quare quo sunt altiores celerius, quo humiliores tardius?* [Why do the higher move more rapidly, and the lower more slowly?]

[267] SALVIATI. Because in a sphere or a circle revolving around its axis, the outer parts move in larger circles and the inner parts in smaller circles, in the same time.

SIMPLICIO. *Quare quae aequinoctiali propiores in maiori, quae remotiores in minori, circulo feruntur?* [Why are those which are nearer the equator carried in larger circles, and those which are further away in smaller circles?]

SALVIATI. To imitate the stellar sphere, where the stars that are nearer to the equator move in larger circles than those which are further away.

SIMPLICIO. *Quare pila eadem sub aequinoctiali tota circa centrum Terrae ambitu maximo, celeritate incredibili, sub polo vero circa centrum proprium gyro nullo, tarditate suprema, volveretur?* [Why does the same ball revolve at incredible speed in a great circle around the centre of the Earth at the equator, but at the pole turns only very slowly around its own centre, not revolving at all?]

SALVIATI. To imitate the fixed stars, which would do the same if the diurnal motion belonged to them.

SIMPLICIO. *Quare eadem res, pila v.g. plumbea, si semel Terram circuivit descripto circulo maximo, eamdem ubique non circummigret secundum circulum maximum, sed translata extra aequinoctialem in circulis minoribus agetur?* [Why does the same object, e.g. a lead ball, if it has once gone around the Earth in a great circle, not go round it everywhere in the same great circle, but rather in smaller circles when it is removed from the plane of the equator?]

SALVIATI. Because that's what some of the fixed stars would do—in fact have done, according to Ptolemy's teaching. Some fixed stars which were once very close to the equator and described very large circles, are now further from it and describe smaller circles.

SAGREDO. There are so many fine things here, I'd consider it a great acquisition if I could only memorize them all. Simplicio, you must lend me this little book of yours, because it must contain a whole host of fascinating and out-of-the-way ideas.

SIMPLICIO. I'll give it to you as a present.

SAGREDO. Oh no, I couldn't deprive you of it. Are these all the questions?

SIMPLICIO. Not at all. Listen to this: *Si latio circularis gravibus et levibus est naturalis, qualis est ea quae fit secundum lineam rectam? nam si naturalis, quomodo et is motus qui circum est, naturalis est, cum specie differat a recto? Si violentus, qui fit ut missile ignitum, sursum evolans, scintillosum caput sursum a Terra, non autem circum, volvatur, etc.?* [If being carried in circular motion is natural to heavy and light bodies, what about rectilinear motion? If it is natural, then how can circular motion be natural, since the two are of different kinds? If it is forced, then why does a fiery arrowhead, flying sparkling up from the Earth, not also revolve in circular motion, etc.?]

SALVIATI. It's already been repeatedly pointed out that circular motion is natural to both the whole and the parts when they are optimally disposed. Rectilinear motion serves to bring into order those parts which are out of their proper place. In fact it would be more correct to say that nothing, whether in its proper place or not, moves in rectilinear motion, but rather in a mixed motion which may even be purely circular; but the only part of this mixed motion which is visible and observable to us is the rectilinear part. The circular part is imperceptible to us, [268] since we ourselves participate in it. This is the answer to the question about rockets; their motion is both upwards and circular, but we don't perceive the circular motion because we are moving with it as well. I don't think this author has ever understood this mixed motion, because he keeps on insisting that rockets go straight upwards and don't have any circular motion at all.

We do not see the circular part of mixed motion, because we participate in it.

SIMPLICIO. *Quare centrum spherae delapsae sub aequatore spiram describit in eius plano, sub aliis parallelis spiram describit in cono? sub polo descendit in axe, lineam gyralem decurrens in superficie cylindrica consignatam?* [Why does the centre of a sphere falling at the equator describe a spiral in the plane of the equator, and at other latitudes a spiral around a cone? Why does it fall along the axis at the pole, tracing a descending line around the surface of a cylinder?]

SALVIATI. Because heavy bodies descend along lines drawn from the circumference of the sphere to its centre, and of these lines the one which ends at the equator describes a circle, and those which end at other latitudes describe conical surfaces, and the axis doesn't trace any other line but remains simply itself.

Now to give you my honest opinion, I don't see how any of these questions provide any grounds for arguing against the motion of the Earth. I'm quite sure that if I were to grant to this author that the Earth is motionless, and I went on to ask him what would happen in all these cases if the Earth were in motion as Copernicus says, he would reply that all these effects would follow which he is now putting up as objections to deny that it is in motion. In other words, he takes necessary consequences and treats them as absurd. So please, if there are any more of these tedious questions, let's quickly get them out of the way.

SIMPLICIO. In this next part he argues against Copernicus and his followers when they say that the motion of the parts when separated from their whole serves only to reunite them with their whole, whereas the circular motion is absolutely natural to their diurnal rotation. Against this he insists that in their opinion, *si tota Terra, una cum aqua, in nihilum redigeretur, nulla grando aut pluvia e nube decideret, sed naturaliter tantum circumferretur; neque ignis ullus aut igneum ascenderet, cum, illorum non improbabili sententia, ignis nullus sit supra.* [if the whole Earth, together with the water, were reduced to nothing, no hail or rain would fall from the clouds, but would simply be carried naturally around; and no fire or burning substance would rise if, as they not implausibly suppose, there is no element of fire above.]

SALVIATI. You've got to admire the prudence of this philosopher, since he's not content to think about what might happen in the ordinary course of nature, but wants to be forewarned in case of things we can be absolutely certain will [269] never happen. I'd like to hear more of his subtle thoughts, so let me grant to him that if the Earth and water were reduced to nothing, hail and rain would cease to fall and fiery substances would no longer rise but would continue going around. What then? What objections will the philosopher raise against me?

SIMPLICIO. His objection is stated in the words which immediately follow: *Quibus tamen experientia et ratio adversatur* [but experience and reason oppose this].

SALVIATI. Well, in that case I must admit defeat, since he has the great advantage of experience which I lack. I've never yet had occasion to see the terrestrial globe and the element of water reduced to nothing, so I haven't been able to observe what hail and rain did in this mini-cataclysm. He does tell us what they did, does he, just for our information?

SIMPLICIO. No, certainly not.

SALVIATI. I'd give anything to be able to meet this person so I could ask him whether, when the Earth disappeared, it also took with it the common centre of gravity, as I assume it did; because in that case, I think the hail and rain must have been left clueless and inert up in the clouds, not knowing what to do with themselves. Or it could be that all the surrounding elements expanded and were drawn into the great empty space left behind when the terrestrial globe disappeared, rushing in to fill it—especially the air which is the most easily distracted. Maybe the more solid bodies, like the many birds which must have been in the air, would draw back more toward the centre of this great empty sphere, since it's reasonable to suppose that substances which contain more matter in a smaller bulk would be assigned a more restricted space, leaving the wider spaces for the more rarefied ones. In the end, when they had died of hunger and been reduced to earth, they would form a new little globe there, together with whatever small amount of water happened to be in the clouds at the time. Or it could be that since these elements are insensitive to light, they wouldn't notice that the Earth had disappeared, and they would continue blindly moving downwards as usual, expecting to encounter the Earth and moving little by little towards the centre, which is where they would end up now if the globe itself wasn't there to stop them.

To give this philosopher, finally, a less indecisive answer, I reply that I know as much about what would happen after the terrestrial globe was wiped out as he would have known about what would happen to it before it was created. I'm sure he will say that he couldn't have imagined any of the [270] things that happened, since only experience has taught him about them. So he must excuse me for not knowing any of the things that he knows about what would happen after this globe was wiped out, since I haven't had the experience that he has.

Tell us now if there is anything else.

SIMPLICIO. There's a figure here showing the terrestrial globe with a large cavity in its centre, filled with air. To show that heavy bodies don't move downwards to be united with the terrestrial globe, as Copernicus says, he places a stone here in the centre, and asks what it would do if it were left to move freely; and he places another on the outside surface of this great cavity, and asks the same question. He says, as regards the first: *Lapis in centro constitutus aut ascendet ad Terram in punctum aliquod, aut non: si secundum, falsum est partes ob solam seiunctionem a toto ad illud moveri; si primum, omnis ratio et experientia renititur, neque gravia in suae gravitatis centro conquiescent. Item, si suspensus lapis liberatus decidat in centrum, separabit se a toto, contra Copernicum; si pendeat, refragatur omnis experientia, cum videamus integros fornices curruere.* [The stone placed in the centre will either rise to some point on the Earth, or not. If not, it is wrong to say that parts move towards the whole solely because they are separated from it. If it does, then this is contrary to both reason and experience, and heavy bodies do not come to rest at their centre of gravity. If the stone which is suspended falls to the centre when it is released, it will be separated from the whole, contrary to what Copernicus says; if it remains suspended, this contradicts all experience, since we see entire vaults collapse.]

Heavy bodies exist prior to the centre of gravity.

If the great mass of heavy bodies was moved, the parts separated from it would follow.

SALVIATI. I'll reply, although I feel at a great disadvantage as I'm arguing against someone who has seen from experience what these stones do in this great cavern, something which I haven't seen myself. I believe that heavy bodies exist before the common centre of gravity, because it's not the centre that attracts matter—the centre being nothing but an indivisible point, and therefore incapable of producing any effect—but rather matter itself, which naturally conspires together to form a common centre, around which parts of equal moment cohere. Hence, I think that if the great mass of heavy bodies was moved to a given place, the parts which had become separated from the whole would follow it and, as long as they were not impeded, would penetrate it until they encountered matter heavier than themselves, at which point they would descend no further. So I think that the whole vault would press down on the cavern full of air, and it would support itself forcibly above all that air only as long as its solidity was strong enough to resist its weight. But I believe that stones which had become detached would descend to the centre and not float above the air. This wouldn't mean that they weren't moving towards the whole, because they would be moving to where all the parts of the whole would cohere if they weren't impeded. [271]

SIMPLICIO. There remains a certain error which he found in a follower of Copernicus, who compared the yearly and diurnal motions of the Earth to those of a cartwheel both on the Earth's surface and on its own axis. His calculations made either the terrestrial globe too big or its orbit too small, given that 365 revolutions of the equator are much less than the circumference of the Earth's orbit.

SALVIATI. I think you must be mistaken, and that this must be the opposite of what the booklet says. The fact is that this author made the globe too small or the orbit too big, not the other way round.

SIMPLICIO. The mistake isn't mine; here's what the booklet says: *Non videt quod vel circulum annuum aequo minorem, vel orbem terreum iusto multo fabricet maiorem* [He fails to see that he makes either the annual circle smaller than it should be, or the terrestrial globe much larger than is proper].

SALVIATI. I have no way of knowing whether the original author made this mistake, since the author of the booklet doesn't name him. But whether this first follower of Copernicus was in error or not, the error made in the booklet is clear and inexcusable, since it passes over such a material error without either noting or correcting it. [The error is attributed here to the author of the booklet, but in fact it is not there.*] But let's excuse it as an oversight rather than anything else.

There is no reason why the circumference of a small circle, rotated a small number of times, cannot measure and describe a line greater than that of a circle, however large.

Besides, if I wasn't already tired of wasting my time on these trivial arguments to very little purpose, I could demonstrate that it's quite possible for a circle no bigger than a cartwheel to describe or measure the circumference of the Earth's orbit, or even a circle a thousand times bigger, not just in 365 revolutions but in less than 20. I say this to show you that there are [272] plenty of subtleties far greater than this one which your author uses to expose Copernicus's error. But please let's pause for breath, and then come on to the other philosopher who is an opponent of Copernicus.

SAGREDO. Indeed I need to pause for breath as well, even though in my case it's only my ears that are tired. If I didn't think I would hear more searching arguments from this other author, I would be tempted to go and take the air in a gondola.

SIMPLICIO. I think you will hear more substantial arguments, because he's both a consummate philosopher and a great mathematician. He has refuted Tycho on the matter of the comets and the new stars.

SALVIATI. Could he be the author of the *Anti-Tycho*?

SIMPLICIO. The very same; but his refutation of the new stars isn't in the *Anti-Tycho*, apart from what I told you before—that he proves they didn't undermine the unchangeability and ingenerability of the heavens. After he wrote the *Anti-Tycho* he discovered how to prove by means of parallax that they're also elemental bodies and are contained within the Moon's orbit; so he wrote another book, *On the Three New Stars, etc.*, where he also included the other arguments against Copernicus. I reproduced for you yesterday what he wrote about these new stars in the *Anti-Tycho*, where he didn't deny that they were heavenly bodies but showed that their appearance didn't affect the unchangeability of the heavens, using the purely philosophical arguments that I gave you. I didn't think to tell you how he then discovered how to remove them from the heavens, since he did so by means of computations and parallaxes, subjects of which I have little or no understanding. I hadn't read that part, but I concentrated on his purely physical arguments against the motion of the Earth.

SALVIATI. I entirely understand. I think that, once we've heard his objections to Copernicus, we should also hear, or at least see, how he proves by means of parallax that these new stars are elemental bodies, given that many highly regarded astronomers place them at a great height among the fixed stars. If this author is able to carry out such a great undertaking as to bring the new stars from the heavens all the way down to the elemental sphere, he deserves to be exalted to the heavens himself, or at least to have his name perpetuated there. So please begin to give us his arguments in opposition to Copernicus, so that we can deal with them as quickly as we can.

[273]

SIMPLICIO. In this case it won't be necessary to read them verbatim, because they're very long-winded. I've read them carefully several times and, as you can see, I've marked the passages containing the nub of the argument in the margin, so it will be enough if I read these. The first begins here: *Et primo, si opinio Copernici recipiatur, criterium naturalis philosophiae, ni prorsus tollatur, vehementer saltem labefactari videtur.* [First, if Copernicus's view is accepted, the criterion of natural philosophy will be severely weakened, if not completely overturned.] All schools of philosophers agree on the criterion that experience and the evidence of the senses are the guides to our investigations, but if Copernicus is right our senses are greatly deceived. We see at close range and with absolute clarity that heavy bodies fall vertically to the ground, without deviating from a straight line by as much as a hairsbreadth; and yet for Copernicus what we see so clearly is a deception, and this motion is not in a straight line at all but a mixture of rectilinear and circular motion.

Copernicus's view violates the criterion of philosophy.

SALVIATI. This is the first argument that Aristotle and Ptolemy and all their followers put forward. We've already answered it at great length and exposed its logical error, making it clear that motion which we have in common with other movable bodies is imperceptible, as if it didn't exist. But since true propositions must have the support of a large number of instances, let me add something more, as a favour to this philosopher. You take his part, Simplicio, and answer my questions for him.

Motion shared in common is as if it did not exist.

First, then, how does the stone affect you so that you perceive its motion when it falls from the top of the tower? It must have some additional or new effect on you when it falls, compared to when it was just resting at the top of the tower; otherwise you wouldn't perceive its descent, or be able to differentiate between when it was in motion and when it was at rest.

A different refutation of the argument based on vertically falling bodies.

SIMPLICIO. I know that it is falling in relation to the tower, because I see it at one moment passing a given point on the tower, then another point lower down, and so on until I see it reach the ground.

SALVIATI. Does that mean that if the stone was dropped from the talons of an eagle in flight and simply fell through the air, which is invisible, and you had no other visible fixed object to relate it to, you wouldn't know that it was in motion?

How the motion of a falling body is perceived.

SIMPLICIO. Yes, I would know, because I would have to [274] raise my head to look up when it was at a great height, and lower my head as it fell. I'd be continually moving my head or my eyes to follow its motion.

SALVIATI. Now you've given the correct answer. So you know the stone is at rest when you can keep it constantly in view without moving your eyes, and you know it's in motion when you have to move your organs of sight—your eyes—in order not to lose sight of it. Therefore you would conclude that an object was motionless as long as you could see it continually from the same viewpoint without moving your eyes.

The motion of the eye allows us to deduce the motion of the object seen.

SIMPLICIO. Yes, I think that would necessarily be the case.

SALVIATI. Now imagine you are on a ship, with your gaze fixed on the top of the mast. Do you think that if the ship was moving, perhaps rapidly, you would have to move your eyes to keep the top of the mast in view and to follow its motion?

SIMPLICIO. I'm certain that I wouldn't have to change anything, not just to keep it in sight, but even if I aimed a gun at it. Whatever motion the ship made, I wouldn't need to move it by a hairsbreadth to keep it aimed at the top of the mast.

SALVIATI. That's because the motion which the ship imparts to the mast it also imparts to you and to your eyes. You don't need to move your eyes at all to keep the top of the mast in view, and therefore it appears to you to be motionless. [The visual ray going from your eye to the mast is like a rope tied between two points on the ship. There are any number of ropes fixed to different points, which remain fixed to these same points whether the ship is in motion or at rest.*]

Now apply this reasoning to the rotation of the Earth and the stone at the top of the tower. You don't perceive its motion because you don't need to move your eyes to follow it; any movement you would have to make is imparted by the Earth to you and the tower alike. If you then add the downward motion which is specific to the stone and is not shared by you, and which is mixed with this circular motion, the circular motion which is common to the stone and to your eye remains imperceptible. Only the downward motion is apparent, because you have to lower your eyes to follow it.

I'd like to be able to propose an experiment to this philosopher to disabuse him of this error. Let him get in a boat, having provided himself with a deep container full of water and a ball of wax or some other material that would sink to the bottom very slowly, so that it falls by barely a *braccio* in a minute. Then make the boat move as fast as possible, so that in a minute it travels more than a hundred *braccia*. He should put the ball gently in the water and let it sink freely, carefully observing its motion. At first he would see it going straight towards that point on the bottom of the container to which it would tend if the boat were stationary; and to his eye and in relation to the container its motion would appear perfectly straight and vertical. Yet it couldn't be denied that its motion would be a compound of straight downwards and circular motion around the element of water. If this is what happens with motion that is not natural, and in materials with which we can experiment both in a state of rest and when they are in motion, and yet our senses appear to be deceived and we still can't discern any difference in the appearances, then how can we expect to discern any when it comes to observing the Earth, which as regards motion and rest has always been in the same state? When do we think we can experiment to see if there is any difference in these effects of local motion when the Earth is in motion or at rest, since it remains eternally in one or other of these two states?

[275]

An experiment to show that motion in common is imperceptible.

SAGREDO. These discussions have helped to settle my stomach, which was somewhat upset by all those fishes and snails. The first point has reminded me how I was disabused of an error which has such an appearance of truth that I doubt whether one person in a thousand would question it. When I was sailing to Syria, I had a very good telescope which had been given to me by our mutual friend who had invented it not many days earlier. I suggested to the sailors that it would be a great aid to navigation if it was used in the ship's crow's nest to spy out distant vessels and identify them. They agreed that it

A subtle investigation of whether a telescope can be used with equal ease at the top of a ship's mast or at its foot.

would be beneficial, but they objected that it would be difficult to use because of the constant rocking of the ship, and especially so at the top of the mast where the pitching is so much greater. They said it would be better to use it at the foot of the mast, where the motion is less than anywhere else on the ship. I won't [276] hide my mistake: I agreed with this view, and for a while I said nothing more in reply. I don't know what it was that prompted me to reflect further on the matter, but finally I realized my folly, excusable though it was, in accepting as true something that was entirely false. It was false, that is, to say that the greater pitching of the ship at the top of the mast than at the foot would make it more difficult to train the telescope on its objective.

SALVIATI. I would have agreed with the sailors, and with your first impression.

SIMPLICIO. So would I, and I still do. I don't think I would change my view however long I thought about it.

SAGREDO. Then for once I can teach a lesson to both of you; and since I find proceeding by question and answer very effective in clarifying matters, I'll adopt that technique—not to mention the pleasure of teasing one's companions by making them say what they didn't know they knew. Let's suppose first that the ships—galleys or other vessels—that we want to discover and recognize are 4, 6, 10, or 20 miles distant, since there's no need for a spyglass to recognize those which are nearer at hand. At a distance of 4 or 6 miles a telescope can easily show the whole vessel, or for that matter a much larger one. Now let me ask you, how many and what kinds of motion affect the crow's nest because of the pitching of the ship?

SALVIATI. Let's imagine the ship is sailing east. Initially, in a completely calm sea, there will be just this forward motion; but when you add the rocking of the waves, there will be a motion which makes the prow and the stern move alternately up and down, so that the crow's nest will lean forwards and back. Other waves will make the ship move sideways, so that the mast rocks to left and right. Others can change the ship's direction and, as if with the rudder, deflect it a little from its easterly course towards either north or south. Yet others may rise up beneath the keel and simply make the ship rise and fall,

Different motions arising from the rocking of the ship.

without deflecting it. I think that these motions are essentially of two kinds: one which changes the angle of the telescope and the other which changes its direction but not its angle, in other words keeping the barrel of the instrument parallel to itself.

Two changes which affect the telescope arising from the pitching of the ship.

[277] SAGREDO. Tell me next what would happen if we first pointed the telescope at that tower at Burano, six miles away, and then moved it through an angle either to right or left, or up or down, just by a nail's breadth. How would this affect the view of the tower?

SALVIATI. The tower would immediately disappear from view, because even such a very small change in the angle here can correspond to hundreds or thousands of *braccia* over there.

SAGREDO. What if we were to re-position the telescope 10 or 12 *braccia* to the right or left, up or down, without changing its angle but keeping the barrel always parallel to itself: what difference would that make to the view of the tower?

SALVIATI. Absolutely none. The change made here and over there must be equal, because the space here and the space there are contained between parallel rays. Since the instrument's field of view is wide enough to encompass many such towers, the tower would not be lost from view at all.

SAGREDO. Coming back now to the ship, we can affirm beyond doubt that moving the telescope 20 or 25 *braccia* to the right or left, up or down, or even forward or back, provided we always keep it parallel to itself, will not deflect the visual ray from the point observed on the object by more than the same 25 *braccia*. Since over a distance of 8 or 10 miles the instrument's field of view embraces a far wider space than a galley or any other ship I may see, this small displacement will not mean that I lose sight of it. It follows that the impediment which makes us lose sight of the object can only come from a change made in the angle, since the displacement of the telescope up or down, or to left or right, caused by the pitching of the ship, can only amount to a few *braccia*.

Suppose now that we have two telescopes, one fixed to the lower part of the ship's mast and the other to the top, not just of the mainmast but of the topmast, where they fly the pennant; and that they are both pointed at a ship 10 miles away. Do you think that any pitching or rolling of the ship and tilting of the mast would produce a greater change in the angle of the telescope at the top of the mast than in the one at the bottom? If a wave makes the bow of the ship rear up, the top of the mast may tilt backwards by 30 or 40 *braccia* more than the base, so that the telescope at the top is moved back by this distance and [278] the one at the bottom by less than a foot; but the change of angle is the same for both. In the same way, a wave striking the side of the ship will move the telescope at the top to the left or right many times more than the one at the bottom, but their angles will either remain unchanged or will both change by the same amount. Since a change of position from right to left, forward or back, or up or down, produces no noticeable impediment to seeing distant objects, whereas a change of angle makes a great difference, we must conclude that there is no greater difficulty in using a telescope at the top of the mast than at the bottom, given that the changes of angle are the same in both places.

SALVIATI. This just shows how carefully one must tread in affirming or denying a proposition. I say again that a confident assertion that it's much more difficult to use a telescope at the top of a mast than at the bottom, because the movement at the top is greater than at the foot, would be enough to persuade anyone that this is the case. So I think it's entirely excusable if some philosophers despair and lose patience with those who won't concede that a cannon ball which they clearly see fall vertically in a straight line really does move in that way, but say that its motion is along an arc, ever more sharply inclined and diagonal.

But let's leave them with their anxiety, and let's hear the other objections our author raises against Copernicus.

SIMPLICIO. He goes on to show how Copernicus's teaching involves denying the evidence of our senses, even the strongest sensations. For instance, we are sensitive to the slightest breeze, and yet we don't feel the force of a constant wind which strikes us at a speed of more than 2529 miles per hour*—this being the distance which the centre of the Earth travels in an hour in the course of its annual orbit. He calculates this carefully, and as he says, if Copernicus is right, *cum Terra movetur circumpositus aër; motus tamen eius, velocior licet ac rapidior celerrimo quocumque*

The annual motion of the Earth ought to cause a constant very strong wind.

vento, a nobis non sentiretur, sed summa tum tranquillitas reputaretur, nisi alius motus accederet. Quid est vero decipi sensum, nisi haec esset deceptio? [The surrounding air is moved with the Earth, and yet its motion would not be felt by us, even though it is more rapid than the strongest wind; indeed we would consider it to be perfectly calm unless some other motion intervened. If this is not deception of the senses, what is?]

SALVIATI. This philosopher must believe that the Earth which Copernicus makes go around the circumference of its orbit, together with its surrounding air, is not the same as the one we live on but some other separate one, because this Earth of ours carries us along with it, at the same speed as the surrounding air. How will we feel it striking us if we're running away from it at the same speed? This gentleman has forgotten that we ourselves, no less than the Earth and the air, are carried around, and therefore we are always in contact with the same part of the air; so it doesn't strike us. [279]

The air does not strike us because we are always in contact with the same part of it.

SIMPLICIO. On the contrary; his very next words are: *Praeterea nos quoque rotamur ex circunductione Terrae, etc.* [Besides, we too are carried around by the revolution of the Earth, etc.]

SALVIATI. In that case I can't either help or excuse him. You'll have to excuse and help him, Simplicio.

SIMPLICIO. For the moment, here on the spot I can't come up with any defence that I find convincing.

SALVIATI. Well, think it over tonight, and then you can defend him in the morning. Meanwhile let's hear his other objections.

SIMPLICIO. He continues with the same argument, showing that Copernicus's view involves denying our own sensations. Now this principle of motion whereby we revolve with the Earth is either intrinsic to us or it is extrinsic, meaning that we are carried along by the Earth. If it's the latter, and since we have no sensation of being carried along, we have to say that our sense of touch is unable to perceive its own object or the impression it makes on our senses. If on the other hand it's intrinsic, then we have no awareness of a local motion deriving from within ourselves, and we will never have any perception of a tendency which is permanently part of ourselves.

Copernicus's view involves denying our sensations.

SALVIATI. The basis of this philosopher's argument, in other words, is this: regardless of whether the principle whereby we move along with the Earth is external or internal, we ought in any case to be aware of it. Since we are not, it is neither external nor internal, and therefore we do not move, and neither does the Earth. My reply is that it can be either external or internal without our being aware of it. If it's external, the example of a boat is more than enough to resolve any difficulties; I say more than enough, because we can make a boat move or remain motionless whenever we like, and make detailed observations of any difference perceptible to the sense of touch which might enable us to say whether it is in motion or not. It's not surprising if this kind of knowledge about the Earth still eludes us, since no one has yet managed to put it to the test; the Earth may have been carrying us along in perpetuity, without us ever having been able to experience what it is like for it to be at rest.

Our motion may be either internal or external, without our being aware of it or feeling it.

[280]

You, Simplicio, if I'm not mistaken, have travelled many times on the boats from Padua to Venice, and if truth be told you will never have felt your participation in the boat's motion unless the boat ran aground or ran into some obstacle and stopped; then you and the other passengers, taken by surprise, would have been in danger of losing your balance. You can be sure that if the Earth were to encounter some obstacle which brought it to a stop, you would be flung up towards the stars, leaving you in no doubt about the impetus you had received. It's true that another of our senses, sight, with the help of reason, can make you aware of the boat's motion, because the trees and buildings in the countryside are separated from the boat and so, when you look at them, they appear to be moving in the opposite direction. But if that was enough to satisfy you of the Earth's motion you have only to look at the stars, which also appear to be moving in the opposite direction, for the same reason. As for not being aware of motion if it's a principle internal to us, there's less reason to be surprised at that: if we don't perceive a motion which is external to us and which is often absent, why should we feel one if it is permanently and immutably present within us?

The motion of a boat is imperceptible to the sense of touch for those who are on board.

The motion of a boat is perceptible to the sense of sight in conjunction with reason.

The motion of the Earth is evident in the stars.

Now, is there anything else in this first objection?

SIMPLICIO. There's just this exclamation: *Ex hac itaque opinione necesse est diffidere nostris sensibus, ut penitus fallacibus vel stupidis in sensibilibus, etiam coniunctissimis, diiudicandis; quam ergo veritatem sperare possumus, a facultate adeo fallaci ortum trahentem?* [So on this view we must necessarily distrust our senses, as being fallacious and obtuse in judging sensory matters, even when they are closely linked to them. For what truth can we hope to find coming from such an unreliable source?]

SALVIATI. I'd rather draw more useful and more certain conclusions from it, by learning to be more cautious and less confident in the first impressions conveyed to us by our senses, because these can easily deceive us. I'd prefer it if this author didn't go to such lengths to persuade us by appealing to our senses that the motion of falling bodies is in a simple straight line and nothing else, or exclaim so indignantly when such a self-evident fact is called into question. This just shows that he thinks those who say that this motion is not rectilinear at all but rather circular must physically see the stone going round in an arc, since he appeals to their senses rather than their reason [281] to settle the matter. But Simplicio, that's not the case at all. I am neutral between these two views, and I'm just acting the part of Copernicus in this role-play of ours; and I have never seen a stone fall in any way except straight downwards, or imagined seeing it otherwise, and I believe it appears to the eyes of everyone else in the same way. So it will be better if we set aside appearances, on which we all agree, and exercise our reason either to confirm that they correspond to reality or to show that they are false.

SAGREDO. This philosopher strikes me as a cut above many of the others who take this view, and if I ever met him I'd like, as a token of my esteem, to remind him of something which he has certainly seen many times. It corresponds closely to the matter we are discussing, and it shows how easily we can be deceived by simple appearances, or rather by sense impressions. Someone walking down a street at night will have the impression that the Moon is keeping pace with them, following them along the eaves of the houses like a cat walking on the tiles, and their eyes would be all too clearly deceived by this appearance if reason didn't intervene.

SIMPLICIO. Indeed there is no lack of examples to assure us that the senses on their own can be deceptive; so let's put these sense impressions to one side and listen to the arguments which follow, which are based *ex rerum natura*, on the nature of things. The first argument is that the Earth cannot, by its nature, move with three completely different motions without going against many manifest axioms. The first axiom is that every effect depends on a cause; the second is that nothing can produce itself, from which it follows that the source of motion and the thing moved cannot be one and the same. This clearly applies to things that are moved by an external mover, but it also follows from the principles above that the same is true of natural motion deriving from an internal principle. Otherwise the mover, as the source of motion, being the cause, and the thing moved being the effect, the cause and effect would be identical. Therefore a body does not move entirely of itself, so that the whole is both the mover and the thing moved; there must be something to differentiate the efficient cause of the motion from what is moved by it. The third axiom is that in things that are subject to the senses, a single thing can produce only one thing; so although the soul in a living creature produces different effects, such as sight, hearing, smell, and generation, it does so by means of different instruments. In short, we can see that different effects in sensible objects derive from differences in their causes.

Arguments ex rerum natura against the motion of the Earth.

Three axioms taken to be self-evident.

[282]

If we now bring these axioms together, it will be plain to see that a simple body like the Earth cannot by its nature move with three very different motions at the same time. Given the assumptions made above, the whole cannot move the whole. Therefore we must distinguish within it three principles for three different motions, as otherwise a single principle would produce several motions. But if the Earth contained three principles of natural motion, in addition to the part which is moved, it would not be a simple body, but one composed of three principles of motion and the part moved. If therefore the Earth is a simple body, it cannot move with three motions. In fact, for the reasons given by Aristotle, it cannot move with any of the motions ascribed to it by Copernicus, since it must move with only one motion towards its centre, as is shown by its parts

A simple body like the Earth cannot move with three different motions.

The Earth cannot move with any of the motions ascribed to it by Copernicus.

which descend at right angles to the spherical surface of the Earth.

SALVIATI. There's a great deal that could be said about the construction of this argument, but I prefer not to examine it at undue length now because it can be answered in just a few words, and what's more in words provided by the very author whom you cite. He says that in a living creature a single principle produces different effects; so my reply to him for now is that in the Earth too, in a similar way, different motions derive from a single principle.

Reply to the arguments ex rerum natura against the motion of the Earth.

SIMPLICIO. That answer certainly won't satisfy the author of this objection; in fact it is completely demolished by what he immediately adds to reinforce his criticism, as you shall hear. He further strengthens his argument with another axiom, namely this: that nature is neither deficient nor excessive in that which is necessary. This is evident to those who study natural phenomena, and especially animals. Since they had to move with many motions, nature has given them many flexible joints, and has joined their parts together in a way appropriate for motion, as at the knees and hips to enable animals to walk and to lie down at will. In humans, in addition, nature has made many flexible joints at the elbow and in the hand, so that they can execute many kinds of motion. From these facts he derives another argument against the threefold motion of the Earth. Either a single continuous body is able to execute different motions without any need of flexible joints, or it is not. If it can do so without them, then nature has made flexible joints in animals in vain, which is contrary to the axiom. But if it cannot, then the Earth, being a single continuous body without any flexible joints, cannot by its nature move with more than one motion. Now you can see how cleverly he refutes your reply, almost as if he had anticipated it.

Fourth axiom against the motion of the Earth.

Animals need flexible joints because of the variety of their motions.

[283]

Another argument against the threefold motion of the Earth.

SALVIATI. Are you serious, or are you speaking ironically?

SIMPLICIO. I'm speaking with the best intelligence I've got.

SALVIATI. In that case you must be confident of having sufficiently sound arguments to hand to be able to defend this philosopher against any other replies made to refute his argument; so please answer me on his behalf, since we can't have him here in person. First of all, you accept as fact that

The flexible joints of animals are not made because of the variety of their motions.

nature gave animals flexible joints and articulations to enable them to execute many different motions. I deny this; in my view animals have flexible joints so that they can move one or more parts of their body while the rest remains motionless. As for their different kinds of motion, I declare that they are all of one kind, namely circular. This is why the ends of all the bones that move are either convex or concave. Some are spherical; these are the bones which have to move in all directions, such as the shoulder joint of a flag-bearer when he waves his banner, or a falconer when he swings the lure to recall a falcon. The same is true of the elbow joint, which enables the hand to turn when it bores a hole with a gimlet. Other joints are cylindrical in form, and turn in only one direction; these are for limbs which bend in only one way, such as the successive finger bones, etc. But without going into more detail, the principle can be stated with a single general statement, namely this: that the motion of a solid body which moves while one of its extremities remains in the same place can only be circular; and since when an animal moves one of its limbs the limb does not separate from the one next to it, this motion must necessarily be circular.

Animal motions are all of the same kind.

The ends of bones which move are all rounded.

Demonstration that the ends of all moving bones must be rounded, and all animal motions must be circular.

[284]

SIMPLICIO. That's not how I understand it. I see animals moving in any number of different ways which are not circular at all—running, jumping, climbing up and down, swimming, and many other ways.

Secondary motions of animals are dependent on their primary motions.

SALVIATI. Yes indeed, but these are all secondary motions deriving from the primary motions of the joints. Jumping or running are motions of the whole body, which are not necessarily circular, but these are the consequence of bending the legs at the knees and the thighs at the hips, motions of the limbs which are circular. The terrestrial globe moves as a whole; there is no question of one part moving while another remains motionless, so there is no need for flexible joints.

Flexible joints are not required for the motion of the Earth.

SIMPLICIO. Taking the philosopher's part, I'll reply that this could well be the case if there was just a single motion, but it's not possible for three completely different motions to be combined in one unarticulated body.

SALVIATI. I'm sure that would be his reply, so I'll counter by attacking from another side, and ask you this: do you think it would be possible by means of articulated joints to adapt the terrestrial globe so that it could participate in three different circular motions? You don't reply? In that case I'll reply on the philosopher's behalf. He would undoubtedly say yes, because otherwise there would have been no point in bringing up the fact that nature makes joints to enable bodies to move with different motions, and saying that the globe cannot have the three motions attributed to it because it lacks flexible joints. If he thought that not even such joints could make the globe capable of such motions, he would simply have said categorically that the globe cannot move with three different motions.

This being the case I would like to ask you, and if possible through you the philosopher who is the author of this argument, to be so good as to explain to me how such joints would need to be arranged to enable the three motions to be easily executed. I grant you four—no, six—months to give your reply. In the meantime it seems to me that a single principle can be the cause of more than one motion in the terrestrial globe, in exactly the same way, as I replied earlier, as a single principle working through different instruments can produce many and diverse motions in animals. There's no need for any articulation, because the motions all belong to the whole and not just to some parts; and since the motions must be circular, there could be no more perfect articulation than a simple sphere.

The question is asked, by means of what kind of flexible joints could the terrestrial globe move with three different motions.

A single principle can be the cause of more than one motion in the Earth.

SIMPLICIO. At most it might be conceded that this could be true of just one motion, but the author and I agree that for three different motions it is not possible. He goes on to reinforce his argument, writing as follows: 'Let us imagine with Copernicus that the Earth moves, of its own accord and by an internal principle, from west to east in the plane of the ecliptic; furthermore, that it turns on its own axis, also by an internal principle, from east to west; and that the third motion deriving from its own inclination is to tilt from north to south and back again.' Given that the Earth is an undivided body not articulated with flexible joints, can our judgement and critical faculties ever embrace the idea that a single natural and undifferentiated principle—a single impulse—could be dispersed into such diverse and virtually contradictory motions? I can't believe that anyone could assert such a thing, unless they had already undertaken to uphold such a position at all costs.

[285]

Another argument against the threefold motion of the Earth.

SALVIATI. Stop there a moment, and find me the place in the book. Let's see: *Fingamus modo cum Copernico, Terram aliqua sua vi et ab indito principio impelli ab occasu ad ortum in eclipticae plano, tum rursus revolvi ab indito etiam principio circa suimet centrum ab ortu in occasum, tertio deflecti rursus suopte nutu a septentrione in austrum et vicissim.**

Grave error by the opponent of Copernicus.

I wondered whether you, Simplicio, had made a mistake in reproducing the author's words; but I see now that he has fallen gravely into error himself, and I'm sorry to say that he has undertaken to attack a position that he hasn't fully understood. These are not the motions that Copernicus attributes to the Earth; I don't know where he got the idea that Copernicus took the annual motion along the ecliptic to be contrary to its motion on its own axis. He can't have read Copernicus's book, because even in the first few chapters he says repeatedly that these motions are both in the same direction, namely from west to east. And even without taking it from others, surely he should have been able to see for himself that if you take the motions of the Sun and the Primum Mobile and attribute them to the Earth, they must necessarily both be in the same direction?

A clever yet simple argument against Copernicus.

SIMPLICIO. Be careful you don't fall into error yourself, along with Copernicus. Isn't the daily motion of the Primum Mobile from east to west, and the Sun's annual motion along the ecliptic in the opposite direction, from west to east? How [286] then can you say that if these same motions are transferred to the Earth they are no longer contraries but in agreement?

SAGREDO. I'm sure Simplicio has shown how this philosopher's error arose, since he must have reasoned in the same way.

The opponent's error exposed by showing how the annual and daily motions, when attributed to the Earth, are in the same direction and not contrary.

SALVIATI. Let's at least take this opportunity to free Simplicio from error. I'm sure he won't have any difficulty in grasping that when he sees the stars rising above the eastern horizon, if this motion did not belong to the stars then the horizon would have to be sinking in the opposite direction, and hence that the Earth was turning on its axis in the opposite direction to the apparent motion of the stars—that is, from west to east, in the order of the signs of the zodiac. As for the annual motion, if the Sun is fixed at the centre of the zodiac and the Earth is moving around its circumference, then for the Sun to appear to move through the zodiac in the order of its signs the Earth must necessarily be moving through them in the same order, since we always see the Sun in the sign opposite to the one where the Earth is found. So, for instance, when the Earth is passing through Aries, the Sun will appear to be passing through Libra; when the Earth is in Taurus, the Sun will appear to be in Scorpio; when the Earth is in Gemini, the Sun will appear in Sagittarius. In other words, both motions are in the same direction, in the order of the signs of the zodiac, the same direction as the Earth's rotation on its axis.

SIMPLICIO. I understand perfectly; I can't think of anything to say to excuse such an error.

SALVIATI. Wait, though, Simplicio; there's another error that's even greater than this. He says that the Earth's daily motion makes it turn on its axis from east to west, yet he doesn't realize that in that case the 24-hour movement of the heavens would appear to us to be from west to east, the very opposite of what we actually observe.

Another graver error shows that the opponent has made little study of Copernicus.

SIMPLICIO. Well, even I wouldn't have made such a grave error as that, and I've only just begun to learn the first elements of spherical astronomy.

SALVIATI. Now you can judge how thoroughly this opponent has studied Copernicus's books if he gets this fundamental hypothesis the wrong way round, since this is the basis for all the points on which Copernicus dissents from the theories of [287] Aristotle and Ptolemy. As for the third motion which he thinks Copernicus assigns to the terrestrial globe, I don't know what motion he means. It's certainly not the one Copernicus attributes to the terrestrial globe along with the other two, namely the annual and diurnal, since that has nothing to do with any declination towards the south or north, but simply serves to maintain the axis of the diurnal rotation continually parallel to itself; so we must conclude that the opponent either didn't understand this or has pretended not to. This serious shortcoming is enough on its own to exonerate us from spending any more time considering his objections, but I want to take them seriously all the same, because to tell the truth they are more deserving of our attention than any number put forward by other insignificant opponents. So, to come back to his

It is doubtful whether the opponent has understood the third motion attributed to the Earth by Copernicus.

objection, it's absolutely not the case that the annual and diurnal motions are opposites; on the contrary, they are both in the same direction, and therefore they can both depend on the same principle. The third motion is a consequence of the annual motion and follows spontaneously from it, so there is no need to invoke any internal or external principle as its cause, as I shall prove in due course.

SAGREDO. I'd like to ask this opponent a question myself, taking natural reason as my guide. He wants to condemn Copernicus if I'm not able immediately to resolve all his doubts and answer all his objections, as if Copernicus's theory was necessarily invalidated by my ignorance. But if he thinks that's a fair way to judge a writer, then he won't find it unreasonable for me to reject Aristotle and Ptolemy if he can't give better answers to the same difficulties that I point out to him in their theories.

The same objection is resolved by examples of similar movements in other heavenly bodies.

He asks me by what principles the terrestrial globe makes its annual motion through the zodiac and its diurnal rotation on its axis around the equator. I reply that they are similar to those by which Saturn moves through the zodiac in 30 years and turns on its axis in the equinoctial plane in a much shorter time, as can be seen from the appearance and disappearance of its collateral globes. Similar again is the principle whereby, as he will readily concede, the Sun traverses the ecliptic in a year, and revolves parallel to the equator in less than a month, as is visibly [288] apparent from the movement of sunspots; and again, it is similar to the principle whereby the Medicean stars traverse the zodiac in 12 years and at the same time revolve in a much shorter period in much smaller circles around Jupiter.

SIMPLICIO. This author would deny all these things as optical illusions produced by the lenses of the telescope.

SAGREDO. Well, that really is asking too much. He says that the unaided eye can't be deceived when it deems falling bodies to move in a straight line, and then says it is deceived in observing these other motions when its power is perfected and increased thirtyfold. Let's tell him, in that case, that the Earth participates in a plurality of motions in a similar way—perhaps in the same way—as a compass that has both a downward motion, as a heavy object, and two circular motions, one horizontal and one vertical along the meridian. Or what else can we say? Tell me, Simplicio: do you think this author would consider that there is a greater difference between rectilinear and circular motion, or between motion and a state of rest?

There is a greater difference between motion and rest than between rectilinear and circular motion.

SIMPLICIO. Without doubt, between motion and a state of rest. This is obvious, because Aristotle says that circular motion is not contrary to rectilinear motion, and in fact the two can be combined, whereas with motion and rest this is impossible.

SAGREDO. In that case it's less implausible to assign to a natural body two internal principles of motion, one rectilinear and one circular, than to give it two principles, also internal, one of motion and one of rest. Both views agree on the natural inclination inherent in the Earth's parts to return to the whole when they are separated from it by force. They differ only in the operation they attribute to the whole; one says that the internal principle of the whole is to be at rest, while the other says it has a principle of circular motion. Now you and this philosopher concede that two principles, one of motion and one of rest, are mutually incompatible, just as their effects are incompatible; whereas this is not the case with rectilinear and circular motion, which have no such resistance to each other.

It is more reasonable to attribute to the Earth two internal principles of motion, rectilinear and circular, than a principle of motion and a principle of rest.

SALVIATI. To this we can add that it's more than likely that the motion of a separated part of the Earth returning to the whole is also circular, as has been said already; so in every respect, as far as the point at issue here is concerned, motion seems more acceptable than rest. So, Simplicio, continue with what remains.

The motion of parts of the Earth returning to the whole may be circular.

[289]

SIMPLICIO. The author reinforces his objection by pointing out another absurdity, which is that the same motions would belong to completely different natures, whereas observation teaches us that different natures act and move in different ways. This is confirmed by reason, because if this were not the case we would have no means of recognizing and differentiating between natures, since we recognize substances by their motions and actions.

Differences of motion enable us to recognize differences of nature.

SAGREDO. I've noticed two or three times in this author's arguments that, in order to prove that something is as he claims, he says that this is how it lends itself to our understanding, or that if it were otherwise we would have no way of knowing this or that

Nature first made things in its own way, and then made human reason able to understand them.

detail, or the principles of philosophy would be undermined. He seems to think that nature first made the human brain, and then arranged things so as to conform to human understanding; but I think it more likely that nature first made things in its own way, and then made human reason with the ability to understand, albeit with great difficulty, something of its secrets.

SALVIATI. I agree. Now tell us, Simplicio, what are these diverse natures to which Copernicus assigns the same motions and actions, contrary to our observation and reason?

Copernicus erroneously assigns the same actions to different natures.

SIMPLICIO. They're stated here: according to Copernicus, water and air—both of which are of a different nature from earth—and everything that exists in these elements must participate in the three motions which he imagines belonging to the terrestrial globe. The author goes on to give a geometrical proof that in Copernicus's view, a cloud suspended in the air which remains for a long time motionless over our heads must necessarily share all three motions of the terrestrial globe. The proof is here; I can't reproduce it from memory, but you can read it for yourself.

SALVIATI. I shan't spend any time reading the proof; in fact I think it was superfluous for the author to include it, since I'm sure that no one who holds that the Earth is in motion will dispute it. So let's take his proof as read and discuss his objection, which doesn't seem to me to prove anything against the Copernican position, since nothing Copernicus says detracts in any way from the motions and actions which allow us to recognize natures, etc. Simplicio, be so kind as to tell me this: can those features which several things have wholly in common serve to enable us to differentiate between the natures of those same things?

Features held in common cannot serve to differentiate between diverse natures.

SIMPLICIO. Certainly not. On the contrary, an identity of [290] actions and features is proof only of an identity of natures.

SALVIATI. Therefore you don't deduce the different natures of air, water, and earth and the other things found in these elements from the actions they all have in common, but from other actions; is that right?

SIMPLICIO. Exactly right.

SALVIATI. In that case, anything which allowed the elements to keep all the motions, actions, and other features that serve to distinguish between their natures, wouldn't deprive us of the ability to distinguish between them, even if it then took away the action which they all have in common, since that serves no purpose in making such a distinction.

SIMPLICIO. I think this reasoning is perfectly sound.

SALVIATI. But is it not your view, and that of this author, of Aristotle and Ptolemy and all their followers, that earth, water, and air are all equally of such a nature as to remain motionless around the centre?

SIMPLICIO. Yes, this is taken as an irrefutable truth.

SALVIATI. In that case your grounds for distinguishing between the nature of these elements and the things belonging to them don't rest on this natural condition which they all have in common; rather their difference must be deduced from other qualities which they don't have in common. It follows that depriving the elements just of this state of rest that is common to them all, and leaving all their other actions unaffected, wouldn't place any obstacle in the way of recognizing their essential difference.

Now this common state of rest is the only thing that Copernicus takes away from them, leaving them weight or lightness, motion whether up or down, slow or fast, rarity or density, or the properties of being hot or cold, dry or moist—everything about them, in fact. So there is no such absurdity in the Copernican position as this author imagines; and sharing an identity of motion signifies neither more nor less than sharing an identity of rest, as far as differentiated or undifferentiated natures are concerned. Now, tell us if there is any other opposing argument.

The fact that the elements share a common motion signifies neither more nor less than their sharing in a common state of rest.

SIMPLICIO. There follows a fourth argument, also based on observation of nature. This is that bodies of the same kind either have motions which agree in kind, or else they agree in [291] a state of rest. On Copernicus's view, however, bodies which agree in kind, and indeed closely resemble each other, would have motions which were quite diverse, in fact diametrically opposed to each other. Stars which are so similar to each other would have such very dissimilar motions that there would be six planets in perpetual revolution while the Sun and the fixed stars would remain perpetually motionless.

Bodies of the same kind have motions which agree in kind.

A further argument against Copernicus.

SALVIATI. The form of this argument strikes me as sound, but I think either its application or its content is faulty, and if the author doesn't modify his assumptions its conclusion will be the direct opposite of what he intends.

The Earth being by nature dark and the Sun and the fixed stars light, are grounds for arguing that the former is in motion and the latter at rest.

The argument goes as follows. Among the heavenly bodies, there are six which are perpetually in motion; these are the six planets. Of the others, namely the Earth, the Sun, and the fixed stars, we don't know which are in motion and which are at rest, given that if the Earth is at rest the Sun and the fixed stars must be in motion, and conversely that the Sun and the fixed stars must be immobile if the Earth is in motion. Since we have no knowledge of which is true, the question is which can be more appropriately deemed to be in motion and which at rest. Natural reason suggests that motion should be attributed to whichever has most in common, in kind and in essence, with those bodies whose motion is not in doubt, and a state of rest to whichever differs most from them; for eternal rest and perpetual motion are such utterly different conditions that a body in constant motion must be utterly different in nature from one that is permanently at rest. Therefore, as long as we are uncertain about which is in motion and which at rest, we must look for some other notable quality that might help us establish which has most in common with the bodies whose motion is not in doubt—the Earth, or the Sun and the fixed stars.

Now nature provides what we are seeking in two striking conditions whose difference is as great as that between motion and rest, namely light and darkness—that is, being naturally brilliant or being dark and totally lacking in light. Bodies shining with an inner and eternal radiance are utterly different in essence from those which lack any kind of light. The Earth is lacking in light, whereas the Sun is radiant with its own light, as are the fixed stars. The six moving planets are wholly lacking in light, like the Earth; so their essence is like that of the Earth and differs from that of the Sun and the fixed stars. Therefore the Earth is in motion, and the Sun and the sphere of the fixed stars are at rest.

[292] SIMPLICIO. The author will firmly deny that the six planets are naturally dark—or else he will affirm the great natural conformity between the six planets and the Sun and the fixed

stars, and their lack of conformity with the Earth, on the basis of other qualities besides darkness and light. In fact, now that I look at it, his fifth argument shows the greatest disparity between the Earth and the heavenly bodies. He writes that it would create great confusion and upheaval in the structure of the universe and between its different parts if, as the Copernican theory maintains, the bilge—where all corruptible matter, water, air, and all the mixed elements accumulate, that is, the Earth—were to be placed between Venus and Mars, that is between heavenly bodies, which are immutable and incorruptible, indeed noble, as Aristotle, Tycho, and others affirm and as everyone, including Copernicus himself, accepts. For Copernicus states that the heavenly bodies are arranged in the best possible order, and that there is no inconstancy in their effect. How much better and more worthy of the natural order—indeed of God the divine architect himself—to separate the pure from the impure, the mortal from the immortal, as all other schools of thought do, thereby teaching that impure and perishable matter is contained within the narrow confines of the orbit of the Moon, while heavenly bodies rise above it in an uninterrupted progression!

Another difference between the Earth and the heavenly bodies, based on purity and impurity.

SALVIATI. It's true that the Copernican system causes upheaval in Aristotle's universe, but we are dealing with our own real and actual universe. When this author, following Aristotle, tries to infer the essential disparity between the Earth and the heavenly bodies from the fact that the latter are incorruptible and the former is corruptible, and concludes from this that motion belongs to the Sun and the fixed stars and immobility to the Earth, he is falling into the logical error of presupposing what he is trying to prove: Aristotle inferred the incorruptibility of the heavenly bodies from their motion, and the point at issue here is whether motion belongs to the heavenly bodies or to the Earth.

Copernicus causes upheaval in Aristotle's universe.

Logical error by the author of the Anti-Tycho.

But we've already spent long enough discussing the emptiness of these rhetorical arguments. What could be more foolish than saying that the Earth and the elements are excluded and separated from the celestial spheres and confined within the sphere of the lunar orbit? Isn't the lunar orbit itself one of the celestial spheres, which they agree has its place right in the

Foolishness of stating that the Earth is not in the heavens.

middle of all the others? It's a novel way of separating the pure from the impure, and the diseased from the healthy, to give those who are infected lodging in the midst of the city! I thought the leper colony was meant to be located as far away [293] as possible. Copernicus did indeed express admiration at the arrangement of the different parts of the universe because God placed the great luminary in the centre of its temple and not to one side, so that its supreme splendour should reach every part of it. As for the placing of the terrestrial globe between Venus and Mars, we shall discuss this shortly, and you can attempt to dislodge it with the arguments of this author. But please let us not encumber the rigour of logical proofs with these flowers of rhetoric, but leave them to orators or rather to poets, who best know how to exalt the most trivial and sometimes even pernicious subjects with their pleasing embellishments. If there's anything else to discuss, let's deal with it as expeditiously as we can.

Argument taken from animals, which need rest even though motion is natural to them.

SIMPLICIO. There remains the sixth and final argument. He considers it highly improbable that a body which is subject to corruption and disintegration could move in perpetual and regular motion, citing the example of animals which, even though they move with the motion that is natural to them, still grow weary and need rest to restore their strength. Such motion has nothing in common with the motion of the Earth, which is immense by comparison; moreover, the Earth has to move with three divergent motions, each taking it in a different direction. Who could argue for such a thing, if they weren't already sworn to uphold it? Copernicus claims that this motion, being natural to the Earth and not forced, is different in its effect from constrained motion, and that while motion resulting from an external impetus does indeed soon dissipate and become unsustainable, that which derives from nature maintains its objects in their optimal state. But this response is invalid, and is destroyed by our reply. An animal, too, is a natural body, not an artificial one, and its movement is natural to it, deriving from its spirit or inner being, unlike constrained motion which derives from an external source to which the object moved makes no contribution. Yet if the animal prolongs its movement it grows weary, and may even die if it pushes itself too far. So you can see how everywhere in nature there are signs contradicting Copernicus's position, and none which support it.

Kepler's argument in favour of Copernicus.

To save me having to take up the role of this opponent again, let me tell you his argument in his dispute with Kepler, who rejected the view of those who considered it inappropriate, indeed impossible, to suppose that the sphere of the fixed stars could be as immense as Copernicus's position required. Kepler argued thus: 'It is more difficult to increase the size of a property out of proportion with the subject to which it belongs, than to enlarge the subject without the property. Therefore it is more plausible to enlarge the sphere of the fixed stars without assuming it is in motion, as Copernicus does, than to increase the immense velocity of the fixed stars as Ptolemy does.' Our author counters this objection by expressing surprise at Kepler's error in supposing that the Ptolemaic hypothesis increases the motion out of proportion to its subject. He argues that on the contrary, the increase is proportionate to the subject, and the velocity of the motion increases as the dimensions of the subject grow. He proves this by imagining a millstone which completes a revolution once every 24 hours, a motion which can be called very slow. If the radius of the millstone is extended as far as the Sun, the velocity of its extremity will equal that of the Sun, and if it is extended as far as the fixed stars it will equal the velocity of the fixed stars, even though the motion at the circumference of the millstone is very slow. Applying this analogy of the millstone to the stellar sphere, let us imagine a point on the radius of that sphere which is as near to its centre as is the radius of the millstone. Then the same motion which is very fast at the stellar sphere is very slow at that point. But it is the size of the body that makes the motion change from very slow to very fast, while still remaining the same motion. So the velocity does not increase out of proportion to the dimensions of the subject, but rather as the subject's dimensions grow, which is quite different from what Kepler supposes.

Objection to Kepler by the author of the Anti-Tycho.

[294]

Circular motion increases in velocity in proportion to the diameter of the circle.

SALVIATI. I don't believe that this author has such a feeble and low opinion of Kepler as to imagine that he doesn't understand that the farthest point on a line drawn from the centre out to the sphere of the stars will move more quickly

The true sense of Kepler's statement explained and defended.

than a point on the same line only two *braccia* from the centre. So he must realize and understand that what Kepler meant to say was that it is less unsuitable, given the norm and the example of other natural bodies, to enlarge a motionless body to an enormous size than to attribute an excessive speed to a body, even one as vast as this. The norm is clearly that velocity decreases as the distance from the centre grows, so that the period of revolution is longer, whereas in a state of rest, which cannot be increased or decreased, the large or small size of the body makes no difference at all. So, if this author's reply aims to counter Kepler's argument, he must believe that for a very small body and a very large one to move [295] in the same period of time the moving principle is the same, since the increase in velocity is the result simply of the increase in size. But this is contrary to the structural rules of nature, according to which smaller spheres complete their revolution in a shorter time; we can observe this in the planets and most clearly of all in the Medicean stars. The period of revolution of Saturn is 30 years, longer than any of the smaller spheres; so to pass from this to a much larger sphere which supposedly makes its revolution in 24 hours is indeed at variance with the norm. If we read this author's reply carefully, therefore, his objection is not to the substance and meaning of Kepler's argument, but rather to the way it is set out and expressed; and even so he is wrong, since he can't deny that he has deliberately pretended to misunderstand Kepler's words in order to accuse him of the grossest ignorance. But it was such a blatant deception that his criticism has not succeeded in diminishing the esteem in which Kepler's teaching is held by the educated public. As for the objection to the perpetual motion of the Earth based on the impossibility of its continuing without growing weary, given that animals which also move with their natural and intrinsic motion grow weary and need to rest their limbs...

The large or small size of a body produces differences in motion, but not when the body is at rest.

The natural order causes smaller spheres to revolve in a shorter time, and larger spheres in a longer time.

Kepler's imagined reply, based on a sarcastic witticism.

SAGREDO. I can imagine Kepler replying that some animals refresh themselves when they are tired by rolling around on the ground, so we need have no fear that the terrestrial globe might grow weary. On the contrary, since it is eternally rotating, we can say that it enjoys a perpetual and tranquil repose.

SALVIATI. You're too sharp and satirical, Sagredo; we are discussing serious matters, so let's leave joking aside.

SAGREDO. Forgive me, Salviati. What I said isn't as wide of the mark as you may think, because movement that provides rest and relief from weariness for a body worn out by a journey can be even more effective in warding off tiredness in the first place; prevention is easier than cure. I'm quite sure that if the motion of animals functioned in the same way as the motion that's attributed to the Earth, they wouldn't grow weary at all. [296] I believe that an animal's body grows weary from using just one part of the body to move both itself and all the rest. So, for instance, in walking only the thighs and the legs are used to carry themselves and everything else, unlike the movement of the heart which appears to be indefatigable because it is moving only itself. Besides, I'm not sure how true it is that an animal's motion is natural rather than constrained; I think the truth is that its vital spirit naturally moves the animal's members with an unnatural motion. For if upward motion is unnatural for heavy bodies, then raising heavy bodies like the legs and thighs to walk can only be done by force, and therefore will cause the mover to grow weary. Going up a staircase takes a heavy body upwards, contrary to its natural tendency, and therefore causes tiredness, because gravity naturally resists such a motion. But if a body is moved with a motion to which it has no resistance, there is no reason for the mover to grow weary or for its strength or power to be diminished; for why should strength be diminished if it is not exercised?

Animals would not grow weary if their motion functioned in the same way as the motion attributed to the terrestrial globe.

The reason why animals grow weary.

The motion of animals should be considered constrained rather than natural.

Strength is not diminished if it is not exercised.

SIMPLICIO. It's the contrary motions with which the terrestrial globe is supposedly moved that are the basis of the author's objection.

SAGREDO. We've already said that the motions are not contrary at all, and that the author is seriously in error on this point. The whole force of his objection rebounds against him, since he claims that the Primum Mobile drags all the lower spheres along in a motion contrary to that which they simultaneously and continually exercise themselves. It's the Primum Mobile, therefore, that should grow weary, since as well as moving itself it has to draw along with it so many other spheres which, moreover, are resisting it with their own contrary

Chiaramonti's objection rebounds against him.

motion. So the author's final conclusion, that when we examine the effects of nature we always find evidence to support the view of Aristotle and Ptolemy and none that does not contradict Copernicus, needs to be considered very carefully. It would be better to say that, since one of these two positions is true and the other is necessarily false, there can never be any reason, experiment, or logical argument in favour of the false view, and none of these can be contrary to the one that is true. So there must inevitably be a great gulf between the reasons and arguments put forward by the proponents of each of these two views, and I will leave it to you, Simplicio, to judge the relative merits of each.

Conclusive arguments can be found for propositions that are true, but not for those that are false.

[297] SALVIATI. I was going to say something in reply to this last of the author's arguments, but with your quick wit, Sagredo, you've taken the words out of my mouth. Nonetheless, even though you have more than adequately replied already, I would still like to add something of what I had in mind.

The author considers it highly improbable that a perishable and corruptible body like the Earth could move perpetually with a regular motion, especially since we see that all animals eventually grow weary and need to rest, and all the more so since the velocity of the Earth's motion would be incomparably greater than that of animals. Now, I fail to understand why he is so perturbed by the velocity of the Earth's motion if he considers the motion of the stellar sphere, which is very many times greater, to be no more disturbing than that of a millstone which makes just one revolution in 24 hours. If the velocity of the Earth's rotation, on the model of that of a millstone, has no greater consequences than that, then the author can cease worrying that the Earth might grow weary, since not even the feeblest and most sluggish of animals—not even a chameleon—would be tired by moving at most five or six *braccia* in 24 hours. If on the other hand he prefers to consider velocity in absolute terms rather than on the model of a millstone, then given the vast distance that the body needs to travel in 24 hours he should be far more reluctant to attribute such motion to the stellar sphere, which has to travel at an incomparably greater velocity than that of the Earth, and also take along thousands of other bodies with it, all of them much bigger than the terrestrial globe.

The sphere of the stars is more at risk of becoming weary than the Earth.

It remains now for us to consider the proofs that lead this author to conclude that the new stars of 1572 and 1604 were sublunar and not celestial, as was commonly believed by astronomers at the time they appeared. This will be no small task; and as the text is new to me and long because of all the calculations it contains, I thought it might be a better use of time if I read as much of it as I can between this evening and tomorrow morning, and then when we meet to continue our discussions tomorrow I will report to you on what I have found. Then, if we have time, we can go on to discuss the annual motion that is attributed to the Earth. Meanwhile if there is anything more you would like to say—especially you, Simplicio—about the matters relating to the diurnal motion which I have talked about at length, we still have a little time left to discuss it.

SIMPLICIO. All that remains for me to say is that I have [298] found our discussions today to be full of acute and ingenious ideas put forward by the advocates of Copernicus in support of the motion of the Earth, but that I don't yet feel persuaded to believe it. In the end what has been said proves only that the arguments for the stability of the Earth are not necessarily conclusive. There hasn't yet been any proof of the opposing view, which necessarily concludes that the Earth is in motion.

SALVIATI. Simplicio, I've never tried to persuade you to change your opinion, and much less would I be so bold as to pronounce definitively on such a great debate. My intention has been, and will be in our future discussions, simply to show you that those who believe that the very rapid motion every 24 hours belongs only to the Earth, and not to the whole universe excluding the Earth, were not blindly convinced that it could and should be so, but carefully listened to and considered the arguments of those who hold the contrary view, and did not dismiss them lightly. Now with this same intention, if you and Sagredo agree, we can go on to consider the other motion attributed to this terrestrial globe first by Aristarchus of Samos and latterly by Nicholas Copernicus. This, as I think you have already heard, is the motion under the zodiac in the space of a year around the Sun, which is fixed and motionless at the centre of the zodiac itself.

SIMPLICIO. This is such a great and noble question that I shall be very curious to hear discussion of it, and I will hope to hear everything that there is to be said on the matter. Then I shall reflect at leisure on what I have heard and have still to hear, and if I gain nothing else I will be able to discuss it on more solid foundations, which will be no small achievement in itself.

SAGREDO. Not to weary Salviati any further, then, let us put an end to today's discussions, and following our usual practice we shall continue our conversation tomorrow, in the hope of hearing great new things.

SIMPLICIO. I'll leave the book on the new stars but take this booklet of conclusions,* so as to re-read the arguments it contains against the annual motion, which is to be the topic of our conversation tomorrow.

THIRD DAY

[299]

SAGREDO. I've been awaiting your arrival so eagerly, in order to hear the new ideas on the annual revolution of our globe, that the hours of this past night and morning have seemed interminable. Not that I have been idle; in fact I've been awake for a good part of the night going over in my mind the discussions we had yesterday, and weighing up the arguments put forward by the advocates of the two contrary theories, that of Aristotle and Ptolemy on the one hand and that of Aristarchus and Copernicus on the other. And I must say that whichever of the two is mistaken is entirely excusable, given the apparent force of the arguments that may have persuaded them—provided, that is, that we limit ourselves to what has been written by these original weighty authors. But the view of the Peripatetics is so ancient that it has had many followers and advocates, whereas the other has very few, first because of its difficulty and then because of its novelty; and among the many supporters of the former, especially among modern authors, there are some who seem to me to support the view which they consider true with arguments that are frankly childish, not to say ridiculous.

SALVIATI. The same thought has occurred to me even more forcefully than to you, having heard so many arguments of this kind that I would be ashamed to repeat them, not for fear of harming the reputation of their authors—we can always avoid naming them—but so as not to debase the honour of the human race. As I've observed this I have concluded that there are some men who reason backwards, first establishing the conclusion in their mind and then, either because it's their own idea or because it comes from someone in whom they have [300] great confidence, imprinting it there so firmly that it's quite impossible to dislodge it. They immediately assent to and applaud whatever arguments they recall or hear from others in support of their settled view, however shallow or simplistic they may be. On the other hand whatever contrary arguments are put to them, however conclusive and well thought out, they

Some, when they argue, first fix in their mind the conclusion that they believe, and then adapt their arguments to it.

receive them not just with disgust but with bitter anger and indignation. Some of them are driven by such fury that they would resort to almost any scheme to suppress and silence their opponent; I've seen some examples of this myself.

SAGREDO. Indeed, such people don't deduce their conclusion from the premises or establish it by reasoned argument, but adapt, or rather distort and overturn, the premises and arguments to their already established and nailed down conclusions. It's not wise to contend with such people, not least because dealing with them is not only unpleasant but dangerous. So we shall continue with our friend Simplicio, whom I have long known to be a man of great integrity and wholly lacking in malice of any kind. What's more, he is very well versed in Peripatetic doctrine, so I can be sure that if he is unable to cite an argument in support of Aristotle's view then it's very unlikely that anyone else can. But here he comes now, all out of breath, the man we've been wanting to see all morning. Simplicio, we've just been maligning you.

SIMPLICIO. Don't blame me for my long delay; the fault lies with Neptune. On this morning's ebb tide the waters withdrew in such a way that the gondola that was bringing me here ran aground just as it had entered a canal not far from here where there is no embankment, and I had to wait more than an hour for the tide to come back in. As I was stuck there, not able to get out of the boat when it unexpectedly ran aground, I started thinking about a detail that I found quite remarkable. As the tide receded and the mud was exposed in many places, you could see the water running rapidly away in little rivulets; and as I was watching and reflecting on this, all of a sudden I saw this motion cease and, without any pause, the same water start to flow back again, and the tide change from retrograde to direct motion, without standing still even for a moment. In all the time I've known Venice I've never happened to see this effect before.

There is no interval of rest in the motion of the water between the outgoing and incoming tide.

[301] SAGREDO. There can't have been many occasions when you found yourself grounded with little rivulets around you, because the smallest incline, when the surface of the open sea rises or falls by as much as the thickness of a sheet of paper, is enough to make the water in such rivulets flow quite a long way. In the

same way, there are places on the sea shore where a rise of 4 or 6 *braccia* in the sea level can cause the water to spread for thousands of acres over the plains.

SIMPLICIO. I understand this very well, but I would still have thought that there would be a perceptible interval of rest between the end of the receding tide and the beginning of its coming in again.

SAGREDO. It may well appear to be so when you observe the tide against a wall or a pile, where the changes in level are vertical, but in reality there is no moment of rest.

SIMPLICIO. But as these are two contrary motions, I thought there must necessarily be an interval of rest between them, following Aristotle's principle that *in puncto regressus mediat quies*: rest intervenes at the point of reversal.*

SAGREDO. I remember this point very well, but I also remember that when I was studying philosophy I was not convinced by Aristotle's proof, and indeed I had many experiences which contradicted it. I could cite them now, but I don't want us to venture into any more ocean depths. We've met here to discuss our subject, if possible without digressing as we have these past two days.

SIMPLICIO. But we may need, if not to digress, at least to extend the discussion considerably. When I got home last night I started re-reading the booklet of conclusions,* and found that it had very conclusive arguments against the annual motion attributed to the Earth; and as I didn't trust myself to be able to reproduce them in detail, I've brought the booklet with me again today.

SAGREDO. Well done. But if we are to continue our discussion as we agreed yesterday, we must first hear what Salviati has to report to us about the book on the new stars;* then we can come on to the annual motion without further interruptions. So, Salviati, what have you got to tell us about these stars? Have they really been brought down from the heavens to these lower regions by the force of the calculations in Simplicio's book?

SALVIATI. I started reading his exposition yesterday evening, [302] and I took another look at it this morning to see whether what I thought I had read in the evening was indeed what it contained, or whether these were illusions or fantasies that had come to me in the night. I regret to say that I did indeed find written and printed in the book what, for the sake of this philosopher's reputation, I would rather not have found there. I can't believe that he is unaware of the worthlessness of his project, both because it is so obvious and because I have heard our friend the Academician speak approvingly of him. I find it too implausible that he should value his own reputation so lightly that to please others he was persuaded to publish a work which he must have known would bring him only the condemnation of intelligent readers.

SAGREDO. You could add that intelligent readers are less than one in a hundred, compared to those who will praise and exalt him above all the greatest intellects there are or ever have been. Someone who was able to defend the Peripatetic immutability of the heavens against a host of astronomers,* and what's more to defeat them with their own weapons! What weight will half a dozen readers in any province who can see the shallowness of his arguments have against the innumerable others who are unable to perceive or understand them, and who believe what they are told and applaud all the more the less they understand? Besides, those few who do understand will decline to make any reply to such poor and inconsequential writing—and with good reason, because those who understand have no need of it, and for those who don't it would be a waste of effort.

SALVIATI. Indeed, the most appropriate punishment for their failure would be silence, if it weren't for other reasons which practically oblige us to react. Apart from anything else, [303] we Italians are allowing ourselves to be seen as ignorant and made a laughing-stock by those from north of the Alps, especially those who have broken away from our religion. I could show you some famous writers who mock our Academician and all Italian mathematicians for allowing the inane writings of someone like Lorenzini* against astronomers to be published and remain unchallenged. But even this could be overlooked, in comparison with the even greater cause for ridicule arising from the hypocrisy of intelligent readers towards the shallow arguments of others who likewise oppose views which they don't understand.

SAGREDO. I could ask for no better example of their arrogance and the unhappy state of someone like Copernicus, obliged to submit to the assaults of those who don't understand even the first principles of the theory for which they have gone to war against him.

SALVIATI. You will be no less amazed at the way they refute the astronomers who affirm that the new stars were above the orbits of the planets, and perhaps even among the fixed stars.

SAGREDO. How have you managed to examine the whole of this book, which is a substantial volume and must contain a large number of proofs, in such a short time?

SALVIATI. I confined myself to his first refutations, where he gives twelve proofs based on the observations of twelve astronomers who all considered the new star that appeared in Cassiopeia in 1572 to be in the firmament of the fixed stars. He proves that on the contrary it was below the sphere of the Moon, by comparing, two by two, its meridian altitudes as taken by different observers located at different latitudes, as I will presently explain. It seems to me from my reading of this first of his procedures that this author was very far from being able to prove anything against astronomers in favour of the Peripatetic philosophers, but rather confirmed their position even more strongly; so I preferred not to examine his other methods with the same level of attention but just glanced at them superficially, as I was sure that these criticisms would be [304] as ineffective as the first. And indeed, as you will see, this whole book can be refuted in very few words, despite all the laborious calculations that it contains.

Method followed by Chiaramonti to refute the astronomers, and by Salviati to refute him.

This, then, is how I shall proceed. This author takes a large number of observations made by twelve or thirteen of his opponents, and in order to defeat them, as I've said, with their own weapons, makes his own calculations based on some of their data and concludes that such stars are below the Moon. Now since I like to proceed by means of question and answer, and as the author himself is not here, I would like Simplicio to reply to the questions I shall put to him as he believes the author would do. So, Simplicio, assuming we are dealing with the star that appeared in Cassiopeia in 1572, tell me if you think it could have been located in different places at the same time, whether in the elemental world, or among the orbits of the planets, or above these among the fixed stars, or even infinitely higher still?

SIMPLICIO. There's no doubt one must say that it was in only one place, and at just one specified distance from the Earth.

SALVIATI. So if the astronomers' observations were accurate, and if this author's calculations were not in error, all of these observations and calculations must necessarily give exactly the same figure for its distance from the Earth, must they not?

SIMPLICIO. I can understand enough to say that it must necessarily be so, and I don't think the author would disagree.

SALVIATI. But if no two of the very many calculations gave the same answer, what would you conclude?

SIMPLICIO. I would conclude that they were all wrong, either because of a mistake in the calculations or because the observations were defective. The most you could say would be that one of the figures, and only one, was correct, but I would have no way of deciding which it was.

SALVIATI. Would you then want to base a dubious conclusion on false foundations, and take it to be true? Clearly not. Now the calculations made by this author are such that not one of them agrees with any of the others, so you can see how reliable they are.

[305] SIMPLICIO. If that's the case, it is indeed a serious defect.

SAGREDO. If I may say something to help Simplicio and the author, I think, Salviati, that your reasoning would indeed be conclusive if the author had been trying to establish the exact distance of this star from the Earth; but I don't believe that was his intention. He meant only to deduce from these observations that the star was sublunar; so if the observations and the calculations based on them consistently show the distance of the star to be less than that of the Moon, that is enough for the author to convict of the grossest ignorance all those astronomers who, through faulty geometry or arithmetic, are unable to draw the correct conclusions from their own observations.

SALVIATI. I think I'd better address myself to you, Sagredo, since you uphold this author's reasoning so perceptively; but first let me make a point so that Simplicio, even though he is not an expert in calculations and geometrical proofs, can at least see how inconclusive the author's arguments are. Both the author and the astronomers against whom he is arguing agree that the new star had no proper motion of its own, but only went around with the diurnal motion of the Primum Mobile. But they disagree over its location, the astronomers placing it in the celestial region, i.e. above the Moon and possibly among the fixed stars, while he considers it to be close to the Earth, i.e. under the arc of the Moon's orbit. Since this new star we are discussing was towards the north and not very far from the pole, so that for us in northern latitudes it never sank below the horizon, it was quite easy with astronomical instruments to take its meridian altitude, both at its minimum beneath the pole and at its maximum above the pole. By comparing these altitudes as observed at different places on the Earth situated at varying distances from the north, i.e. at varying polar elevations, it was possible to deduce the star's distance from the Earth.

[306] If the new star was in the firmament among the fixed stars, then its meridian altitudes taken at different polar elevations would have to differ by the same amount as between the polar elevations themselves. So for example, if the elevation of the star above the horizon was 30 degrees at a place where the polar elevation was, say, 45 degrees, then the star's elevation should be greater by 4 or 5 degrees further north where the polar elevation was also greater by 4 or 5 degrees. If, on the other hand, the distance of the star from the Earth was very small by comparison with that of the firmament, then its meridian altitudes should have increased significantly more than the polar elevation as the pole was approached. This greater increase—the amount by which the increase in the star's elevation exceeds that of the polar altitude—is called a difference in parallax, and from this it is easy, following a clear and reliable method, to calculate the distance of the star from the centre of the Earth.

If the new star is in the firmament of the fixed stars, the difference between its minimum and maximum elevations will not be greater than between the polar elevations.

Now this author takes the observations made by thirteen astronomers* at different polar elevations, and selects a small number of these as he chooses. He then uses twelve pairings to calculate that the height of the new star was always lower than that of the Moon. But he only manages to do this by making the assumption, which I find quite sickening, that anyone into whose hands his book falls will be grossly ignorant. I wonder how the astronomers he inveighs against are able to keep silent, especially Kepler, whom he particularly chooses to attack and who is not usually slow to speak out, unless he decided that it was beneath him.

So that you can see for yourselves, I've copied out on this sheet the conclusions he draws from his twelve comparisons.

1. The first is from two observations, one by Maurolico and the other by Hainzel, from which it is deduced that the star's distance is less than 3 terrestrial radii from the centre, the difference of parallax being 4 degrees, 42 minutes and 30 seconds.. 3 radii;

2. from the observations of Hainzel and [307] Schuler, with a parallax of 8 minutes and 30 seconds, he deduces that its distance from the centre is more than.. 25 radii;

3. from the observations of Tycho and Hainzel, with a parallax of 10 minutes, he deduces the distance from the centre to be slightly less than.. 19 radii;

4. from the observations of Tycho and the Landgrave, with a parallax of 14 minutes, he gives the distance from the centre as around.............. 10 radii;

5. from the observations of Hainzel and Gemma, with a parallax of 42 minutes and 30 seconds, the distance is implied as around............. 4 radii;

6. from the observations of the Landgrave and Camerarius, with a parallax of 8 minutes, the distance is found to be around......................... 4 radii;

7. from the observations of Tycho and Hagek, with a parallax of 6 minutes, he gives the distance as.. 32 radii;

8. from the observations of Hagek and Ursinus, with a parallax of 43 minutes, he gives the star's distance from the surface of the Earth as........ 1/2 a radius;

9. from the observations of the Landgrave and Busch, with a parallax of 15 minutes, he gives the

distance from the surface of the Earth as .... $^1/_{48}$ of a radius;

10. from the observations of Maurolico and Muñoz, with a parallax of 4 degrees and 30 minutes, he gives the distance from the surface of the Earth as .... $^1/_5$ of a radius;

11. from the observations of Muñoz and Gemma, with a parallax of 55 minutes, we have a distance from the centre of around .... 13 radii;

12. from the observations of Muñoz and Ursinus, with a parallax of 1 degree and 36 minutes, the distance from the centre is found to be less than .... 7 radii. [308]

These are twelve collations that the author has chosen to make, among the very many, as he says, that could have been made by combining the observations of these thirteen observers; and I think we can assume that he chose the twelve which are most favourable to prove his case.

SAGREDO. I would like to know whether among the many measurements the author did not choose there were any that were unfavourable to his case, in other words those from which calculations would show that the new star was above the Moon. It seems reasonable to me at first sight to suppose that there were, since the figures we have seen differ so widely among themselves, some giving the star's distance from the Earth as four, six, ten, a hundred, a thousand, and fifteen hundred times greater than others. So I can well suspect that there were some among the figures he did not calculate that would favour the opposite view, all the more so because I don't think the astronomers who made these observations were lacking in intelligence and expertise in these computations, which are hardly the most difficult calculations in the world. In fact, given that even among these twelve comparisons there are some which show the star as only a few miles from the Earth and others which show it as just a small distance below the Moon, I would consider it nothing less than miraculous if there were not some, favouring the opposite view, that put the star at least 20 *braccia* beyond the lunar orbit. And it would be more extraordinary still if all these astronomers were so blind as not to realize such a glaring mistake.

SALVIATI. Prepare to be amazed when you hear the excesses of confidence in one's own authority and the stupidity of others to which people can be led when they want to argue and show themselves more intelligent than others. The measurements the author chose to ignore include some which show the new star to be not just above the Moon but even above the fixed stars—and not just a few but the majority, as you can see from this other sheet of paper where I've noted them down.

[309] SAGREDO. What does the author say about them? Did he not consider them?

SALVIATI. Yes he did, but he says the calculations that show the star at an infinite distance are based on erroneous observations, and that they contradict each other.

SIMPLICIO. That strikes me as a very weak answer, because the other side could equally well say that the observations showing the star to be in the region of the elements are erroneous.

SALVIATI. Simplicio, you would be amazed and also outraged if I could show you the clever artifice—not that it's really so very clever—by which this author tries to ingratiate himself with you, flattering your ears and inflating your ambition, and hides his cunning by playing on your credulity and that of the other pure philosophers. He claims to have convinced and silenced these mere astronomers who presumed to question the absolute immutability of the Aristotelian heavens, and what's more to have silenced and convinced them by means of their own weapons. I'll do my best to show you, and Sagredo must excuse Simplicio and me if we try his patience while I attempt to make clear at perhaps excessive length—excessive, that is, for his quick understanding—what should certainly not remain unrecognized and hidden from him.

SAGREDO. Not only will you not try my patience, but I will relish hearing what you have to say. I only wish that all Peripatetic philosophers were able to listen, so that they would realize how much they are obliged to this defender of theirs.

SALVIATI. Tell me then, Simplicio, is it clear to you how, if the new star is located in the meridian circle towards the north, a person travelling half a day's journey northwards would see the star's elevation above the horizon increase by the same amount as the pole star, assuming that its true location was among the fixed stars? Whereas if its location was significantly lower, i.e. closer to the Earth, its elevation would increase by more than that of the pole star, and the increase would be greater the closer it was to the Earth?

SIMPLICIO. I think this is quite clear, and I'll try to show it in a mathematical diagram: [310]

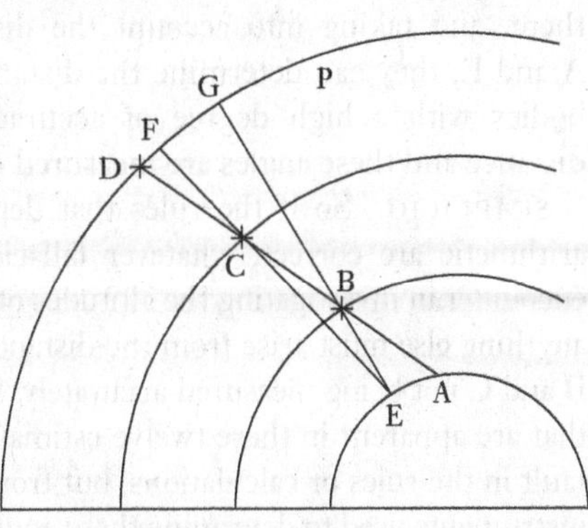

On this large circle I will mark the pole star as point P, and on these two lower circles I will mark two stars, B and C, observed from point A on Earth. These two stars are seen along the same line ABC prolonged until it meets a fixed star D. If I then walk along the Earth's surface to point E, the two stars will appear to separate from the fixed star D and approach the pole star P. The lower star B, moving more, will appear to me at G, and star C, moving less, will appear at F. But the fixed star D will have remained at the same distance from the pole star.

SALVIATI. I can see that you've understood very well. I think you understand, too, that as star B is lower than star C, the angle formed by the visual rays from the two places A and E meeting at C—that is, the angle ACE—is narrower, or if you like, more acute, than the angle formed at B by the rays AB and EB.

SIMPLICIO. Yes, that's clear to see.

SALVIATI. And because the Earth is tiny and almost imperceptible in comparison with the firmament, and therefore the distance AE that we can walk on Earth is also tiny compared to the immense length of the lines EG and EF from the Earth to the firmament, you can understand that star C could be raised so much higher from the Earth that the angle formed at it by the rays from these same points A and E would become extremely acute, so much so that it would be absolutely imperceptible and effectively nil.

SIMPLICIO. This too I understand perfectly.

SALVIATI. Now you need also to know, Simplicio, that astronomers and mathematicians have found infallible geometrical and mathematical rules by means of which, using the sizes of these angles at B and C and the difference between them, and taking into account the distance between points A and E, they can determine the distance of the most remote bodies with a high degree of accuracy, provided that this distance and these angles are measured correctly.

SIMPLICIO. So if the rules that depend on geometry and arithmetic are correct, whatever fallacies and errors we may encounter in investigating the altitudes of new stars or comets or anything else must arise from the distance AE and the angles at [311] B and C not being measured accurately. So all the discrepancies that are apparent in these twelve estimates derive not from any fault in the rules or calculations, but from errors made with the instruments used to determine these angles and distances.

SALVIATI. That's true, and there's no problem about that. Now note carefully that as the star moves further away, from B to C, making the angle ever more acute, the ray EBG moves progressively further away from that part of the ray ABD which is below the angle. This can be seen in the line ECF, the lower part of which EC is further away from the part AC than is EB. But however far away it moves, the lines AD and EF can never completely separate, as they must finally meet at the star; they could only be said to separate and become parallel if the distance was extended to infinity, and this cannot happen. However—and this is the point to note—we have already seen that the distance from the Earth to the firmament can be considered infinite in relation to the small size of the Earth, and therefore the angle contained by the rays extended from points A and E to terminate at a fixed star can be reckoned as nil, and such rays considered as parallel lines. It follows that we can only affirm that the new star was in the firmament if a comparison between observations made at different places produces calculations showing the angle to be imperceptible and the lines to be parallel. If however the angle is of a significant size, it follows that the new star is lower than the fixed stars and, if the angle ABE is greater than would be formed at the centre of the Moon, that it is lower even than the Moon.

SIMPLICIO. Is the distance of the Moon then not so great as to make such an angle on the Moon appear imperceptible?

SALVIATI. No, it's not; and indeed the angle is perceptible not only at the Moon but also at the Sun.

SIMPLICIO. In that case the angle could be observable in the new star, and this would not mean it was necessarily lower than the Sun, let alone the Moon.

SALVIATI. Yes, it could be, and in fact it is in this case, as you will see in due course. But first I shall prepare the ground so that, even though you're not expert in astronomical calculations, you can understand and see for yourself how this author was more [312] interested in pleasing the Peripatetics by covering up and dissembling various points than in presenting the unvarnished truth. So let's proceed further. I think you've clearly understood from what has been said so far that the distance of the new star can never be so great that the angle we've mentioned several times would disappear altogether and the rays from observers at points A and E would become parallel. This being so, you must also understand perfectly that if any observations produced calculations showing that this angle was nil or the lines were truly parallel, these observations must be faulty, at least in some small degree. If, on the other hand, the calculations showed these lines diverging not just to the point of equidistance—that is, having become parallel—but having gone beyond that point, so that they were further apart where they were above the angle than below it, then we would have to conclude beyond a doubt that those observations were made even less accurately and were quite erroneous, as they lead to a conclusion that is manifestly impossible. Now you must take it from me when I state as a categorical truth that when two straight lines are drawn from two points on another straight line, if the angles included between them on the third line are greater than two right angles the lines will be further apart at the top than at the bottom. If the angles were equal to two right angles the lines would be parallel; and if they were less than two right angles the lines would converge and, if prolonged, would necessarily form a triangle.

SIMPLICIO. I know this without needing to take it on trust from you; I'm not so devoid of geometrical knowledge that I don't recognize a proposition that I've read a thousand times in Aristotle, namely that the three angles in a triangle are always equal to two right angles. So in my diagram, I entirely understand that in the triangle ABE, assuming that EA is a straight line, the sum of the three angles A, E, and B equals two right angles, and therefore that the angles E and A together are less than two right angles by the amount of angle B. Moreover, if the lines AB and EB are widened, while keeping [313] them fixed at points A and E, until the angle contained by them in the direction of B vanishes, the two angles beneath will remain equal to two right angles, and the lines will be parallel. If they were widened further, the angles at E and A would become greater than two right angles.

SALVIATI. You are a veritable Archimedes, and you have spared me the need to expend any more words explaining that whenever the calculations result in the two angles A and E being greater than two right angles, the observations must necessarily be wrong. This is what I wanted to be certain that you understood, and which I wasn't sure I could explain in a way that could be fully grasped by a pure Peripatetic philosopher. So now let's proceed to what follows.

To recapitulate the point you agreed just now, given that the new star could not be in different places but must be in just one place, it followed that whenever the calculations based on these astronomers' observations showed it not to be in the same place there must have been an error in the observations, that is in taking the elevation of the pole star, or the altitude of the new star, or both. Now, as very few of the many comparisons between pairs of observations agreed in showing the star to be in the same place, it is only these few that may be considered as being correct; all the others are definitely incorrect.

SAGREDO. So we must give more credence to these very few than to all the others put together. You say there are very few that agree with each other. I can see two of these twelve comparisons, the fifth and the sixth, that both put the star's distance from the centre of the Earth at 4 radii; so it is more likely that the new star was in the region of the elements rather than the celestial one.

SALVIATI. Not so. If you read carefully you will see that they don't state the distance to have been exactly 4 radii, but 'around 4 radii'; and you will see that the two distances differed by several hundred miles. Look: the fifth comparison, here, is 13,389 miles, while the sixth is 13,100; so the fifth is greater by almost 300 miles.

SAGREDO. Which then are the few that agree in locating the star in the same place?

SALVIATI. Unfortunately for this author, there are five which all place it in the firmament, as you can see from this other note where I have recorded numerous other comparisons. But I'm willing to concede to the author more than perhaps he would ask, and that is that every combination of these observations contains an element of error. I think this is inevitable: every comparison requires four observations, namely two different polar elevations and two different altitudes of the new star, made by different observers in different places and with different instruments; and anyone with any experience of such measurements will say it's not possible for all four of these to be made without some error creeping in. We know that the same observer in taking a single polar elevation in the same place and with the same instrument, as I've done myself hundreds of times, will find variations of a minute or so and sometimes much more. You can find examples of this in several places in this very book. So I ask you, Simplicio, bearing all this in mind: do you think this author takes these thirteen observers to have been alert, intelligent men who were expert in handling these instruments, or does he consider them clumsy and inexperienced?

[314]

Astronomical instruments are easily subject to error.

SIMPLICIO. He must surely have considered them highly prudent and intelligent; if he thought they were not equal to this task, he would risk having his book rejected as worthless because it was based on suppositions that were full of errors. Besides, he would be treating us as simpletons if he thought he could persuade us that a false proposition was true on the basis of their incompetence.

SALVIATI. In that case, since these observers are competent and yet have made errors which we need to correct if we are to derive as much information as possible from their observations, we should make the smallest emendations and corrections to them that we can, consistent with bringing their observations from the realm of impossibility into that of possibility. So for instance, if we can moderate an obvious error and a patent impossibility in one of their observations by adding or subtracting two or three minutes and thereby make it possible, we should not try to correct it by adding or subtracting fifteen, twenty, or fifty minutes.

SIMPLICIO. I don't think the author would dissent from this, because given that they are careful and expert observers, we should expect them to be more likely to err by a little than by much.

[315] SALVIATI. Now note this next point. Of the places where the new star might be located, some are manifestly impossible, while others are possible. It is absolutely impossible that it was at an infinite distance beyond the fixed stars, because there is no such place in the universe, and even if there were a star placed there would be invisible to us. It is also impossible that it went creeping along the Earth's surface, and much less that it was inside the terrestrial globe. The possible locations are those which are the subject of debate, as it is not contrary to reason to say that a visible object with the appearance of a star could be either above the Moon or below it. Now the attempts to establish its true location by means of observations and calculations, with as much certainty as human diligence can achieve, have mostly placed it at a more than infinite distance above the firmament, or else very close to the surface of the Earth and in some cases even under its surface. Of those which don't place it in an impossible position, there are none that agree among themselves; so we must necessarily conclude that all the observations are wrong, and if we want to derive any benefit from all this effort we must resort to making corrections and emending all the observations.

SIMPLICIO. The author will say that the observations which place the star in an impossible location should be discounted as infinitely erroneous and wrong, and that we should accept only those that locate it in places that are not impossible. Only among these should we try, by means of the most probable and most numerous comparisons, if not to fix its exact location in [316] terms of its true distance from the centre of the Earth, at least to establish whether it was among the elements or among the celestial bodies.

SALVIATI. Your argument is the same as the author makes to defend his position, but he does so in a way that's very unfair to his opponent. It's this above all that has amazed me at his excessive confidence, not only in his own authority but in the blindness and lack of perception of astronomers. So let me speak for the astronomers, and you can reply for the author. First I ask you whether astronomers, when they are observing with their instruments to find, say, the elevation of a star above the horizon, can be in error both in overstating and understating its position. In other words, can they err sometimes by judging that the star is higher than it really is and sometimes that it is lower, or are their errors always of the same kind, either by always giving readings that are too high and never too low, or vice versa?

SIMPLICIO. I don't doubt that they can be equally liable to err in either direction.

SALVIATI. I think the author would give the same reply. So applying now to their calculations these two contrary kinds of error, into either of which the observers of the new star could have fallen, one kind will show the star to be higher than it really is, and the other will show it to be lower. Since we have agreed that all the calculations are wrong, on what grounds does this author want us to accept that those showing the star to be near the Earth are closer to the truth than those showing it to be excessively far away?

SIMPLICIO. From what I have understood from our discussions so far, the author doesn't exclude the observations and comparisons that show the star to be more distant than the Moon, or even than the Sun; he only excludes those that would show it to be, as you yourself have said, more than an infinite distance away. As even you reject such a distance as impossible, he also discounts these observations as infinitely shown to be false and impossible. It seems to me, therefore, that if you want to convince him you must produce calculations which are more precise, or more numerous, or made by more diligent observers, which place the star at such and such a distance above the Moon or above the Sun—at a place, that is, where it is possible for it to be—just as he has produced these twelve which all place the star below the Moon, in places which exist in the universe and where it is possible for it to be. [317]

SALVIATI. Ah, here is the misunderstanding that you, Simplicio, have in common with the author—you in one respect, the author in another. From what you say I can see you have made the assumption that the greatly exaggerated figures for the distance of the star grow in proportion to the errors the observers have made with their instruments, and conversely that we can deduce the magnitude of the observation errors from the scale of the exaggerations. So, when you hear that a given observation shows the distance of the star to be infinite, you assume that the observation error was also infinite and hence incapable of being corrected, and therefore that it can be discounted. But, my dear Simplicio, this doesn't follow. I can excuse you for not having understood how this works, as you are not expert in these exercises, but I can't provide a similar cover for the author's mistake. He pretends not to understand it, and has assumed that we readers won't understand it either, and so hopes to exploit our ignorance to increase the acceptance of his theory among the uncomprehending masses.

To correct your error, therefore, and to enlighten those who are more credulous than well informed, I must tell you that it can happen—and in the majority of cases does happen—that an observation showing a star to be, say, at the distance of Saturn, can with the addition or subtraction of just one minute to the elevation taken by the instrument send the star to an infinite distance, and so move it from a possible location to an impossible one. Conversely, calculations based on these observations that show the star at an infinite distance can, with the addition or reduction of just one minute, bring it back to a location that is possible. And what I say about correcting the measurement by a minute can also apply to a correction of half a minute, a sixth of a minute, or less. So you must keep firmly in mind that at very remote distances, such as, e.g., the distance of Saturn or the fixed stars, the smallest errors made by the observer with his instrument will change the location from finite and possible to infinite and impossible. The same is not [318] true of distances that are sublunar or close to the Earth. Here it can happen that an observation giving the star's distance as, say, four terrestrial radii, could be increased or decreased not just by a minute but by ten or a hundred minutes or more without affecting the calculation enough to make the star infinitely distant or even further than the Moon. You will see from this that the magnitude of what we might call instrumental errors should be reckoned not by the result of the calculation but by the number of degrees and minutes actually counted on the instrument. The observations we should consider most accurate and least subject to error are those which require the addition or subtraction of the fewest minutes to restore the star to a possible location. Then, among possible locations we should consider the true position to have been that in which the largest number of distances, calculated on the most accurate observations, concur.

SIMPLICIO. I don't fully understand what you have said, and I can't comprehend how it can be that in very large distances a greater exaggeration can arise from an error of a single minute than from an error of ten or a hundred minutes in small distances. But I'd be glad to have it explained to me.

SALVIATI. You can see it in practice if not in theory from this short summary that I've worked out from all the combinations and some of the estimates that the author omitted. I've written them down on this same sheet of paper.

SIMPLICIO. You must have spent all the time since yesterday, which is not more than eighteen hours, on these calculations, without stopping to eat or sleep.

SALVIATI. On the contrary, I've refreshed myself in both ways. I can make such calculations very quickly; to tell the truth, I'm more than a little surprised that this author is so long-winded and includes so many computations that are quite unnecessary for the problem he wants to solve. To explain this fully, I have copied out on this piece of paper all the observations made by thirteen astronomers that are recorded by the author, setting down the polar elevations and the meridian altitudes of the new star, both the minimum readings below the pole and the maximum readings above it. They will also show briefly how all the observations of the astronomers recorded by the author show that it is more probable that the new star is above the Moon and all the planets, and that it is among the fixed stars or even above them. Here are the figures.

[319]

Tycho.

Altitude of the pole	55° 58′	
Altitude of the star	84° 0′	maximum
	27° 57′	minimum
These are the figures in his first publication; in the second the minimum is	27° 45′	

Hainzel.

Polar altitude	48° 22′
Altitude of the star	76° 34′
	76° 33′ 45″
	76° 35″
	20° 9′ 40″
	20° 9′ 30″
	20° 9′ 20″

Peucer and Schuler.

Polar altitude	51° 54′
Altitude of the star	79° 56′
	23° 33′

The Landgrave.

Polar altitude	51° 18′
Altitude of the star	79° 30′

Camerarius.

Polar altitude	52° 24′
Altitude of the star	80° 30′
	80° 27′
	80° 26′
	24° 28′
	24° 20′
	24° 17′

[320]

Hagek.

Polar altitude	48° 22′
Altitude of the star	20° 15′

Ursinus.

Polar altitude	49° 24′
The star	79°
	22°

Muñoz.

Polar altitude	39° 30′
The star	67° 30′
	11° 30′

Gemma.

Polar altitude	50° 50′
The star	79° 45′

Maurolico.

Polar altitude	38° 30′
The star	62°

Busch.

Polar altitude	51° 10′
The star	79° 20′
	22° 40′

Reinhold.

Polar altitude	51° 18′
The star	79° 30′
	23° 2′

Now to explain my whole line of reasoning, let us start with these five calculations that the author ignored, perhaps because they weakened his case, as they all show the star to be many terrestrial radii above the Moon. Here is the first, based on the observations of the Landgrave of Hesse and Tycho, who the author admits are among the most accurate observers. In this first example I'll explain the procedure that I followed in the investigation and this will serve you for all the others, as they all follow the same rule, varying only in the data they use. These are the number of degrees of polar elevation and of altitude of the new star above the horizon and whose distance from the centre of the Earth we are trying to establish, expressed as radii of the terrestrial globe. The distance in miles is immaterial, so calculating this and also the distance between the places where the observations were made is a waste of time and effort, and I don't know why the author did so—especially as he ends by converting miles back into radii of the terrestrial globe.

SIMPLICIO. Perhaps it was because he wanted to discover [321] the distance of the star in such smaller units and fractions of units, establishing it to within a few inches. We lay readers who don't understand your arithmetical rules find it astonishing when you report your conclusions as something like this: 'Therefore the distance of the comet or the new star from the centre of the Earth was three hundred and seventy-three thousand, eight hundred and seven miles and two hundred and eleven four thousand and ninety sevenths (373,807 211/4097)'. Seeing such precision in your recording of these details makes us believe it's impossible that you, who make your calculations to the nearest inch, could end up misleading us by a hundred miles.

SALVIATI. This argument and excuse of yours would be valid if it were the case that a *braccio* more or less was significant in a distance of thousands of miles, and if we could be so certain of the assumptions we take as correct that we could be sure of finally arriving at an incontrovertible truth. But you can see from the author's twelve computations that the distances of the star that they yield differ from each other, and are therefore wide of the truth, by hundreds and thousands of miles. Now if I am quite sure that the answer I am seeking must inevitably be hundreds of miles wide of the mark, why should I labour over my calculations for fear of erring by an inch? But let's finally come to the operation, which I perform as follows.

Tycho, as you can see in the note, observed the star at a polar altitude of 55° 58′, and the polar altitude of the Landgrave was 51° 18′. The height of the star at the meridian as taken by Tycho was 27° 45′, and by the Landgrave was 23° 3′. We can set these altitudes down as follows:

	Tycho	Pole 55° 58′	Star 27° 45′
	Landgrave	Pole 51° 18′	Star 23° 3′
Then I subtract the smaller numbers from the larger, giving the differences as follows:		4° 40′	4° 42′
	Parallax	2 minutes	

The difference between the polar elevations, 4° 40′, being smaller than between the elevations of the new star, 4° 42′, there is a difference in parallax of 2 minutes. [322]

Having established these figures, I take the author's diagram here, where B is the Landgrave's location, D that of Tycho, C the location of the new star, and A the centre of the Earth. ABE is the vertical line from the Landgrave's location, ADF the vertical line from Tycho's, and the angle BCD the difference in parallax.

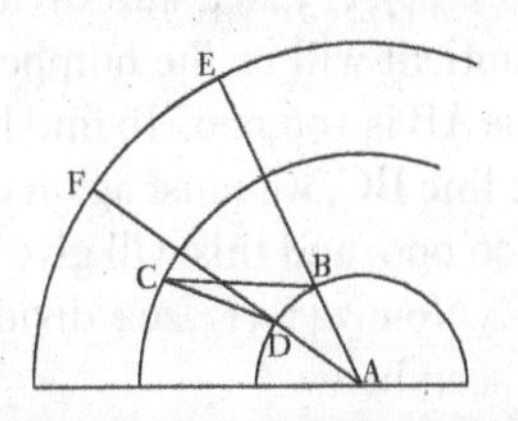

Angles BAD	4°40′	chord 8142 units where the radius AB is 100,000
BDF	92° 20′	
BDC	154° 45′	sine 42657
BCD	0° 2′	sine 58

58   42657   8142
8142
85314
170628
42657
341256
59
58 | 3473 | 13294
571
5

Now as the angle BAD, contained between the two verticals, is equal to the difference in the polar altitudes, it will be 4° 40′; I will make a separate note of this here. Next I find the chord of this angle, using this table of arcs and chords; it is 8142 parts where the radius AB is 100,000, and I note this next to it. Then I can easily find the angle BDC, because half of the angle BAD, namely 2° 20′, added to a right angle, gives the angle BDF as 92° 20′. Add to this the angle CDF, which is the deviation from the vertical of the greater altitude of the star, namely 62° 15′, and this gives the size of angle BDC* as [323] 154° 45′. I make a note of this too, together with its sine, which the table shows is 42,657, and under this I note the angle of parallax BCD, which is 0° 2′, together with its sine which is 58.

Now in the triangle BCD, the side DB is to the side BC as the sine of the opposite angle BCD is to the opposite angle BDC.

Therefore if the line BD was 58, BC would be 42,657. Since the chord DB is 8142 when the radius AB is 100,000, and we are seeking to find out how many such parts is the line BC, we can use the golden rule to say: if when BD is 58 BC is 42,657, how much would BC be if DB were 8142? I multiply the second term by the third, which gives 347,313,294, and divide this by the first, which is 58. The quotient will be the number of parts in the line BC when the radius AB is 100,000. To find how many radii BA are contained in the line BC, we must again divide the quotient we have found by 100,000, and this will give the number of radii contained in BC. Now 347,313,294 divided by 58 gives 5,988,160¼, as may be seen below:*

```
      5988160¼
58 | 347313294
      5717941
      54 3
```

which divided by 100,000 gives us 59 $^{88160}/_{100000}$

```
1 | 00000 | 59 | 88160
```

However, we can shorten this procedure considerably if we divide the first product above, 347,313,294, by the product of the multiplication of the two numbers, 58 and 100,000, thus:

```
                59
58 | 00000 | 3473 | 13294
                571
                 5
```

and this likewise gives 59 $^{5113294}/_{5800000}$.

[324] This is the number of radii contained in the line BC, and adding one for the line AB, we will have a little less than 61 radii for the two lines ABC. Therefore the distance in a straight line from the centre A to the star C will be more than 60 radii, which puts it above the Moon by more than 27 radii, by Ptolemy's reckoning, and more than 8 by that of Copernicus, who the author tells us reckons the Moon to be 52 radii from the centre of the Earth.*

By this similar computation, based on the observations of Camerarius and Muñoz, I find the star to be at a similar distance, again more than 60 radii. Here are the observations, followed by the calculations.

Polar altitude of:	Camerarius	52° 24′
	Muñoz	39° 30′
Difference in polar altitude		12° 54′

Altitude of the star	Camerarius	24° 28′	
	Muñoz	11° 30′	
Difference in altitude of the star		12° 58′	12° 58′
			12° 54′
Difference in parallax (angle BCD)			0° 4′

Angles	BAD	12° 54′	its chord 22466
	BDC	161° 59′	sine 30930
	BCD	0° 4′	sine 116

Golden rule [325]

```
          22466
     116  30930  22466
     -----------------
          673980
        202194
        67398
    --------------
        59          Distance BC = 59 radii, almost 60
116 | 6948 | 73380
      1144
      10
```

The next computation is based on the observations of Tycho and Muñoz, and its calculations show the star to have been 478 radii or more from the centre of the Earth.

Polar altitude	Tycho	55° 58′	
	Muñoz	39° 30′	
Difference in polar altitude		16° 28′	
Altitude of the star	Tycho	84° 0′	
	Muñoz	67° 30′	
Difference in altitude of the star		16° 30′	16° 30′
			16° 28′
Difference in parallax (angle BCD)			0° 2′

Angles			
	BAD	16° 28′	its chord 28640
	BDC	104° 14′	sine 96930
	BCD	0° 2′	sine 58

Golden rule

58 96930 28640
28640
3877200
58158
77544
19386
478
58 | 27760 | 75200
4506
53

[326] This next computation gives the star as more than 358 radii distant from the centre.

Polar altitude	Peucer	51° 54′	
	Muñoz	39° 30′	
		12° 24′	
Altitude of the star	Peucer	79° 56′	
	Muñoz	67° 30′	
		12° 26′	12° 26′
			12° 24′
			0° 2′

Angles	BAD	12° 24′	chord 21600
	BDC	106° 16′	sine 95996
	BCD	0° 2′	sine 58

Golden rule

```
 58  95996  21600
            21600
        ---------
         57597600
            95996
       -----------
          191992
           357
  58 | 20735 | 13600
          3339
           42
```

Here is another computation that shows the star to be more than 716 radii distant from the centre.

Polar altitude	Landgrave	51° 18′
	Hainzel	48° 22′
		2° 56′
Altitude of the star	Landgrave	79° 30′ 0″
	Hainzel	76° 33′ 45″
		2° 56′ 15″

```
 2° 56′ 15″
 2° 56′ 0″
 ---------
 0° 0′ 15″
```

Angles	BAD	2° 56′	chord 5120
	BDC	101° 58′	sine 97845
	BCD	0° 0′ 15″	sine 7

Golden rule [327]

```
  7   97845  5120
              5120
        ----------
         1956900
           97845
          489225
      -----------
           715
   7 | 5009 | 66400
          134
```

As you can see, these are five computations which place the star a long way above the Moon. Now I would like you to consider the point I mentioned a little earlier, namely that at great distances a change, or rather a correction, of a very few minutes alters the star's position by a very large amount. In the first investigation, for example, where the calculation showed the star to be 60 radii distant from the centre with a parallax of two minutes, a correction of two minutes or even less would be all that was needed for anyone wanting to maintain that it was in the firmament, because in that case the parallax would disappear or would be so small as to place the star at an immense distance, such as everyone accepts that of the firmament to be. In the second computation, a correction of less than four minutes has the same effect. In the third and fourth, as in the first, two minutes are enough to locate the star even above the fixed stars. In the last example above, a quarter of a minute, or fifteen seconds, gives the same result. But the same is not true for sublunar altitudes; imagine any distance you please and try to adjust the author's computations so that they all produce the same defined distance, and you will see how much greater are the emendations you would have to make.

SAGREDO. It could only help us to understand more fully if we could see some examples of what you say.

SALVIATI. Choose any defined sublunar distance you like for the location of the star, and with very little trouble we shall be able to establish whether corrections similar to those we have made above, which we saw were sufficient to place it among the fixed stars, will bring it to your chosen place.

[328] SAGREDO. To take the distance that most helps the author's case, let us suppose that it is the largest of all those found in the course of his twelve investigations. Since the dispute is between astronomers who say that the star is higher than the Moon and the author who places it below, then even the smallest difference that shows it to be below the Moon will give him the victory.

SALVIATI. Let us then take the seventh computation, based on the observations of Tycho and Thaddeus Hagek, from which the author reckons the distance of the star from the centre as 32 radii, as this is the location that most favours his case. What's more, to give him every possible advantage, let us place the star at the distance most disadvantageous to the astronomers, which is to locate it even beyond the firmament. With these suppositions, let us go through his other eleven investigations to see what corrections would be needed to bring the star up to a distance of 32 radii. Let us start with the first, based on the observations of Hainzel and Maurolico, from which the author finds the distance to be around 3 radii from the centre, with a parallax of 4° 42′ 30″. Let us see whether reducing this to just 20 minutes brings the star up to 32 radii. Here is the calculation, and it's very brief and exact. I multiply the sine of the angle BDC by the chord BD, and divide the result (ignoring the last five figures) by the sine of the parallax. The result is 28½ radii; so not even a correction of 4° 22′ 30″, subtracted from 4° 42′ 30″, brings the star up to an elevation of 32 radii. For Simplicio's information, the correction that would be needed is 262½ minutes.

Polar altitude	Hainzel	48° 22′	
	Maurolico	38° 30′	
		9° 52′	
Altitude of the star	Hainzel	76° 34′ 30″	
	Maurolico	62° 0′ 0″	
		14° 34′ 30″	14° 34′ 30″
			9° 52′ 0″
	Parallax		4° 42′ 30″

BAD	9° 52′	chord 17200
BDC	108° 21′ 30″	sine 94910
BCD	0° 20′	sine 582

[329]

94910
17200
18982000
66437
9491
28

582 | 16324 | 52000
4688
2

In the second calculation, based on the observations of Hainzel and Schuler, with a parallax of 0° 8′ 30″, the star is found to be at a height of around 25 radii, as can be seen in the following calculation.

BD	chord 6166	97987
		6166
BDC	sine 97987	587922
BCD	sine 247	587922
		97987
		587922

24
247 | 6041 | 87842
1103
11

If we reduce the parallax of 8′ 30″ to 7′, whose sine is 204, the star is raised to around 30 radii; so a correction of 1′ 30″ is not sufficient.

29
204 6041 87842
1965
12

Now let us see what correction is needed for the third computation, based on the observations of Hainzel and Tycho, which puts the star at around 19 radii high, with a parallax of 10 minutes. The usual angles, with their sines and chord, are those found by the author, and they make the distance of the star around 19 radii high, as the author's own calculation shows; so in order to raise it the parallax must be reduced, following the [330] rule that he himself observes in the ninth investigation. Let us then suppose the parallax to be 6 minutes, the sine of which is 175. If we make the division the distance of the star will still be found to be less than 31 radii; so a correction of 4 minutes is too little for the author's needs.

Angles	BAD	7° 36′	chord 13254
	BDC	155° 52′	sine 40886
	BCD	0° 10′	sine 291

```
            13254
            40886
        ---------
            79524
           106032
          106032
          53016
      -----------
          18                     30
  291 | 5419 | 03044     175 | 5419
      | 2501 |                |   16
         18
```

We can apply the same rule to the fourth and the remaining computations, using the author's own chords and sines. In the fourth example the parallax is 14 minutes and the height established is less than 10 radii. If the parallax is reduced from 14 minutes to 4 minutes, the star is not raised even to 31 radii; so a correction of 10 minutes out of 14 is still not enough.

BD chord 8142
BDC sine 43235
BCD sine 407

```
          43235
           8142
      ---------
          86470
         172940
          43235
        345880
      ---------
            30
  116 | 3520 | 19370
           4
```

In the author's fifth example the sines and chords are: [331]

BD chord 4034*
BDC sine 97998
BCD sine 1236

```
          97998
           4034
      ---------
         391992
        293994
      391992
      ---------
            27
  145 | 3953 | 23932
          1058
            3
```

The parallax is 42′ 30″, which gives the height of the star as around 4 radii. Correcting the parallax by reducing it from 42′ 30″ to just 5′ is not enough to raise the star even to 28 radii; so an emendation of 37′ 30″ is still too little.

In the sixth example the chord, sines and parallax are these:

BD		chord 1920
BDC		sine 40248
BCD	0° 8′	sine 233

```
      40248
       1920
   --------
     804960
    362232
    40248
   --------
    26
29 | 772 | 76160
    198
     1
```

The altitude of the star is found to be around 4 radii above the Earth; let us see where it is placed if the parallax is brought down from 8 minutes to just one minute. The calculation above shows that the star is raised only to around 27 radii; so a correction of 7 minutes out of 8 is not enough.

[332] In the eighth case the chord, sines and parallax are:

BD	chord 1804
BDC	sine 36643*
BCD	sine 29

```
      36643
       1804
   --------
     146572
    293144
    36643
   --------
    22
29 | 661 | 03972
    83
     2
```

From this the author calculates the height of the star as 1½ radii, with a parallax of 43 minutes. Reducing this to 1 minute still shows the star as less than 24 radii distant; so a correction of 42 minutes is not enough.

Let us look now at the ninth example. The chord and sines are below, and the parallax is 15 minutes. From this the author calculates that the star's distance from the surface of the Earth was less than one forty-seventh of a radius. This, however, resulted from an error of calculation. It actually comes out, as we shall see below, as more than one fifth; it is in fact around 90/436, which is more than a fifth.

BD	chord 232	39046
BDC	sine 39046	232
BCD	sine 436	78092
		117138
		78092
		436 | 90 | 58672

The author rightly adds that to correct the observations it is not enough to reduce the parallactic difference to a minute or even to an eighth of a minute. I will go further and say that not even a tenth of a minute will bring the star's altitude to 32 radii, since the sine of one tenth of a minute, i.e. 6 seconds, is 3. If following our rule we divide this into 90, or rather if we divide 9058762 by 300000, this comes to 30 58672/100000, or slightly more than 30½ radii.* [333]

The tenth example gives the star's height as one fifth of a radius, with these angles and sines and a parallax of 4° 30′. Even reducing this from 4° 30′ to 2′ does not raise the star to 29 radii.

BD		chord 1746	1746
BDC		sine 92050	92050
BCD	4° 30′	sine 7846	87300
			3492
			15714
			27
			58 | 1607 | 19300
			441
			4

In the eleventh investigation the author places the star around 13 radii distant, with a parallax of 55 minutes. Let us see how high it would raise the star if the parallax were reduced to 20 minutes. The calculation below raises it to a little less than 33 radii, so the correction would be just under 35 minutes out of 55.

BD		chord 19748	96166
BDC		sine 96166	19748
BCD	0° 55′	sine 1600	769328
			384664
			673162
			865494
			96166
			32
			582 | 18990 | 86168
			1536
			36

[334] The twelfth example has a parallax of 1° 36′, and makes the star no higher than 6 radii. Reducing the parallax to 20′ would take the star to a distance of less than 30 radii; so a correction of 1° 16′ is not enough.

BD		chord 17258	17258
BDC		sine 96150	96150
BCD	1° 36′	sine 2792	862900
			17258
			103548
			155322
			28
			582 | 16593 | 56700
			4957
			29

Corrections to the parallax in the author's ten investigations to bring the star to an altitude of 32 radii.						
Degrees	Minutes	Seconds		Degrees	Minutes	Seconds
4	22	30	out of	4	42	30
	4		out of	0	10	
	10		out of	0	14	
	37		out of	0	42	30
	7		out of	0	8	
	42		out of	0	43	
	14	50	out of	0	15	
4	28		out of	4	30	

	35	out of	0	55	
1	16	out of	1	36	
	216			296	60
	540			540	9
	756			836	540

From this it can be seen that to reduce the altitude of the star to 32 radii requires the subtraction of 756 from the total parallax of 836, reducing it to 80, and that even this correction is not sufficient.

This shows, as I have noted above, that if the author decided [335] to take the distance of 32 radii as the true height of the new star, he would have to subtract a total of more than 756 minutes from the sum total of the other ten investigations to bring the parallax down sufficiently to confirm this location (I say ten because his second example gives a very high result and needs a correction of only two minutes to reduce it to 32 radii). The five examples that I have calculated, on the other hand, showing the star to be above the Moon, need a correction of only 10¼ minutes to place it in the firmament. If you add to these another five investigations which place the star definitely in the firmament without the need for any corrections, we will have ten investigations which agree in locating it in the firmament, subject only to correcting five of them, as we have seen, by 10¼ minutes. Contrast this with the author's other ten investigations, which to bring the star's height to 32 radii require an emendation of 756 minutes out of 836—that is, it is necessary to subtract 756 from the total of 836, and even this correction is not sufficient.

The following are the five investigations which immediately and without the need for any correction show the star with no parallax and therefore in the firmament, indeed in its remotest parts and as high as the pole itself.

Polar altitude	Camerarius	52° 24′
	Peucer	51° 54′
		0° 30′
Altitude of the star	Camerarius	80° 26′
	Peucer	79° 56′
		0° 30′

Polar altitude	Landgrave	51° 18′
	Hainzel	48° 22′
		2° 56′
Altitude of the star	Landgrave	79° 30′
	Hainzel	76° 34′
		2° 56′
[336] Polar altitude	Tycho	55° 58′
	Peucer	51° 54
		4° 4′
Altitude of the star	Tycho	84° 00′
	Peucer	79° 56′
		4° 4′
Polar altitude	Reinhold	51° 18′
	Hainzel	48° 22′
		2° 56′
Altitude of the star	Reinhold	79° 30′
	Hainzel	76° 34′
		2° 56′
Polar altitude	Camerarius	52° 24′
	Hagek	48° 22′
		4° 2′
Altitude of the star	Camerarius	24° 17′
	Hagek	20° 15′
		4° 2′

Of the other pairings that can be made between the observations of all these astronomers, those that raise the star to an infinite height are far more numerous—about thirty more, in fact—than those where the calculations place it lower than the Moon. And since, as we have agreed, it is reasonable to believe that the observers are more likely to have made a small error than a large one, it is clear that the corrections needed to bring the star down from an infinite distance will bring it first, and with a smaller correction, to the firmament than to a position below the Moon; so these observations all favour the view of those who place it among the fixed stars. Moreover, the corrections needed in these cases are much less than in those where the star has to be raised from a position improbably close to the Earth to a height more favourable to the author, as can be seen in the examples above. There are three where the position is impossible, since they place the star less than one radius from the centre of the Earth, so that it would have to circulate in some way under the Earth. These are the comparisons where the polar altitude taken by one of the observers is higher than the other, while the altitude of the star taken by the former is less than that by the latter. This is the case with the following combinations.

The first is the comparison between the Landgrave and Gemma. Here the polar altitude given by the Landgrave, 51° 18′, is greater than that given by Gemma, which is 50° 50′, while the Landgrave's altitude of the star, at 79° 30′, is less than Gemma's, 79° 45′.

Polar altitude	Landgrave	51° 18′
	Gemma	50° 50′
Altitude of the star	Landgrave	79° 30′
	Gemma	79° 45′

[337]

The other two are the following.

Polar altitude	Busch	51° 10′
	Gemma	50° 50′
Altitude of the star	Busch	79° 20′
	Gemma	79° 45′
Polar altitude	Reinhold	51° 18′
	Gemma	50° 50′

Altitude of the star	Reinhold	79° 30′
	Gemma	79° 45′

From what I have shown you so far, you can see how much this first method proposed by the author for establishing the distance of the star undermines his case, and how much more clearly and probably it points to the star being among the most distant of the fixed stars.

SIMPLICIO. So far it seems to me that the ineffectiveness of the author's proofs has been exposed very clearly. I see, though, that all this takes up only a few pages of his book, so it may be that he has other arguments which will be more conclusive than these first ones.

SALVIATI. On the contrary, they can't fail to be even weaker, if the arguments we have seen thus far are any guide to the rest. It's clear that the uncertainty and inconclusiveness of these results derive from the observers' errors in the use of their instruments, and from taking the measurements of the polar altitude and the altitude of the star as accurate when they can easily all be wrong. Yet astronomers have had centuries to establish polar altitudes at their leisure, and moreover the meridian altitudes of a star are easiest to observe, as they are clearly defined and allow time for the observer to follow them, since they don't move perceptibly in a short time as they do when they are remote from the meridian. This [338] being the case, what confidence can we have in calculations based on observations that are more numerous, harder to carry out, more fleeting in their variations, and what's more, made using instruments that are clumsier and more unreliable? To judge by the quick look I have had at the proofs which follow, the calculations are based on the altitudes of stars taken in different vertical circles, known by their Arabic name as azimuths.* These observations are made using an instrument that is simultaneously adjustable in a horizontal circle as well as a vertical one, so that at the same moment as taking the altitude the observer must also measure the distance of the vertical point in which the star is from the meridian. The procedure must then be repeated after a significant time interval, making a precise note of the time elapsed, using a timepiece* or other observations of the stars. The author then takes this mass of observations and compares them with another set made by another observer, in another country, using a different instrument and at a different time. From these he tries to deduce what the altitude of the star and its horizontal latitudes would have been at the same time and hour as the first observations, and it is on this adjustment that he finally bases his calculation. Now I leave it to you to judge how much confidence we can have in the result of such investigations. In any case, I have no doubt that if anyone wanted to cudgel their brains over such lengthy computations they would find that, just as with the earlier examples, those supporting the author's position would be outnumbered by those in favour of the opposing view; but I don't think it is worth going to such trouble over a question which is not among our primary concerns.

SAGREDO. I agree with you about that. But if this whole question is fraught with so much confusion, uncertainty, and error, what are the grounds for the confidence with which so many astronomers have asserted categorically that the new star was in the highest heavens?

SALVIATI. There are two kinds of observations, both of them very simple, very easy, and very secure, either of which is more than enough to establish its being located in the firmament, or at least a very long way above the Moon. The first is that its distance from the pole did not change, or changed only minimally, whether it was at its highest point on the meridian or at its lowest. The second is that it constantly maintained the same distance from its neighbours among the fixed stars, and in particular from the eleventh star in Cassiopeia, from which it was not more than one and a half [339] degrees distant. Both of these facts show beyond doubt that it had no parallax at all, or one so small that a very rapid calculation will confirm that it was at a very great distance from the Earth.

SAGREDO. Surely the author understands these points, so if he's seen them how does he defend himself against them?

SALVIATI. When someone who can't find any effective remedy for their mistakes comes out with ridiculous excuses, we sometimes say they are clutching at straws. But this author is clutching not at straws but at gossamer threads, as you will clearly see if we examine the two points I have just mentioned. First, I have written down the following brief calculations to show what we can learn from the polar distances of each of these observers. For you to understand these fully, I must first point out that whenever a new star or any other phenomenon is close to the Earth, and is turning in the diurnal motion about the pole, it will appear more distant from the pole when it is in the lower part of the meridian than when it is in the upper part. This diagram will show this. Here point T indicates the centre of the Earth, O the location of the observer. The arc of the firmament is marked VPC, and the celestial pole P. As the phenomenon moves through the circle FS it is seen sometimes below the pole, along the line OFC, and sometimes above it, along the line OSD, so that it appears in the firmament at points D and C. But its true positions in relation to the centre of the Earth are B and A, which are equidistant from the pole. So it is already clear that the apparent location of the phenomenon S, namely point D, is nearer to the pole than the other apparent location at point C, seen along the line OFC. This is the first point to note. The second point you should note is that the apparent lower distance from the pole exceeds the apparent upper distance, also from the pole, by more than the lower parallax of the phenomenon: that is, the arc CP (the apparent lower distance) exceeds the arc PD (the apparent upper distance) by more than the arc CA (which is the lower parallax). This is easily deduced: the arc CP exceeds PD by more than it does PB, PB being greater than PD; but PB is

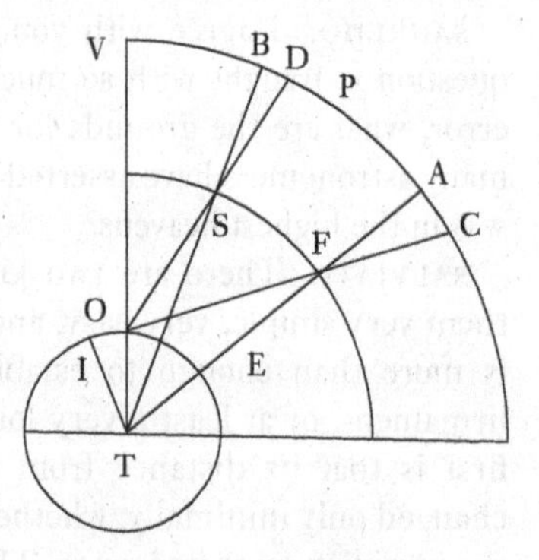

equal to PA, and CP exceeds PA by the arc CA. Therefore the arc CP exceeds the arc PD by more than the arc CA, which is [340] the parallax of the phenomenon located at F; and this is what we wanted to establish.

To give every possible advantage to the author, let us suppose that the parallax of the star at F represents the whole amount by which the arc CP (the distance below the pole) exceeds the arc PD (the distance above the pole). Now I come to examine what we learn from the observations of all the astronomers cited by the author; and they all without exception work against him and contrary to his purpose. Let us begin with these observations of Busch, who found the distance of the star from the pole to be 28° 10′ when it was above the pole, and 28° 30′ when it was below. The difference, then, is 20 minutes, which for the author's benefit let us take to be entirely the parallax of the star at F, that is, the angle TFO. The distance to the zenith, that is the arc CV, is then 67° 20′. With these two values found, let the line CO be extended and a perpendicular TI dropped on to it. Now consider the triangle TOI, in which angle I is a right angle, and angle IOT is known, being opposite to the angle VOC which is the distance of the star from the zenith. Moreover, in the triangle TIF which is also a right triangle, the angle F is known, as we have taken it to be the parallax. So let us now make a separate note of the two angles IOT and IFT, and also note their sines.

Now in the triangle IOT, if the whole sine TO is 100,000 units, the sine TI is 92,276 units; moreover in the triangle IFT, if the whole sine TF is 100,000 units the sine TI is 582 units. So to find the number of units in TF if TO is 100,000, we use the golden rule to ask: where TI is 582, TF is 100,000; what then would be the value of TF if TI were 92,276? We multiply 92,276 by 100,000, which gives 9,227,600,000; then we divide this by 582, which as we can see gives 15,854,982; and this is the number of units in TF where TO is 100,000. Hence, to find how many times the line TO is found in TF, we shall divide 15,854,982 by 100,000, which gives close to 158 and a half; and this is the number of radii making up the distance of star F from the centre T. Seeing that the product of multiplying 92,276 by 100,000 is first divided by 582, and then the quotient is divided by 100,000, we can make the calculation shorter by omitting the multiplication of 92,276 by 100,000 and simply divide the sine 92,276 by the sine 582, which immediately gives the same result, as can be seen below: 92,276 divided by 582 [341] also gives close to 158 and a half. The point to remember, then, is that the division of TI as the sine of angle TOI by TI as the sine of the angle IFT gives us the distance we are seeking, namely TF, in terms of the radius TO.

Angles	IOT	67° 20′	sine 92276
	IFT	0° 20′	sine 582
TI	TF	TI	TF
582	100000	92276	0

15854982
582 | 9227600000
3407002246
49297867
325414

100000 | 158 | 54982

158
582 | 92276
34070
492
3

Now see what Peucer's observations tell us. His lower distance from the pole is 28° 21′, and the upper distance is 28° 2′, so the difference is 19 minutes, and the distance from the zenith is 66° 27′. From these figures we can deduce the star's distance from the centre to be just under 166 radii.

Angles	IAC	66° 27′	sine 91672
	IEC	0° 19′	sine 553

165 427/553
553 | 91672
36397
312
4

Here is what Tycho's observations show, choosing those that are more favourable to our opponent: lower distance from the pole 28° 13′, upper distance 28° 2′, treating the difference of 11 minutes as entirely belonging to the parallax. The distance from the zenith is 66° 15′. The calculation below finds the star's distance from the centre to be 276 9/16 radii.

Angles	IAC	62° 15′	sine 88500		276 9/16
	IEC	0° 11′	sine 320	320	88500
					2418
					21

[342]

The observations of Reinhold, which are the following, give the star's distance from the centre as 793 radii.

Angles	IAC	66° 58′	sine 92026		793 38/116
	IEC	0° 4′	sine 116	116	92026
					10888
					33

From the following observation by the Landgrave the star's distance from the centre can be deduced as 1057 radii.

Angles	IAC	66° 57′	sine 92012		1057 53/87
	IEC	0° 3′	sine 87	87	92012
					5663
					5

Taking two of the observations by Camerarius that are most favourable to the author's case, the distance of the star from the centre is found to be 3143 radii.

Angles	IAC	65° 43′	sine 91152		3143
	IEC	0° 1′	sine 29	29	91152
					4295
					1

The observations of Muñoz do not give any parallax, and therefore place the new star among the most distant of the fixed stars. Hainzel's show it at an infinite distance in space, but with an emendation of half a minute bring it back among the fixed stars; those of Ursinus yield the same result with a correction of 12 minutes. The other astronomers do not give distances above and below the pole, and so no conclu- [343] sion can be drawn from them. So you can see how all the observations of every astronomer agree, contrary to the author's case, in placing the star in the highest regions of the heavens.

SAGREDO. How then does he defend himself in the face of such plain evidence against him?

SALVIATI. By clutching at one of his weakest threads. He says that the parallax is reduced because of refraction, which counteracts its effect by raising the apparent position of the phenomenon whereas parallax lowers it. Now to give you an idea of the validity of this pathetic argument, even if the effect of refraction was as great as some astronomers have recently claimed, the maximum difference it would make in falsely raising the appearance of a phenomenon that was already 23 or 24 degrees above the horizon would be to reduce its parallax by about 3 minutes. The effect of this adjustment would be minimal in bringing the star's position down below the Moon; in fact in some cases it would be less than the concession we made in granting that the whole of the difference between the distances below and above the pole was due to parallax. This has a much clearer and more evident effect than that of refraction, the scale of which I have reason to doubt. In any case, my question to the author is this: does he believe that the astronomers whose observations he uses were aware of these effects of refraction and took account of them, or not? If they did know about them and took them into account, then it is reasonable to assume that they made allowance for them when they gave the true elevation of the star, making the appropriate adjustment to the degrees of elevation shown by their instruments to cancel out the distortion caused by refraction so that the figures they gave were accurate and correct, and not those that were apparent and wrong. If however he thinks that these authors did not give any consideration to these effects of refraction, then he must admit that their estimate of every measurement where accuracy depends on adjusting for refraction is equally in error. This includes finding exact figures for the elevation of the pole, as these are commonly derived from two meridian elevations of one of the permanently visible fixed stars; and these elevations will be altered by refraction in just the same [344] way as that of the new star. Hence the polar elevation deduced from them will also be defective, sharing the same flaw as this author attributes to the elevations proposed for the new star: both will be shown as higher than their true position, with the same degree of error. But it's an error that doesn't affect our present discussion at all, since all we need to know is the difference between the two distances of the new star from the pole, when it is below the pole and when it is above it; and it's clear that these distances will be the same whether the distortion caused by refraction applies to both the star and the pole or whether both have been corrected. The author's argument would carry some weight—although very little—if he had assured us that the polar elevations had been established accurately and corrected for the error due to refraction, and that the same astronomers had failed to guard against this error when they established the elevation of the new star; but he has given us no such assurance, and I doubt whether he could, or whether, as is more credible, the observers neglected to take such a precaution.

SAGREDO. This seems to me more than enough to rule out this objection. Tell me how he gets round the fact that the new star always maintained the same distance from the neighbouring fixed stars around it.

SALVIATI. He does so in the same way, by grasping at two threads that are even more tenuous than the first. One is also linked to refraction, but with even less justification: he says that as the effect of refraction makes the new star appear higher than its true location, this casts doubt on its apparent distances from the neighbouring fixed stars around it. I can't help admiring how he pretends not to realize that this same refraction will have the same effect on the new star as on its existing neighbour, raising them both equally, so that the distance between them remains unaffected. His other defence is even more unfortunate, not to say ridiculous, and it is based on the error that can arise in the observer's use of the sextant—this being the instrument used to measure the interval between two stars. He says that as the observer cannot align the centre of the pupil of his eye with the centre of the sextant, but holds the instrument above the centre by the distance between the pupil and somewhere on the cheekbone where he rests its top, the angle perceived by the eye is more [345] acute than the angle formed by the arms of the sextant. He says that the angle between the rays will vary depending on whether one is observing stars at a low elevation above the horizon or when they are at a great height; and he says that this angle will be different if the observer raises the instrument without moving his head. But if when the observer raised the sextant he bent his neck back so that his head was raised along with the instrument, the angle would remain unchanged; so the author's objection is based on the assumption that observers, in using their instrument, failed to raise their head as they should, something which does not seem very likely. Yet even supposing that this was indeed what happened, I leave you to judge what the difference would be between the acute angles of two isosceles triangles, one of which had sides of four *braccia* in length and the other sides of four *braccia* minus the thickness of a lentil—for the difference in the length of the two visual rays can definitely not be any greater than this, if in one case the line from the centre of the pupil falls vertically onto the plane of the arm of the sextant (a line which cannot be longer than a thumb's breadth), and in the other it falls onto this plane not vertically but obliquely, making a slightly acute angle to the sextant's scale, because the observer has raised the instrument without also raising his head. But to free this author once and for all from these pathetic excuses, he needs to know—since he clearly has not had much experience in the use of astronomical instruments—that the arms of a sextant or quadrant are fitted with two sights, one at the midpoint of the arm and the other at its outer extremity, which are raised an inch or more above the surface of the arm, and that one aligns one's visual ray with the top of these sights, keeping the instrument a hand's breadth or two or more away from the eye, so that neither the pupil, nor the cheekbone, nor any other part of the body is in [346] contact with or leaning against the instrument. What's more, the instrument itself is not held or raised by hand, especially if it is one of the large pieces that are generally used, which weigh tens or hundreds or even thousands of pounds and rest on a very firm base; so this whole objection falls away.

These are the specious arguments put forward by this author, which even if they were of solid steel wouldn't be able to raise his case by a hundredth of a minute, and yet he persuades himself that they will make us believe he has overcome a difference of more than a hundred minutes. I'm referring to the fact that no perceptible difference has been observed in the distance between the new star and any fixed star at any point in their circulation, a difference which should have been plain to see even with the naked eye, without the aid of any instrument, if it had been in the vicinity of the Moon. This would especially be the case with the eleventh star in Cassiopeia, its neighbour only one and a half degrees away from it, where the variation should have been more than two lunar diameters, as the most intelligent astronomers of the time realized.

SAGREDO. I feel as if I'm watching a poor farmer whose hoped-for crops have all been beaten down and destroyed by a storm, and who goes, stooping and weary, picking up such meagre remains as would not suffice to feed a chicken for a single day.

SALVIATI. Indeed, this author has set out with very inadequate weapons to resist those who attack the immutability of the heavens, and has tried with very fragile chains to draw the new star in Cassiopeia down from the highest heaven to these low elemental regions. Now, since I think we have demonstrated very clearly the great difference between the arguments of the astronomers and this author who opposes them, it will be good for us to leave this aspect and return to our main subject. The next topic to be considered is the annual motion commonly attributed to the Sun, but then taken away from the Sun and transferred to the Earth, first by Aristarchus of Samos and then by Copernicus. I see that Simplicio has come stoutly armed against this position, notably with the rapier and shield of the booklet of conclusions or mathematical disquisitions. So it will be good to begin by putting forward the opposing arguments in this book.

SIMPLICIO. I would prefer, if you don't mind, to keep these until the end, as they are the most recently discovered.

[347] SALVIATI. Then we must continue with the procedure we have followed so far, with you setting out in order the contrary arguments of Aristotle and other ancient writers. I will do the same, so that no argument is left without being thoroughly examined and discussed; and Sagredo, with his quick thinking, will intervene with his thoughts as they occur to him.

SAGREDO. I will indeed, taking liberties as usual; and as you command me to do so you will also be bound to excuse me.

SALVIATI. Far from excusing you, I shall be obliged to thank you for the favour you do us. Now let Simplicio begin to put forward the difficulties which make him unable to believe that the Earth moves around a fixed centre in the same way as the other planets.

SIMPLICIO. The first and greatest difficulty is the contradiction and incompatibility of being both at the centre and remote from it. If the terrestrial globe does indeed move in the course of a year around the circumference of a circle, namely the zodiac, then it cannot possibly at the same time be at the centre of the zodiac. But the Earth's location at the centre has been proved in many ways by Aristotle, Ptolemy, and others.

SALVIATI. Your logic is impeccable: there's no doubt that if we are to prove that the Earth moves around the circumference of a circle, we must first prove that it is not at the centre of the same circle. So we must now establish whether the Earth is at this centre, as you say it is, or whether it revolves around it, as I say it does. First, however, we need to be clear whether you and I have the same concept of this centre or not. So please state what and where you understand this centre to be.

SIMPLICIO. I understand it to be the centre of the universe, of the world, of the stellar sphere, of the heavens.

SALVIATI. I could reasonably question whether such a centre exists in nature, since neither you nor anyone else has ever proved whether the world is finite and has a shape, or is infinite and boundless.* But conceding for now that it is finite and has a bounded spherical shape, and therefore has a centre, let us see how credible it is to say that the Earth, rather than any other body, is located at this centre.

No one has ever proved whether the world is finite or infinite.

[348] SIMPLICIO. Aristotle demonstrates in a hundred ways that the world is finite, bounded, and spherical.*

SALVIATI. These hundred ways all come down to just one, which itself is invalid. If I deny his presupposition that the universe is in motion, all his demonstrations fall to the ground, because he only proves that the universe is finite and bounded on the assumption that it is in motion. But so as not to multiply our points of disagreement, let's concede for now that the world is finite and spherical, and that it has a centre. Since its spherical shape and its centre are posited on its being in motion, it seems reasonable to begin our investigation into the location of this centre with the circular motion of the celestial bodies. Aristotle himself, indeed, reasoned and concluded in this way, saying that the centre of the universe was the centre around which all the celestial spheres rotate, which he believed to be also the location of the terrestrial globe. So tell me, Simplicio: if Aristotle had found himself obliged by evident experience to modify his account of the disposition and order of the universe, and to admit that one of these two propositions had been mistaken, which do you think he would choose? Would it be locating the Earth at the centre of the universe, or saying that the celestial spheres revolve around this centre?

Aristotle's demonstrations proving that the universe is finite fall to the ground if it is denied that it is in motion.

Aristotle defines the centre of the universe as the point around which all the celestial spheres rotate.

The question of which of two propositions contradicting his teaching Aristotle would accept, if obliged to choose one of them.

SIMPLICIO. If this were to arise, I think that the Peripatetics...

SALVIATI. I'm not asking about the Peripatetics; I'm asking about Aristotle himself. I know very well how the Peripatetics would respond. Humble and reverent servants of Aristotle as they are, they would deny all the experiences and all the observations in the world; they would refuse even to look at them, for fear of having to accept them.* They would maintain that the world is as Aristotle described it, not as nature shows it to be; for if you took away Aristotle's authority, what other basis would they have for arguing? So please tell me what you think Aristotle himself would do.

SIMPLICIO. I must admit that I can't decide which of these two he would regard as the lesser difficulty.

SALVIATI. Please don't use this word 'difficulty' to describe what may prove to be true; the difficulty was wanting to place the Earth at the centre of the celestial revolutions. But as you don't know which way Aristotle would decide, and as I consider him to be a man of great intelligence, let us proceed by considering which of the two choices seems the more reasonable, and assume that this is the one he would have chosen.

[349] Going back, then, to the beginning of our discussion, let us assume for Aristotle's sake that the world is spherical and has a circular motion—although the only physical evidence that we have of its size are the fixed stars. It must therefore have a centre, like anything that is spherical in shape and is in circular motion. Since, moreover, we can be sure that within the stellar sphere there are many other spheres with their own stars, one inside the other, which also move with a circular motion, we must ask whether it is more reasonable to believe that all these inner spheres move around this same centre of the world, or around another centre which is remote from it. So, Simplicio, let us have your opinion on this point.

It is more appropriate that the container and what is contained should move around the same centre than around different centres.

SIMPLICIO. Provided we need go no further than this single proposition, and we could be sure of not encountering some other difficulty, I would say that it is much more reasonable to suppose that the container and what it contains all move around a common centre, than around different centres.

If the centre of the world is also the centre around which the planets move, the Sun and not the Earth is located there.

SALVIATI. If, then, it is true that the centre of the world is the same as the centre around which the celestial bodies, i.e. the planets, move, it is quite certain that it is not the Earth but the Sun which is located at the centre of the world. So on this first basic general point, we can say that the centre is occupied by the Sun, and the Earth is as remote from the centre as it is from the Sun.

SIMPLICIO. On what do you base your argument that the Sun and not the Earth is at the centre of the revolution of the planets?

Observations from which it can be deduced that the Sun and not the Earth is at the centre of the revolution of the heavens.

SALVIATI. On observations which are absolutely clear, and therefore conclusive. The most palpable of those which show that the Sun and not the Earth is at the centre, is that the planets all appear at varying distances from the Earth. These variations are so great that Venus, for example, is six times further away from us at its furthest point than at its closest, and Mars soars almost eight times higher at one point than at another. So you see that Aristotle was a little deceived in thinking that they were always equally distant from us.

SIMPLICIO. What then are the indications that they revolve around the Sun?

SALVIATI. In the case of the three outer planets, Mars, Jupiter, and Saturn, this can be deduced from the fact that they are always close to the Earth when they are in opposition to the Sun, and remote from us when they approach conjunction with the Sun. The extent of this variation in distance is such that [350] Mars appears fully sixty times larger when it is close to us than when it is furthest away. In the case of Venus and Mercury, we can be certain that they revolve around the Sun because they never move very far away from it, and because they appear now above and now below the Sun, as the changing shape of Venus conclusively proves. As for the Moon, it is clear that it cannot be separated from the Earth in any way, for reasons which we shall see in more detail as we proceed.

The changing shape of Venus indicates that its motion is around the Sun.

The Moon cannot be separated from the Earth.

SAGREDO. I look forward to hearing even more marvellous things arising from this annual motion of the Earth than were those depending on its diurnal rotation.

SALVIATI. You won't be disappointed. As far as the diurnal rotation was concerned, the only effect it had or could have on the celestial bodies was to make them all appear to move very fast in the opposite direction. But this annual motion, when it is combined with the individual motions of all the planets, produces a host of extraordinary effects which so far have perplexed all the world's greatest intellects. But to return to our first general principles, I repeat that the centre of the celestial revolutions of the five planets—Saturn, Jupiter, Mars, Venus, and Mercury—is the Sun; and it will be the centre of the Earth's motion as well, if we succeed in locating the Earth in the heavens. As for the Moon, it has a circular motion around the Earth, from which, as I've said, it is impossible for it to be separated; but this doesn't prevent it from also going around the Sun, along with the Earth in its annual motion.

The annual motion of the Earth, combined with the motions of the other planets, produces extraordinary appearances.

SIMPLICIO. I'm not yet entirely sure about this structure. Perhaps it will be easier to understand and discuss if we make a drawing of it.

SALVIATI. So we shall; in fact, so that you are both more convinced and more amazed, I would like you to draw it yourself. You will see that although you don't think you understand it, you actually understand it very well; and you will be able to describe it perfectly just by answering my questions. So take a sheet of paper and a pair of compasses, and let this blank page be the immense expanse of the universe, in which you will distribute and arrange the various parts as reason dictates to you.

Drawing of the system of the universe, based on appearances.

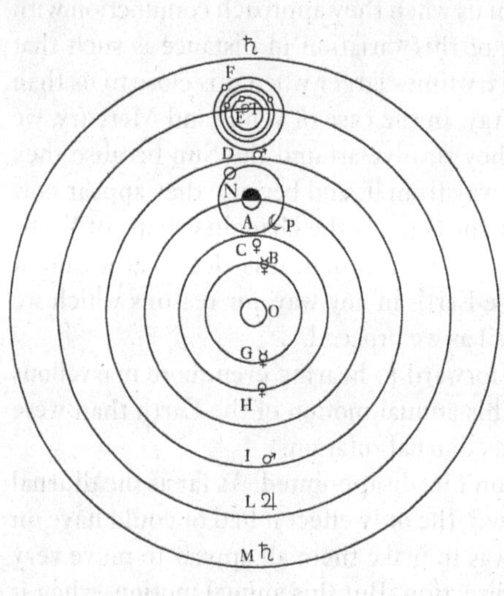

First, since you hold firmly that the Earth is located in this universe without needing me to tell you, mark a point wherever you like to show where you think it is located, and assign a letter to it.

[351] SIMPLICIO. Let this be the location of the terrestrial globe, marked A.

SALVIATI. Good. Secondly, you know very well that the Earth is not located within the body of the Sun or even close to it, but is some distance away from it. So place the Sun wherever you like, as distant from the Earth as you think fit, and mark it with a letter as well.

SIMPLICIO. There it is: let this be the location of the solar body, marked O.

SALVIATI. Now that we've established these two, we need to consider how to place Venus in such a way that its position and motion conform to what the evidence of our senses shows us. So call to mind what you know about what happens with this star, either from our earlier discussions or from your own observations, and place it where you think it should be.

SIMPLICIO. I will assume that the appearances you have described, which I have also read in my booklet of proofs, are correct: namely, that this star never recedes from the Sun beyond a fixed interval of about forty degrees, so that not only does it never reach opposition to the Sun, but not even a quadrature nor as a sextile aspect. Furthermore, it appears almost forty times larger at one time than at another. It is at its largest when it is retrograde and moving towards evening conjunction with the Sun, and at its smallest when it is moving forwards towards morning conjunction. Moreover, when it appears at its largest it has a horned shape, and at its smallest it appears perfectly round. Since all these appearances are correct, it seems to me inescapable to say that this star revolves in a circle around the Sun. This circle cannot be said to encompass or contain the Earth within it, nor can it be either below the Sun, i.e. between the Sun and the Earth, or above the Sun. It cannot enclose the Earth, because then Venus would sometimes appear in opposition to the Sun; it cannot be below the Sun, because then it would appear crescent-shaped at the time of both conjunctions; and it cannot be above the Sun, because then it would always appear round and never horned. So I shall mark its place with this circle CH around the Sun, but without enclosing the Earth.

Venus is very large at its evening conjunction, and very small at the morning one.

Necessary conclusion that Venus must revolve around the Sun.

[352]

SALVIATI. Now that you've found a place for Venus, you need to think about Mercury. As you know, it always stays close around the Sun, moving away from it much less than Venus. So decide where you're going to place it.

SIMPLICIO. There is no doubt that, as it imitates Venus, there is plenty of space for it in a smaller circle, inside the orbit of Venus, and also moving around the Sun. A conclusive reason for this, and especially for its closeness to the Sun, is its brightness, which exceeds that of Venus and the other planets. So on this basis we can mark its circle here, with the letters BG.

Mercury is concluded to revolve around the Sun, inside the orbit of Venus.

SALVIATI. Now, where shall we place Mars?

SIMPLICIO. Mars comes into opposition with the Sun, so its orbit must enclose the Earth. But I see that it must embrace the Sun as well, because if it passed below the Sun and not above it, it would appear horned when it is in conjunction with

Mars must include the Earth within its orbit, and also the Sun.

When Mars is at opposition to the Sun it appears sixty times larger than at conjunction.

Jupiter and Saturn also circulate around the Earth and the Sun.

The approach and recession of the three outer planets is double the distance of the Sun.

The apparent difference in size is less for Saturn than for Jupiter, and less for Jupiter than for Mars; the reason for this.

The Moon's orbit encloses the Earth, but not the Sun.

the Sun, in the same way as Venus and the Moon, whereas in fact it always appears round. So its orbit must enclose the Sun as well as the Earth. Since I remember you saying that when Mars is in opposition to the Sun it is sixty times larger than when it approaches its conjunction, I think that these appearances will correspond very well with an orbit around the Sun which also encloses the Earth, which I will draw here and mark with the letters DI. D is the point where Mars is closest to the Earth and is in opposition to the Sun; when it is at point I it is in conjunction with the Sun and at its greatest distance from the Earth. And since we can observe the same appearances in Jupiter and Saturn, although with Jupiter showing much less variation than Mars, and Saturn even less than Jupiter, I think we can also satisfactorily place these two planets in two circles around the Sun. I will mark this first one, for Jupiter, EL, and another, higher, for Saturn, FM.

SALVIATI. You have distinguished yourself so far. Now since, as you see, the distance by which the three outer planets approach and recede is measured by double the distance between the Earth and the Sun, this produces a greater variation in the distance of Mars than of Jupiter, because Mars's orbit DI [353] is smaller than Jupiter's orbit EL. Similarly the variation in the distance of Saturn is less than that of Jupiter, because Jupiter's orbit EL is smaller than Saturn's orbit FM. This all corresponds exactly with the appearances.

Now it remains for you to decide the place you will assign to the Moon.

SIMPLICIO. Following the same procedure, which seems to me absolutely conclusive, we can see that the Moon comes into conjunction with and opposition to the Sun, and hence we must say that its orbit encloses the Earth. But there is no evidence that it encloses the Sun, because if it did it would not appear as a crescent when it approaches conjunction with the Sun, but would always be round and full of light. Moreover, it would not be able to eclipse the Sun from us, as it often does, by coming between us and the Sun. Therefore we must assign it an orbit around the Earth, which I shall mark here NP, where P is the point at which it appears to us from the Earth, A, to be in conjunction with the Sun, so that sometimes it eclipses the

Sun. When it is at N we see it in opposition to the Sun; in this position it can sometimes fall under the Earth's shadow and be obscured.

SALVIATI. And now, Simplicio, what are we to do with the fixed stars? Should we show them scattered across the vast abyss of the universe, at varying distances from any defined point, or should we place them on the surface of a sphere, so that each of them is equally distant from its centre?

SIMPLICIO. I would rather take a middle path and assign them to a sphere around a specified centre, made up of two spherical surfaces, one of them very high and concave and the other lower and convex. I would place the innumerable host of fixed stars between these two surfaces, at varying heights. This could be defined as the sphere of the universe, which contains within it the orbits of all the planets we have already drawn.

Probable position of the fixed stars.

How we should conceive the sphere of the universe.

SALVIATI. So now, Simplicio, we have arranged the celestial bodies exactly as Copernicus placed them, and you have done this with your own hand. What's more, you have assigned to each its proper motion, except for the Sun, the Earth, and the fixed stars. You have given Mercury and Venus a circular motion around the Sun which does not enclose the Earth. You have made the three outer planets, Mars, Jupiter, and Saturn, also revolve around the Sun, enclosing the Earth within their orbits. And you have shown the Moon with no motion other than its orbit around the Earth, without enclosing the Sun. In all of these motions, too, you are in agreement with Copernicus.

[354]

It remains now for us to decide how to distribute three things among the Sun, the Earth, and the sphere of the fixed stars. These are a state of rest, which appears to belong to the Earth; the annual motion under the zodiac, which appears to belong to the Sun; and the diurnal motion, which appears to belong to the fixed stars, with the rest of the universe apart from the Earth participating in it. Now since it is the case that all the planetary spheres—Mercury, Venus, Mars, Jupiter, and Saturn—revolve around the Sun as their centre, it seems reasonable to assume that it is the Sun, rather than the Earth, that is at rest; for in a movable sphere it is reasonable to suppose that the centre is at rest rather than some other point

A state of rest, annual motion, and diurnal motion to be distributed among the Sun, the Earth, and the firmament of fixed stars.

In a moving sphere, it seems more reasonable for the centre to be at rest than any other part.

If the annual motion is attributed to the Earth, the diurnal motion should be assigned to it as well.

remote from the centre. The Earth is placed between two moving bodies, namely Venus and Mars, one of which completes its revolution in nine months and the other in two years; so we can appropriately attribute the annual motion to the Earth, and leave the state of rest to the Sun. This being the case, it necessarily follows that the diurnal motion, too, must belong to the Earth. Otherwise, the Sun being at rest, if the Earth did not turn on its axis but had only its annual motion around the Sun, our year would consist of just one day and one night of six months each, as has been said on another occasion.* See, too, how neatly this removes the need for the precipitous motion of the universe every 24 hours, and how the fixed stars, which are themselves so many Suns, enjoy perpetual rest in the same way as our own Sun. Finally, see how easily this first sketch is able to account for so many great appearances in the celestial bodies.

SAGREDO. I see this very well. But just as you take this simplicity as grounds for the strong probability of this system, others may draw the opposite conclusion. Since this model was put forward in ancient times by the Pythagoreans, and since it fits the appearances so well, they might reasonably wonder why it has attracted so few supporters in the course of thousands of years, and why Aristotle himself rejected it. And more recently Copernicus himself has met with the same fate.

Utterly childish arguments which suffice to confirm the ignorant in their belief that the Earth is at rest.

SALVIATI. Sagredo, you wouldn't be so surprised that this view has found so few followers if you had encountered, as I have very many times, the kind of stupidities that suffice to make the common people stubborn and resistant to even listening, much less assenting, to these novel ideas. I don't think [355] we should attach much weight to the views of those who say that since they can't have breakfast one morning in Constantinople and dinner the same evening in Japan, this is conclusive proof to confirm their fixed conviction that the Earth is at rest. Or they say that the Earth is so heavy that it cannot possibly climb up above the Sun and then come hurtling down again. There is an infinite number of such people, and there is no point in paying any attention to their foolish ideas or trying to win their assent. When we are dealing with subtle and delicate arguments, we don't need the company of those who can't see beyond generic definitions and are unable to make any kind of distinctions. Besides, what impression can you expect to make with all the proofs in the world on minds which are so obtuse that they can't recognize their own sheer folly?

My amazement is quite different from yours, Sagredo: you wonder that the view of the Pythagoreans has so few followers, but I am astonished how anyone hitherto has embraced and followed it. My admiration for their originality of mind knows no bounds. Having accepted this view and deemed it to be true, the force of their intellect has so overridden their senses that they have been able to follow what reason dictated to them rather than the manifest evidence of their senses to the contrary. You have already examined the arguments against the diurnal rotation of the Earth and we have seen how plausible they are, and the fact that they were accepted as conclusive by Ptolemy, Aristotle, and all their followers is a strong testimony to their effectiveness. But the experiences which openly contradict the Earth's annual movement seem even more plainly to argue against it; so I say again that I cannot find words to express my admiration for the way in which Aristarchus and Copernicus were able to let their reason conquer their senses and command their assent.

The view of Copernicus shown to be improbable.

In Aristarchus and Copernicus, reason and argument prevailed over the manifest evidence of the senses.

SAGREDO. Should we then expect yet more forceful arguments against this annual motion?

SALVIATI. Indeed we should, and based on such strong sense evidence that, if reason had not been accompanied by a superior and more excellent sense perception than our natural everyday senses, I strongly suspect that I too would have been much more resistant to the Copernican system. But I was enlightened by a light much brighter than any ordinary lamp.* [356]

SAGREDO. Come now, Salviati, let's get down to brass tacks, as the saying goes. I feel as if we are wasting every word we expend on anything else.

SALVIATI. I'm here at your service.

SIMPLICIO. [Please, gentlemen, let me put my mind at rest. I am all at sea because of a point Salviati touched on just now, and I will be able to understand your speculations more clearly once the waves have been calmed. For it's hard to see reflections clearly in an undulating mirror, as the Latin poet* so elegantly put it when he said 'I have just seen myself on the shore when the sea was calm and windless'.

SALVIATI. You are quite right; so tell us your doubts.

SIMPLICIO. In your last point you dismissed as equally obtuse those who deny the Earth's diurnal motion because they don't see themselves being transported to Persia or Japan, and those who oppose its annual motion because of their reluctance to accept that the vast and weighty mass of the terrestrial globe could rise and sink as it would have to do if it were revolving with such a motion around the Sun. I am not embarrassed to be numbered among such simpletons because I have the same reluctance in my mind as regards the second point, against the annual motion, especially when I see the resistance to being moved, even on a level surface, not just of a mountain but of a rock which is only a small part of a mountainside. So please don't be scornful of such objections but answer them, not only for me but for others who find them irrefutable; because I think it's hard to expect anyone, however simple they may be, to recognize and confess to being simple just because they hear someone calling them so.

[357] SAGREDO. On the contrary, the simpler they are the more impervious they are to any suggestion that they might be wrong. But this seems to me a good moment for us to resolve this and other objections of a similar kind, not only to satisfy Simplicio but, no less important, because it is clear that there is no lack of men who are very well versed in general philosophy and other sciences, but who through ignorance of astronomy, mathematics, and any other discipline that sharpens the intellect to arrive at the truth, are persuaded by such vacuous arguments. I think poor Copernicus deserves to be pitied for this, as he could never be sure that the criticism of his theories would not fall into the hands of people who lack the capacity to follow his reasoning, subtle as it is and therefore hard to understand, but who are persuaded by such vain appearances as these that it is mistaken, and so keep on declaring that his theories are erroneous and false. So, if they lack the capacity to follow his more abstruse arguments, it is as well to make them recognize the ineffectiveness of these objections, so that this recognition might make them

more cautious in their judgement and less ready to condemn the theory which they now believe to be erroneous. I will therefore mention two other objections that I heard not long ago from highly literate people against the diurnal motion, and then we can come on to the annual motion.

The first objection was that, if it were true that it is not the Sun and the other stars that rise above the eastern horizon, but rather that the eastern part of the Earth sinks below it while they remain motionless, then within a few hours the mountains situated in the east would necessarily decline downwards because of the rotation of the globe, to the point where instead of having to climb up to reach their summit we would have to go downhill in order to get up there. The second was that, if the diurnal motion belonged to the Earth, it would have to be so [358] rapid that someone at the bottom of a well would be able to see a star directly overhead for only a fleeting moment, as it would be visible to them only for the very short time it took for the passage of two or three *braccia* of the Earth's circumference, this being the width of the well. Yet we see from experience that the apparent passage of such a star across the well takes some considerable time; and this necessarily proves that the mouth of the well cannot possibly move at the speed that would be required for the diurnal rotation, and consequently that the Earth is immobile.

SIMPLICIO. I find the second of these two arguments quite conclusive; but as for the first, I think I could resolve it for myself. It seems to me that for the terrestrial globe to move a mountain towards the east in the course of its rotation on its axis, is the same as the mountain being uprooted from its place and dragged across the surface of the Earth while the globe remained motionless; and I see no difference between moving a mountain over the surface of the Earth and sailing a ship over the surface of the sea. Hence, if the objection of the mountain were valid, it would equally follow that when the ship in the course of its voyage had travelled many degrees distant from our ports, a sailor going up the mast would no longer climb upwards but would move on the level or even downwards; and this does not happen. I have never heard any sailor, even among those who have circumnavigated the globe, say that there was any difference in this or any of the other tasks carried out on board a ship as a result of its being in one place rather than another.

SALVIATI. You argue very well. If it had ever occurred to the author of this objection that if the Earth was rotating, the nearby mountain to his east would in the space of two hours have been moved to where, say, Mount Olympus or Mount Carmel are now, he would have understood that his own line of argument obliged him to believe and acknowledge that reaching the summit of these mountains would *de facto* have meant [359] descending. These are the kind of thinkers who deny the existence of the Antipodes because it's not possible to walk with one's head down and one's feet attached to the ceiling. From concepts which are true and which they understand perfectly, they are not able to deduce the simplest solutions to resolve their doubts. In this case, they understand very well that to gravitate and descend is to tend towards the centre of the globe and that to ascend is to move away from the centre; but it is beyond them to grasp that people at the Antipodes to us can stand and walk about with no trouble at all because they do just as we do, keeping their feet towards the centre of the Earth and their head towards the sky.

SAGREDO. Yet we know that those who are highly competent in other disciplines are out of their depth with such ideas. This is all the more reason why, as I was saying just now, it would be good to deal with all the objections, even the feeblest. So let us respond to the argument about the well.

SALVIATI. This second argument does indeed have some appearance of being conclusive. Nonetheless I am certain that if we could question the author to whom it occurred and asked him to explain exactly the effect that he thought should follow, and which seems to him not to follow, if it were true that the diurnal rotation belonged to the Earth, then, as I've said, I think he would get entangled in explaining his difficulty and its consequences no less than if he disentangled himself by thinking about it.

SIMPLICIO. I'm sure that he would, since to tell the truth I am in the same state of confusion myself. On a first reading the argument strikes me as sound, but on the other hand I can see dimly that if the reasoning is valid then the immense speed at which the star would appear to move if the Earth was in motion would apply equally if the motion belonged to the star—indeed much more, since its motion would have to be many thousand times more rapid than that of the Earth. But then again, the star's disappearing from sight simply by passing the mouth of the well, which is just two or three *braccia* across, would have to be so fleeting as to be imperceptible, since the Earth and the well would pass [360] through more than two million *braccia* in an hour. And yet such a star is visible from the bottom of this same well for quite a long interval of time. So I would very much like to have this question clarified for me.

SALVIATI. This makes me even more certain that the author of this objection is confused, since you yourself, Simplicio, have not fully grasped what you mean to say. I think this comes chiefly from your overlooking a distinction which is of prime importance in this matter. Tell me, therefore, in making this observation of the star's passing over the mouth of the well, do you make any distinction between the greater or lesser depth of the well—whether the observer, that is, is more or less distant from the mouth? Because I didn't hear you make any reference to this.

SIMPLICIO. Truthfully, I didn't think of it, but your question makes me realize that it is indeed an essential distinction. I can already begin to understand that in determining the duration of the star's transit, the depth of the well can make as much difference as its width.

SALVIATI. I suspect, rather, that its width makes no difference at all, or only very little.

SIMPLICIO. But surely travelling across a width of ten *braccia* takes ten times longer than to pass over one *braccio*? I'm certain that a boat ten *braccia* long will pass before my sight sooner than a galley a hundred *braccia* long.

SALVIATI. We are still persisting in the age-old idea that we only move as much as our legs carry us. What you say is true, my dear Simplicio, as long as the object that you see is moving while you as the observer are standing still; but if you are in the well while both you and the well are carried along by the rotation of the Earth, do you not see that the mouth of the well will not pass over you in an hour, or a thousand hours, or in all eternity? The effect on you of the motion or non-motion of the Earth in such a case will not be apparent in the mouth of the [361] well, but only in some other separate object that does not share in the same state, either of motion or of rest.

SIMPLICIO. That's understood; but supposing I am in the well and am carried along with it by the diurnal motion, and the star that I see is motionless, only the mouth of the well allows me to see the star passing overhead; so if the opening is not more than three *braccia* wide, out of the many millions of *braccia* of the rest of the Earth's surface which blocks my view, how can the time when I can see the star be a perceptible part of the time when it is hidden?

SALVIATI. You are still falling back into the same misunderstanding, and it's clear that you need someone to help you out of it. It's not the width of the well that determines the length of time the star appears to you; if it was you would see the star all the time, since the mouth of the well makes its passage visible to you all the time. The length of time depends on the extent of the motionless heaven that you are able to see through the opening of the well.

SIMPLICIO. But isn't the part of the sky that is visible to me the same fraction of the whole celestial sphere as the mouth of the well is of the whole terrestrial globe?

SALVIATI. I would like you to answer that question for yourself; so tell me, is the mouth of the well always the same fraction of the surface of the Earth?

SIMPLICIO. Yes, without doubt it's always the same.

SALVIATI. And what about the part of the heaven that is seen by the person in the well: is that always the same fraction of the whole celestial sphere?

SIMPLICIO. Some of the confusion is clearing from my mind, and I'm beginning to understand what you said a short while ago, that the depth of the well is a significant factor in this. There's no doubt that the further the observer's eye is from the mouth of the well, the smaller the part of the heaven that will be visible, and consequently the more rapidly it will pass over and become invisible to the observer at the bottom of the well.

SALVIATI. But is there any point in the well from which one can see exactly the same fraction of the celestial sphere as the mouth of the well is in relation to the surface of the Earth?

SIMPLICIO. I think that if the well went down to the centre of the Earth, perhaps from there one would see a part of the sky corresponding to the part of the Earth represented by the well. But as one moved away from the centre and came up towards the surface, the part of the heaven that was visible would steadily increase. [362]

SALVIATI. And finally, when one's eye was level with the mouth of the well, half of the heaven or very slightly less would be visible, and—assuming one was at the equator—it would pass over in twelve hours.]

I've already sketched out the form of the Copernican system for you. The first very powerful attack against its validity is mounted by Mars itself. If it were true that the variation in Mars's distance from the Earth at its nearest and farthest points was twice the distance of the Earth from the Sun, its appearance when it is closest to us would need to be more than sixty times larger than when it is furthest away. Yet we see no such difference in its apparent size; rather, when it is in opposition to the Sun and nearest to the Earth it appears no more than four or five times larger than when it is close to conjunction and becomes hidden by the rays of the Sun. Another and even greater difficulty is presented by Venus. If, as Copernicus asserts, its revolution around the Sun takes it alternately above and below the Sun, moving away from us and towards us by as much as the diameter of the circle which it describes, then it ought to appear to us almost forty times larger when it is below the Sun and nearest to us than when it is above the Sun and near conjunction with it; and yet the difference is almost imperceptible.

Mars mounts a fierce attack against the Copernican system.

The appearances of Venus do not agree with the Copernican system.

Venus presents another difficulty against Copernicus.

There is a further difficulty. If, as seems reasonable, the body of Venus is dark in itself and shines only with light from the Sun, as does the Moon, then it ought to appear crescent shaped when it is below the Sun, in the same way as the Moon does when it is near the Sun; yet no such appearance occurs. For this reason Copernicus affirmed that either it shone with its own light or it was formed of a material that could absorb the light

Copernicus states that Venus either shines with its own light or is made of a transparent material.

Copernicus says nothing about the small variation in the size of Venus and Mars.

from the Sun and allow it to pass right through, so that it always appeared to shine brightly. In this way he accounted for the fact that the shape of Venus does not change, but he said nothing about the small variation in its size. As for Mars, he said much less than was needed, I think because he could not to his own satisfaction save an appearance that so undermined his position, which he nonetheless maintained to be true because he was persuaded by so many other considerations.

The Moon is a major disruption in the order of the planets.

Moreover, to say that all the planets, including the Earth, revolve around the Sun as their centre, and that the Moon alone is an exception to this order, having its own motion around the Earth as well as an annual motion around the Sun along with the Earth and the whole elemental sphere, seems to disrupt the whole order to such an extent as to make it improbable and false. It is because of these objections that I am amazed that Aristarchus and Copernicus, who cannot fail to have observed them and yet were unable to resolve them, were still so confident in what reason dictated to them in the light of other remarkable observations that they confidently affirmed that the structure of the universe could not [363] be in any other form than the one they had described. There are other very serious and yet intriguing difficulties which are not so easy for the average mind to resolve but which Copernicus understood and explained, which we shall defer until later, after we have dealt with other objections that others have put forward which seem to contradict his position.

Response to the first three objections to the Copernican system.

We come now to explain and reply to the three major objections brought forward so far. Of the first two, I declare that not only do they not contradict the Copernican system, but they strongly and unequivocally support it; for Mars and Venus do indeed appear with the variations in size that the system assigns to them, and Venus does appear crescent-shaped when it is below the Sun, changing its shape in exactly the same way as the Moon.

SAGREDO. How is it then that this was not apparent to Copernicus but is clear to you?

SALVIATI. These things can only be apprehended through the sense of sight, and nature has not endowed humankind with such perfect sight as would allow them to perceive such differences; indeed the organ of sight itself is an impediment to seeing them. But now that in our own time it has pleased God to allow human ingenuity to discover such a marvellous invention as to perfect our sight by multiplying it 4, 6, 10, 20, 30, and 40 times, an infinite number of objects that were invisible, because of either their remoteness or their very small size, have now been made perfectly visible by means of the telescope.

SAGREDO. But Venus and Mars are not objects that are invisible because of their remoteness or their small size; on the contrary, we can see them perfectly with our unaided natural vision. Why then can we not discern the differences in their size and shape?

The reason why Venus and Mars do not appear to vary in size as they should.

SALVIATI. This is due in large part to the impediment in the eye itself, as I suggested just now. The eye does not represent distant shining objects to us in their pure, unadorned form, but rather we see them surrounded by adventitious and alien rays, which are so long and dense that the bare body of an object appears to us 10, 20, 100, or a thousand times larger than it would if the radiant halo which does not belong to it were removed.

SAGREDO. I recall now that I have read something about this, but I don't remember whether it was in our mutual friend's *Letters on the Sunspots* or his *Assayer.** But I think it could only be helpful, both to refresh my memory and to inform Simplicio, who may not have read these works, if you could explain this point to us more fully, as I think it is very necessary for an understanding of the point we are discussing here. [364]

The working of the telescope considered fallacious by the Peripatetics.

SIMPLICIO. Indeed, everything that Salviati is telling us now is quite new to me. To tell the truth, I have not been curious to read these books, nor have I hitherto given much credence to this newly introduced spyglass. I have rather followed in the footsteps of my fellow Peripatetic philosophers, and have considered what others have admired as marvellous achievements to be deceptive illusions produced by the glasses. If I have been wrong about this I shall be glad to be corrected, and I shall listen all the more attentively because my interest is aroused by the other novelties you have described to us.

SALVIATI. The confidence these gentlemen have in their own astuteness is as unjustified as their lack of respect for the judgement of others. It is quite something for them to consider themselves better qualified to judge this instrument, having never tried it, than those who have experience of using it a thousand times and continue to do so every day. But please let us leave these obdurate individuals, as even to criticize them is to honour them with more attention than they deserve. Returning to our subject, then, I say that shining objects appear to our eyes surrounded by extra rays and therefore seem much bigger than they would if we saw their bodies devoid of this radiation. This is because their light is refracted in the liquid on the surface of our pupils, or because it is reflected on the edge of our eyelids and the rays are reflected onto the pupils, or for some other reason; and this enlargement is proportionately greater the smaller the luminous object. So if we suppose that the extra rays extend for, say, four inches, and that these surround a circle four inches in diameter, the apparent size of the circle would increase ninefold; however—

Shining objects appear surrounded by adventitious rays.

The reason why luminous bodies appear more enlarged the smaller they are.

SIMPLICIO. I think you meant to say threefold, because if you add four inches on either side of a circle whose diameter is also four inches you have multiplied it by three, not increased it by nine.

Areas of surfaces increase in proportion to the square of their lines.

SALVIATI. A little geometry, Simplicio. It's true that the diameter is increased three times, but the surface area, which is what we are discussing here, is increased nine times. The areas of circles are to each other as the squares of their diameters; so a circle with a diameter of four inches has to one of twelve inches the same ratio as the square of four squared has to the square of twelve, that is 16 to 144. Therefore it will be nine times larger, not three. So, to continue, if we add this same halo of four inches to a circle with a diameter of only two inches, the diameter of the halo will be ten inches, and its area will be to that of the bare circle as 100 to 4, these being the squares of 10 and 2; so the enlargement would be 25 times greater. Finally, adding a halo of four inches to a small circle of just one inch in diameter would increase its area 81 times, and so on, the increase being in ever greater proportion to the object as the object itself becomes smaller. [365]

SAGREDO. I must say that the difficulty which has troubled Simplicio did not trouble me, but there are some other points on which I would welcome more clarification. In particular, I would be glad to know on what grounds you say that this enlargement is always the same in all visible objects.

The more brightly objects shine, the more they appear to be enlarged.

An easy experiment to demonstrate how secondary rays make stars appear enlarged.

Jupiter appears enlarged less than Sirius.

SALVIATI. I've already partly explained this when I said that this enlargement applies only to shining objects and not to those that are dark; so let me add the rest now. The shining objects whose light shines most brightly are those that produce a greater and stronger reflection on the pupil of our eye, and therefore their appearance is enlarged much more than those which shine less brightly. And so as not to dwell at any more length on this point, let us come to see what the true source of knowledge teaches us, by observing the star of Jupiter this evening when it is quite dark. We shall see it shining brightly, and very large. Then let us look at it through a tube, or close our fist and put it to our eye so that we look through the narrow space between the palm of our hand and our fingers, or indeed through a hole made with a fine needle in a sheet of paper. We shall see Jupiter shorn of its rays, and so small that we will think it is less than a sixtieth of its size when we saw it with the naked eye as a great torch. Then we can look at Sirius, [366] the Dog Star, the largest and most resplendent of the fixed stars, which to the naked eye appears scarcely any smaller than Jupiter; but if in the same way we take away its halo its body will appear so small that it will seem less than a twentieth of the size of Jupiter, and indeed anyone with less than perfect vision will have difficulty seeing it at all. From this it is reasonable to conclude that this star, which shines much more brightly than Jupiter, also has a greater effect on its surrounding radiation than does Jupiter. The radiation around the Sun and Moon is negligible, because their large size means that they take up so much of the space in our field of vision that it leaves no room for secondary rays, so we see their bodies shorn and clearly defined. We can confirm this same truth, namely that bodies that shine with a brighter light have much more surrounding radiation than those whose light is fainter, with another experiment that I have carried out a number of times. I have seen Jupiter and Venus together on several

The Sun and Moon are enlarged only a little.

An experiment showing that more brightly shining bodies have much more radiation than those that are less bright.

occasions, 25 or 30 degrees from the Sun, and when it was quite dark Venus appeared a good eight or even ten times larger than Jupiter, as long as they were observed with the naked eye; but when they were observed through the telescope the true size of the body of Jupiter was seen to be four times or more larger than that of Venus. Yet Venus shone incomparably more brightly than the very pale light of Jupiter, for the simple reason that Jupiter is at a very great distance from the Sun and from us, whereas Venus is close to us and to the Sun.

The telescope is the best means of seeing stars devoid of their surrounding halo.

Once these points are established, it should not be hard to understand how it comes about that when Mars is in opposition to the Sun, and therefore seven times or more closer to the Earth than when it is near to conjunction, it appears only four or five times larger in the former state than in the latter, when we ought to see it as more than fifty times larger. The cause is simply its surrounding radiation, and when we strip away its secondary rays we will find it to be enlarged in precisely the right proportions. And the telescope is the best, indeed the only means of taking away its halo. By enlarging its body 900 or a thousand times, it enables us to see Mars unadorned and as clearly delineated as the Moon, and with the difference between its two positions varying in exactly the right proportions.

A second reason for the small apparent enlargement of Venus.

[367]

As for Venus, it ought to appear almost forty times larger at its evening conjunction when it is below the Sun than at its opposite conjunction in the morning, and yet we perceive it not even doubled in size. The reason for this is that, in addition to the effect of its surrounding radiation, it is crescent-shaped, and its horns are not only thin but receive the Sun's light obliquely and therefore very faintly; so its light, being meagre and weak, produces less bright and less extensive irradiation than when we see it with the whole of its hemisphere illuminated. But seen through the telescope its horns are as clearly and sharply defined as those of the Moon, and appear as part of a huge circle, almost forty times larger than its body when it is above the Sun, towards the end of its morning appearance.

SAGREDO. How pleased you would be, Copernicus, if you could see this part of your system confirmed by such clear experimental evidence!

SALVIATI. Yes, but how much less would have been the fame of his great intellect among those who understand! Because, as I've said before, he persisted in affirming what reason led him to conclude even though the evidence of the senses pointed to the opposite. I never cease to wonder that he was willing constantly to insist that Venus revolves around the Sun and is six times further away from us at one time than at another, and yet it always looks the same to us when it should have appeared forty times larger.

Copernicus was convinced by reason despite the evidence of the senses.

SAGREDO. I think the differences in size in the appearance of Jupiter, Saturn, and Mercury must also correspond exactly to their varying distances from the Earth.

SALVIATI. I have observed exactly this in the two outer planets for the past twenty-two years. It's not possible to make useful observations of Mercury because it can only be seen when it is at its greatest elongation from the Sun, when the differences in its distance from the Earth are not perceptible. Hence these differences are unobservable, as are its changes in shape, but they must certainly take place as they do in Venus. When we do see it, it should appear as a semicircle, in the same way as Venus does at its maximum elongation; but its body is so small, and it shines so brightly because of its proximity to the Sun, that the telescope is not powerful enough to shave off its halo to make it appear completely cropped.

Mercury does not lend itself to clear observations.

It remains for us to dispose of what appeared to be a serious problem regarding the Earth's motion, namely that while all the planets turn around the Sun it alone is not solitary like all the others, but is accompanied by the Moon and the whole elemental sphere in its annual motion around the Sun. Here we must again applaud the exceptional perceptiveness of Copernicus, and at the same time lament his misfortune in not being alive in our own time to see how the apparent absurdity of the Earth's motion in concert with the Moon is removed. For we can now see Jupiter, like another Earth, revolving around the Sun in twelve years accompanied by not one Moon but by four, together with whatever may be contained within the orbit of the four Medicean stars.

Removal of the difficulty arising from the Earth's revolving around the Sun not on its own, but accompanied by the Moon.

[368]

SAGREDO. On what grounds do you call the four planets of Jupiter Moons?

The Medicean stars are like four Moons around Jupiter.

SALVIATI. Because that is how they would appear to anyone observing them from Jupiter. They are dark in themselves, receiving their light from the Sun, as is apparent from the fact that they are eclipsed when they enter the cone of Jupiter's shadow. As the only part of them that is illuminated is the hemisphere facing the Sun, they appear to us to be always wholly bright, since we are outside their orbits and closer than they are to the Sun. But to anyone who was on Jupiter they would appear wholly illuminated only when they were in the upper part of their circles; when they were in the lower part, that is when they were between Jupiter and the Sun, they would appear on Jupiter to be crescent-shaped. In short, they would appear to the Jovians with the same changes of shape as the Moon does to us Terrestrials. So you can see now how marvellously these three notes, which at first seemed so dissonant, harmonize with the Copernican system. I think this will allow Simplicio to see the degree of probability with which we can conclude that it is not the Earth but the Sun that is at the centre of the revolution of the planets. And since the Earth is now located among the bodies whose motion around the Sun is not in doubt—that is, above Mercury and Venus and below Saturn, Jupiter, and Mars—is it not also highly probable, indeed necessary, to grant that the Earth too is in motion around it?

SIMPLICIO. These phenomena are so large and so evident that it's inconceivable Ptolemy and his followers were not aware of them; and if they were, they must have found a way of accounting for such evident appearances which was relevant and convincing, since they were so widely accepted for such a long time.

The main aim of astronomers is to provide an explanation of the appearances.

[369]

SALVIATI. You make your point well. But you must understand that the main aim of pure astronomers is simply to provide an explanation for the appearances of the heavenly bodies, and to adapt their structures and combinations of circles to these appearances and the motions of the stars, so that the motions according to their calculations agree with the appearances themselves. They are not unduly concerned if they admit some incongruities which in reality create difficulties in other respects. In fact Copernicus himself writes that in his first studies he restored the science of astronomy on the basis of Ptolemy's own suppositions, correcting the planetary movements so that his computations corresponded more accurately with the appearances and the appearances with his calculations—always however taking the planets separately one at a time. But he goes on to say that when he tried to bring the individual models together into an overall structure, the result was a monster, an impossible creature made up of limbs which were out of all proportion and incompatible with each other, so that however satisfactory it might be from the pure mathematical astronomer's point of view, it could not satisfy or put at rest the mind of a physical astronomer. He understood very well that if the celestial appearances could be saved by making assumptions that in nature were false, it would be much better to do so using suppositions that were true; so he set to work researching whether any of the ancients cited had proposed a different structure of the world from that commonly received from Ptolemy. When he found that some Pythagoreans had attributed the diurnal rotation, in particular, to the Earth, and others the annual motion as well, he began to examine under these two new suppositions the appearances and peculiarities of the planetary motions, all of which he had readily to hand; and finding that the parts and the whole corresponded with perfect ease he embraced this new system, putting an end to his quest.

Copernicus restored astronomy on the assumptions of Ptolemy.

What prompted Copernicus to establish his system.

SIMPLICIO. What inconsistencies in the Ptolemaic system are so great that there are not even greater ones in the Copernican?

SALVIATI. The infections are in Ptolemy; the remedies for them are in Copernicus. First of all, surely every school of thought will regard it as a great inconsistency that a body having a natural circular motion should move irregularly around its own centre* and regularly around another point. Yet such disparate motions are part of Ptolemy's model, whereas in Copernicus' the motions of all bodies are constant around their own centre. With Ptolemy it is necessary to assign contrary motions to the celestial bodies, [370] making them all move from east to west and at the same time from west to east; whereas with Copernicus all the celestial revolutions are in one direction, from west to east. And how are we to account for the apparent motion of the planets, which is so inconsistent

Inconsistencies in the Ptolemaic system.

that they appear not just to move sometimes more rapidly and sometimes more slowly, but sometimes to stop altogether, and then subsequently to turn back and move for a long spell in the opposite direction? To save this appearance Ptolemy introduced huge epicycles, adapting them one by one to each planet, with some rules describing motions that are impossible; these can all be eliminated with one simple motion of the Earth. And would you not consider it a great absurdity, Simplicio, if in Ptolemy's model, where each planet has its own assigned orbit, one above the other, it was necessary to say that Mars often fell so far from its place above the sphere of the Sun that it broke into the Sun's orbit and came closer to the Earth than the Sun itself, and then shortly afterwards rose an immense distance above it? Yet this too, along with other irregularities, is remedied purely by the simple annual motion of the Earth.

SAGREDO. These stations, retrograde motions and direct motions have always struck me as highly improbable, and I would like to understand better how they work in the Copernican system.

A very strong argument in favour of Copernicus is that he removes the need for stations and retrograde motions of the planets.

The annual motion of the Earth alone causes the great irregularities in the motion of the five planets.

Demonstration that the irregularities in the three outer planets derive from the annual motion of the Earth.

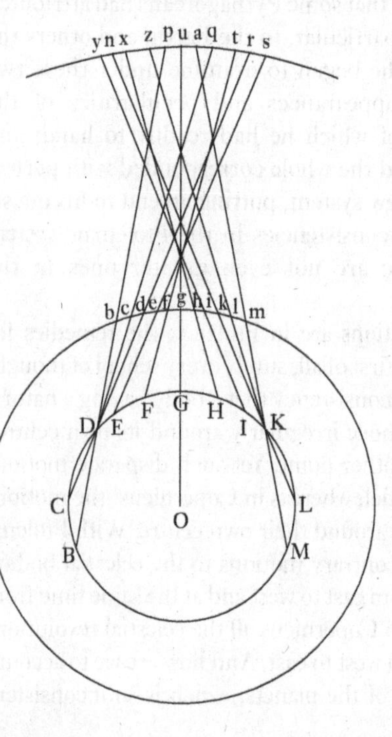

SALVIATI. Sagredo, this account of how they work should be enough on its own to persuade anyone who wasn't totally stubborn and unteachable to accept the theory in its entirety. I tell you, then, that the annual motion of the Earth between Mars and Venus alone is the cause of the apparent irregularities in the motions of all five planets, without any change to their orbital period—thirty years for Saturn, twelve for Jupiter, two for Mars, nine months for Venus, and about eighty days for Mercury. This figure will explain the whole matter fully and easily. Suppose the Sun to be located in the centre O, around which we shall draw the orbit described by the Earth in its annual motion, BGM. Then the circle described by, for example, Jupiter, around the Sun in twelve years, will be *bgm*, and by the zodiac in the sphere of the fixed stars it will be *yus*. Now let us take a number of equal arcs in the Earth's annual orbit, BC,CD, DE, EF, FG, GH, HI, IK, KL, LM; and other arcs in the circle of Jupiter through which it passes at the same times as the Earth passes [371] through its arcs, which we mark *bc*, *cd*, *de*, *ef*, *fg*, *gh*, *hi*, *ik*, *kl*, *lm*. Each of these will be proportionately smaller than the arcs in the Earth's orbit, because the motion of Jupiter below the zodiac is slower than the annual motion of the Earth. So supposing that when the Earth is at B Jupiter is at *b*, it will appear to us to be at *p* in the zodiac, looking along the straight line B*bp*. Then let the Earth move from B to C, and Jupiter from *b* to *c* in the same period of time; Jupiter will appear to us to have reached *q* in the zodiac in direct motion, according to the order of the signs, from *p* to *q*. Then when the Earth moves to D and Jupiter to *d*, it will appear to us in the zodiac at *r*. From E, Jupiter passing on to *e* will be seen in the zodiac at *s*, still moving in direct motion. But when the Earth, beginning to come directly between Jupiter and the Sun, reaches F, Jupiter at *f* will appear to us at *t*, and so will have begun apparently to move backwards through the zodiac. In the time in which the Earth has passed through the arc EF Jupiter will have slowed down between points *s* and *t*, and will appear to us to be almost stationary. Then when the Earth reaches G, Jupiter, in opposition to the Sun, will be at *g* and will be seen in the zodiac at *u*, having apparently moved backwards through the whole of the zodiac arc *tu*; but in reality it will have continued to follow its constant course, not just in its own circle but in that of the zodiac as well, in relation to the zodiac's centre and to the Sun which is at its centre. As the Earth and Jupiter continue their motions, when the Earth reaches H Jupiter will be at *h* and will appear to have moved a long way backwards in the zodiac, through the whole of the arc *ux*; and when the Earth comes to I and Jupiter to *i*, it will appeared to have moved only the small distance *xy* and will appear to be stationary there. Moving on, when the Earth is at K and Jupiter is at *k*, in the zodiac it will have passed through the arc *yn* in direct motion. When the Earth in its course comes to L Jupiter at *l* will be seen at point *z*; and finally when Jupiter is at *m*, seen from the Earth at M it will [372] have moved on to *a*, again in direct motion. So the whole of Jupiter's apparent retrograde motion through the zodiac will correspond to the arc *sy*, while in its own circle it will have passed through the arc *ei*, and the Earth in its own circle through the arc EI.

Retrogradations are more frequent in Saturn, less so in Jupiter, and less again in Mars; the reason for this.

Retrogradations of Venus and Mercury demonstrated by Apollonius and by Copernicus.

What we have said here about Jupiter can be taken to apply to Saturn and Mars as well. With Saturn these retrogradations are rather more frequent than with Jupiter, because Saturn's motion is slower so that the Earth catches up with it more quickly. With Mars they are rarer, because its motion is more rapid and therefore it takes longer for the Earth to catch up with it. Venus and Mercury too, whose orbits are contained within that of the Earth, appear with their stations and retrogradations not because these correspond to their real motions but because of the annual motion of the Earth, as Copernicus, following Apollonius of Perga,* clearly proves in book 5 of his *On the Revolutions*, chapter 35.

The annual motion of the Earth is perfectly suited to explain the irregular motions of the five planets.

The Sun itself testifies to the annual motion of the Earth.

So you see, gentlemen, how neatly and simply the annual motion, assuming that it belongs to the Earth, lends itself to explain the apparent irregularities that are to be observed in the motions of the five planets, Saturn, Jupiter, Mars, Venus, and Mercury, removing all the irregularities and establishing their motions as constant and regular; and Nicholas Copernicus was the first to make plain the reason for this marvellous effect. But there is another effect, no less astonishing than this, which compels our human understanding to acknowledge this annual revolution and attribute it to our terrestrial globe, with a knot perhaps even more difficult to untie. A new and unheard-of conjecture is brought to us by the Sun itself, which clearly did not want to be alone in not attesting to such an important conclusion, but rather wanted to share in it as the most significant witness of all. So listen now to this tremendous new wonder.

The Lincean Academician is the first discoverer of the sunspots and all the other novelties in the heavens.

The first to discover and observe the sunspots,* and indeed all the other novelties in the heavens, was our friend the Lincean Academician. He discovered them in the year 1610, when he was still lecturing in mathematics at the University of Padua, and he spoke of them there and in Venice with various people, some of whom are still alive. A year later he showed them to many gentlemen in Rome, as he states in the first of his letters to Mark Welser, a Magistrate in Augsburg. He was the first to affirm, contrary to the views of those who were too timid and too defensive of the immutability of the heavens, that the sunspots were material objects that formed and dissolved in short periods of time; that as regards their location, they were contiguous with the surface of the Sun; and that they revolved around the Sun, or rather completed their rotations carried along by the Sun's globe itself, which turns on its own axis in the space of a little less than a month. At first he considered this motion of the Sun to be around an axis perpendicular to the plane of the ecliptic, since the arcs described by these spots across the body of the Sun appeared to our eyes as straight lines parallel to the plane of the ecliptic. However, they are subject to various other erratic and irregular movements which partly disrupt their course, making them change their position in a random and completely disordered way. Sometimes many of them collide together and then separate again; some divide and change their shape, often in bizarre ways. Such unpredictable changes partly disrupted the primary periodic course of these spots, but they did not make our friend change his mind and believe that these deviations had some fixed and essential cause; rather he persisted in his belief that all the variations in their appearance were the result of these accidental changes. In just the same way, someone observing the motion of our clouds from some distant region would see them moving, with a very rapid and constant motion, as they are carried along every twenty-four hours by the diurnal spinning of the Earth (assuming that this motion

History of the Academician's progress in the observation of the sunspots over a long period of time.

[373]

belongs to the Earth) in circles parallel to the equator; but their course would be partly disrupted by the incidental movements caused by the winds that drive them randomly in all directions.

It happened at that time that signor Welser sent him some letters about these sunspots, written by someone under the assumed name of Apelles, pressing him to give his honest opinion of the letters and asking him for his own view about the nature of the spots. Our friend obliged him with three letters in which he first showed the weakness of Apelles' ideas, and secondly explained his own views, adding that when Apelles had had time to reflect further he would undoubtedly come to the same view himself, as indeed he did. Then, since our Academician considered (and others with an understanding of natural phenomena agreed with him) that he had investigated and explained enough on this matter in these three letters, if not to exhaust everything that human curiosity could seek or desire, at least as much as it was possible for human reason to establish, he laid aside his observations for some time while he [374] was occupied with other studies, although he did make the occasional isolated observation to oblige a friend. Some years later when we were at my villa, Le Selve, at my request and encouraged by a long period of brilliant clear skies, we observed the entire transit of one very large, dense and isolated spot, systematically writing down its position day by day when the Sun was at the meridian. When we realized that the course of its transit, far from being straight, was somewhat curved, we decided to make further observations at regular intervals. The stimulus to do so was all the stronger because of an idea that suddenly came to my guest, and which he expressed to me with these words: 'Filippo, I think the way is opening up for us to reach a deduction of great consequence. If the axis around which the Sun rotates is not perpendicular to the ecliptic but is inclined to it, as the curved course that we have just observed leads me to think, then we shall be able to make a conjecture about the positions of the Sun and the Earth on firmer and more conclusive grounds than any observation hitherto has been able to provide.' My interest aroused by this bold promise, I pressed him to explain his thinking to me; this

An idea which suddenly occurred to the Lincean Academician on the important implication of the motion of the sunspots.

was his reply: 'If the annual motion belongs to the Earth, around the Sun along the ecliptic, and if the Sun, located at the centre of the ecliptic, rotates there not around the axis of the ecliptic—which would also be the axis of the annual motion of the Earth—but rather at an incline to it, then we will necessarily see erratic alterations in the apparent motions of the sunspots. Assuming that the axis of the Sun's rotation remains constant and unchanging at the same inclination and towards the same point in the universe, then several consequences will necessarily follow for how they appear to us, carried along as we are by the terrestrial globe in its annual motion around the Sun. First, the passage of the sunspots will indeed appear to us sometimes to be in a straight line, but this will happen only twice a year; at all other times they will appear to move along visibly curved arcs. Secondly, these arcs will appear to us in one half of the year curved in the opposite way to their appearance in the other half; that is, for six months the arcs will curve towards the upper part of the Sun's surface, and for the other six months towards the lower part. Third, as the sunspots are born, so to speak, and first appear to our sight on the left side of the Sun's surface and move across to set and disappear on the right side, for six months the points where they first appear in the east will be lower than the opposite points where they disappear. For the other six months the opposite will happen; that is, the spots will rise at more elevated points and will move downwards to disappear at lower points. There will be only two days in the year when their rising and setting points will be in equilibrium. From those moments of balance the inclination of their courses will steadily increase, becoming more pronounced each day, until in three months they will reach their maximum obliquity. From that point they will start to diminish and in the same period of time will return to equilibrium again. Moreover—and this is the fourth marvel—the day when they are at their maximum obliquity will be the same day as their course is seen as a straight line, and the day of equilibrium will be when the curve of their transit is seen to be most pronounced. At other times, as the inclination diminishes and they move towards equilibrium,

Surprising changes to be observed in the motion of the sunspots foreseen by the Academician, if the annual motion belongs to the Earth.

[375]

the curvature of their journey, in contrast, will steadily increase.'

SAGREDO. My dear Salviati, I know it's bad manners to interrupt you when you are explaining; but I think it's just as bad to allow you to go on at greater length when you are wasting your breath, because to be honest I'm not able to form a clear idea of even one of the conclusions you've been describing. But from the general somewhat confused impression I have gained they seem to me to be matters of great consequence, and so I would like to have some understanding of them.

SALVIATI. My experience was the same as yours when my guest described them to me purely verbally. He helped me to understand by taking a material instrument, just a simple astronomical sphere, but using the circles for another purpose than that for which they are commonly used. Since we don't have a sphere here I will manage by drawing the circles on paper as need arises. So to represent the first phenomenon that I mentioned, namely that there could be only two occasions in the year when the passage of the spots appears to us to be made along straight lines, let us mark O here as the centre of the Earth's orbit, or let us say the ecliptic, and likewise also the centre of the globe of the Sun itself. Given the great distance between the Earth and the Sun, we may suppose that we here on Earth can see only half of its sphere. So let us draw this circle ABCD around the centre O, to represent the outer limit dividing the hemisphere of the Sun that is visible to us from the other half that is hidden from us. Now our eye, no less than the centre of the Earth, is understood to be in the plane of the ecliptic, as is the centre of the Sun. So if we represent the body of the Sun as being bisected by this plane, the line bisecting it will appear to us as a straight line, BOD; and if we drop onto it a vertical line

The first phenomenon to be noted in the motion of the sunspots, as a consequence of which all the others can be explained.

[376]

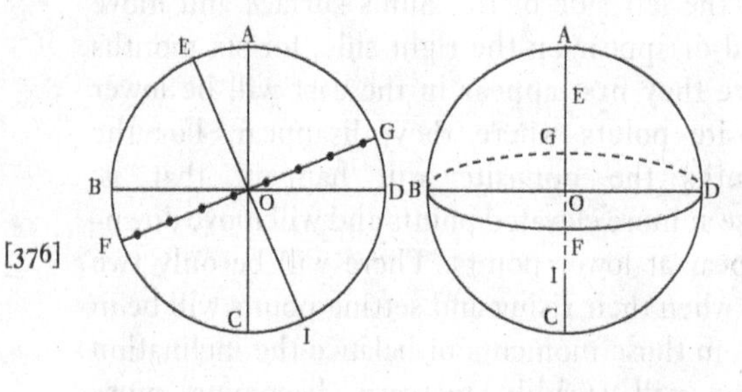

AOC this will be the axis of the ecliptic and of the terrestrial globe's annual motion. Now suppose that the Sun's body rotates, without any change at its centre, not around the axis AOC (which is vertical to the plane of the ecliptic) but around an axis at an inclination to it, such as EOI here. Suppose, too, that this axis remains permanently fixed and immutable at the same inclination and directed towards the same points in the firmament and in the universe. As the solar globe rotates, every point on its surface apart from the poles describes the circumference of a circle, larger or smaller depending on its greater or lesser distance from the poles. So if we take point F which is equidistant from the poles and mark the diameter FOG, this will be vertical to the axis EI, and will be the diameter of the great circle described around the poles E and I. Now if the Earth, and we along with it, are at a point on the ecliptic where the hemisphere of the Sun that is visible to us is bounded by the circle ABCD, and the hemisphere in the course of its normal passage through the poles A and C passes also through E and I, it is clear that the great circle whose diameter is FG will be vertical to the circle ABCD. The ray from our eye to the centre O is also perpendicular to the circle ABCD; so it will also fall in the plane of the circle whose diameter is FG, and therefore that circle's circumference will appear to us as a straight line, which is identical to FG. Hence any sunspot that appeared at point F, as it was carried along by the Sun's rotation, would trace a circumference on the surface of the Sun that would appear to us as a straight line. So its passage would appear to be straight, as would that of any other [377] spots tracing smaller circles in the same revolution, as they would all be parallel to the great circle, and our eye is at an immense distance from them.

Consider now the Earth's position six months later, when it will have passed through half the ecliptic and will be facing the hemisphere of the Sun that is currently hidden from us. The limit of the part that is visible to us will be the same circle ABCD, and it will still pass through the poles E and I; so you will understand that the same will be true of the course of the sunspots, that is, they will all appear to be in straight lines. But since this happens only when the circle limiting our sight passes through the poles E and I, and since this circle is constantly changing because of the Earth's annual motion, its passing through the fixed poles E and I lasts only for a moment, and consequently it is only for a moment that the motion of the sunspots appears as a straight line. It will be clear from what has been said so far that when the spots appear and begin their motion at point F and proceed towards point G, they move from the left and upwards towards the right; but when the Earth is in the diametrically opposite position they will appear at point G, which will be on the observer's left, but their course will be downwards towards the right at F.

Now let us imagine that the Earth has moved by a quarter of a circle from its present position. Again in this figure we mark the circle limiting our vision ABCD, with its axis AC as before, through which would pass the plane of our meridian. The axis of the Sun's rotation would also be in this plane, with one of its poles facing us, as it is in the hemisphere that we can see; we shall mark this as point E. The other pole will be in the hemisphere that is hidden from us; I mark this as point I. So if the axis EI is at an inclination with its upper part E inclined towards us, the great circle described by the rotation of the Sun will be BFDG; and the half of this which is visible to us, that is BFD, will no longer appear as a straight line, because the poles E and I are not in the circumference ABCD. Rather it will appear as a curved line with its convex side towards the lower part C; and the same will clearly be true of all the smaller circles parallel to the great circle BFD. It is clear too that when the Earth is at a point diametrically opposed to this, so that the other hemisphere of the Sun, currently hidden from us, is visible, we will see the part DGB of the same great circle, curved with its convex side towards the top part A; and in [378] this position the course of the sunspots will be first through the arc BFD and then through the arc DGB, and their first appearance at point B and final disappearance at point D will be balanced, neither being more or less elevated than the other.

If we now place the Earth at a point on the ecliptic such that neither the horizon ABCD nor the meridian AC passes through the poles of the axis EI, as in this third figure, with the visible pole E falling between the arc of the horizon AB and the section of the meridian AC, the diameter of the great circle will be FOG, with FNG as the visible semicircle and GSF the one that is hidden. Of these, the former will be curved with its convex side N towards the bottom part, and the latter curves with its highest point S towards the upper part of the Sun. The points where the sunspots appear and disappear, i.e. the points F and G, will not be balanced as B and D were; rather, F will be lower and G higher, but with a less marked difference than in the first figure. The arc FNG, similarly, will be curved, but not to the same extent as the earlier arc BFD. In this position, therefore, the courses of the sunspots will be curved, and will be upwards from F on the left to G on the right. And if the Earth is at a point diametrically opposite this, so that the visible hemisphere of the Sun is the one that is currently hidden, and bounded by the same horizon ABCD, then it is clear that the course of the sunspots will be along the arc GSF, starting at the higher point G which however will be on the observer's left, and moving downwards to terminate at point F on the right.

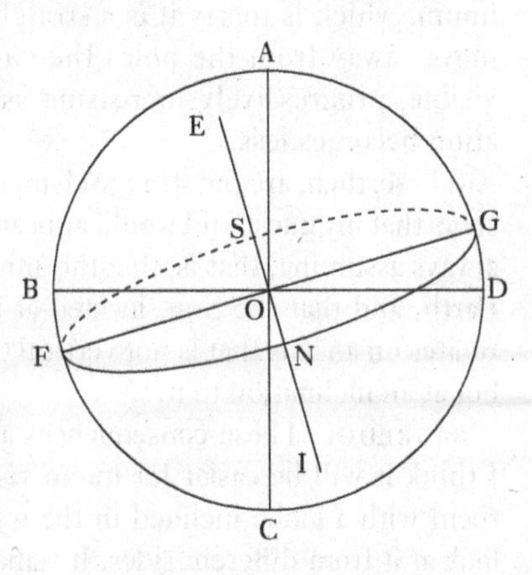

From what I have explained so far, it should not be difficult to understand how all the changes in the apparent courses of the sunspots derive from the passage of the line marking the limits of the solar hemispheres through the poles of the Sun's rotation, or at a greater or lesser distance from them. So the further the poles are from this line, the more the courses of the sunspots will be curved and the less they will be oblique; and at their maximum divergence, which is when these poles intersect with the meridian, the curvature is at its greatest but the inclination is at its minimum, which is to say it is in equilibrium, as the second figure shows. But when the poles fall on this line, as shown in the first figure, the inclination is at its greatest but the curvature is at its minimum, which is to say it is a straight line. Then, as this line moves away from the poles the curvature starts to become visible, progressively increasing as the obliquity or inclination becomes less.

[379] These, then, are the strange changes in the course of the sunspots that my guest said would appear from one time to another, always assuming, that is, that the annual motion belongs to the Earth, and that the Sun, located at the centre of the ecliptic, rotates on an axis that is not vertical to the plane of this ecliptic but at an inclination to it.

SAGREDO. These consequences are quite clear to me now. I think it will be easier for me to visualize them if I compare them with a globe inclined in the way you describe, and then look at it from different sides. It remains now for you to tell us how the consequences you envisaged turned out in practice.

The observed events corresponded to the predictions.

SALVIATI. It turned out that as we continued making very careful observations over many months, noting the passage of various sunspots at different times of the year with the greatest possible accuracy, what happened corresponded exactly with our predictions.

SAGREDO. Well, Simplicio, if what Salviati says is true—and it wouldn't be right to cast doubt on his words—the Ptolemaics and Aristotelians will need sound arguments, strong theories, and well-founded experiments to counter such a weighty conclusion, if they are to prevent their theory from collapsing altogether.

Although ascribing the annual motion to the Earth corresponds to the appearances of the sunspots, it does not follow that, conversely, the annual motion of the Earth can be inferred from the appearances of the spots.

SIMPLICIO. Just a minute, my dear sir; you may not yet be at the point you think you have reached. I haven't entirely mastered the substance of what Salviati has said, but my logic teaches me that, as regards the form of his argument, it by no means obliges me to conclude in favour of the Copernican theory that the Sun is at rest at the centre of the zodiac and that the Earth is in motion within the zodiac's circumference. It may be true that, given this rotation of the Sun and this revolution of the Earth, such and such changes will necessarily be observed in the course of the sunspots; but it does not follow that, conversely, observing these changes in the sunspots necessarily leads to the conclusion that the Earth moves around the circumference of the zodiac and the Sun is at its centre. Who can assure me that such movements of the sunspots cannot also be observed if the Sun is in motion along the ecliptic and the inhabitants of the Earth are fixed at its centre? So until [380] you can demonstrate to me that these appearances cannot be explained if the Sun is in motion and the Earth is fixed, I will not change my opinion and my belief that it is the Sun that moves and the Earth is at rest.

SAGREDO. Simplicio defends himself strongly, and he makes a very acute counter-argument in defence of Aristotle and Ptolemy. If the truth be told, I think he has gained much skill in arguing rigorously from his conversations with Salviati, even though they have only been for a short time. I understand that this effect has been noticed in others as well. As for investigating and judging whether the apparent variations in the motions of the sunspots can be adequately explained assuming that the Earth is fixed and the Sun is in motion, I look forward to hearing Salviati expound his thinking to us. I'm sure he has given the matter some thought and has mastered whatever there is to be said about it.

Pure Peripatetic philosophers will mock the sunspots and their appearances as illusions produced by the lenses in the telescope.

SALVIATI. I have indeed given it some thought on many occasions, and have also discussed it with my friend and guest. As for what philosophers and astronomers will argue in support of the ancient system, we can be sure of what one group of them will say. We can be sure, that is, that the true pure Peripatetic philosophers will mock those who spend their time on what in their view are foolish trifles. They will dismiss all these appearances as vain illusions produced by the lenses, and thus will effortlessly free themselves from the need to think any further about them. As for philosophers who are also astronomers, we have thought very carefully about the arguments that could be brought forward about this, and we have not found any solution that could satisfy both the course of the sunspots and the course of our reasoning. I shall set out for you the thoughts that occurred to us, and you must decide what weight you will give to them.

If the Earth is fixed at the centre of the zodiac, four different motions must be attributed to the Sun, as is explained at length.

[381]

If the apparent motions of the sunspots are as was described above, and if the Earth is fixed at the centre of the ecliptic on whose circumference the centre of the Sun is placed, then all the variations that are observed in these movements must be caused by motions in the Sun itself. In the first place, it must revolve upon itself carrying the sunspots along with it, since we have assumed, or rather proved, that they are in contact with the solar surface. Secondly, it must be affirmed that the axis of the Sun's rotation is not parallel to the axis of the ecliptic, in other words that it is not perpendicular to the plane of the ecliptic, because if it were then the courses of the sunspots would appear to us as straight lines parallel to the ecliptic. Its axis, then, is tilted, because their courses mostly appear as curved lines. In the third place, it will be necessary to say that the inclination of this axis is not fixed and continuously directed towards the same point in the universe; on the contrary, it changes direction from one moment to the next. The reason for this is that if its inclination were continuously directed towards the same point, the courses of the sunspots would never appear to change. In whatever form they first appeared—straight or curved, curving upwards or downwards, ascending or descending—they would always appear the same. Therefore this axis must be said to be variable. Sometimes it will be in the plane of the circle marking the outer limit of the visible hemisphere; this will be when the passages of the sunspots appear to be made in straight lines at their greatest inclination, as happens twice a year. At other times it will be in the plane of the meridian of the observer, so that one of its poles falls within the solar hemisphere that is visible to us and the other in the hemisphere that is hidden. Both of these poles will be at a distance from the extremities, or poles, of another axis, which must necessarily be assigned to the Sun, and which is parallel to the axis of the ecliptic; this distance will correspond to the inclination of the axis of the rotation of the sunspots. In addition, the pole falling in the visible hemisphere will be in the upper part of the solar disc at one time, and in the lower part at another. This must be the case because when the paths of the sunspots are level and at their maximum curvature, the convex side of the curve is towards the lower part on one occasion and towards the upper part on the other. Then, since these appearances are all in constant change—the degrees of inclination and curvature being now greater and now smaller, and the inclination sometimes reduced to perfect equilibrium and the curvature to a perfect straight line—it is necessary to suppose that the axis of the monthly revolution of the sunspots has a revolution of its own, making its poles describe two circles around the poles of another axis which, as I have said, must necessarily be assigned to the Sun. The radius of these circles must correspond to the degree of inclination of that axis, and the period of its rotation must be a year, since that is the frequency with which all these appear- [382] ances and variations in the paths of the sunspots are repeated. It is clear from the fact that the maximum inclinations and the maximum curvatures are always the same, that the rotation of this axis is around the poles of another axis parallel to that of the ecliptic, and not around any other points.

All this means that to maintain the Earth at rest at the centre we have to attribute to the Sun two motions around its own centre, on two different axes, one completing its revolution in a year and the other in less than a month. This assumption strikes me as very difficult to make, in fact virtually impossible. It means also attributing to the body of the Sun two other motions around the Earth, also on different axes, one around the ecliptic in a year and the other forming spirals or circles parallel to the equinoctial once a day. There is no reason why the third motion to be attributed to the Sun (not the one of almost a month that carries the sunspots, but the other involving the axis and the poles of the monthly motion) should complete its period in a year, reflecting the annual motion around the ecliptic, rather than in twenty-four hours, reflecting the diurnal motion around the poles of the equinoctial.

I know that what I say is somewhat obscure now, but it will become clear when we discuss the third annual motion that Copernicus assigns to the Earth.* But for now, if it is possible to reduce these four incompatible motions—all of which must of necessity be attributed to the body of the Sun itself—to a single very simple motion of the Sun around one unchangeable axis, without adding anything to the motions that for many other reasons have been assigned to the terrestrial globe, and moreover to save the appearances of so many erratic variations in the motions of the sunspots, then I think that is an offer that should not be refused. And this, Simplicio, is as much as our friend and I could think of that could be put forward by the Copernicans and the Ptolemaics to explain these appearances while maintaining their respective theories. It is for you to make use of it as your judgement decides.

SIMPLICIO. I recognize that I am not qualified to pronounce on such an important decision. As for my own view, I shall remain neutral, while hoping that the time may come when we are enlightened by a higher contemplation than this human reasoning of ours, when the veil will be removed from our minds and the darkness that clouds our understanding will be taken away.

[383] SAGREDO. Simplicio's excellent and pious counsel deserves to be adopted and followed by everyone. Coming as it does from the supreme wisdom* and the highest authority, it alone can be embraced with confidence. But I will be a little bolder than he is and will declare that, as far as human reason is permitted to penetrate and staying within the bounds of conjecture and probable arguments, I have never encountered anything that has inspired my mind to greater wonder than these two theories. Of all the ingenious arguments I have ever heard, none, apart from pure geometrical and mathematical proofs, has been so inescapable in its reasoning. Based on the stations and retrogradations of the five planets, in one case, and on the wandering movements of the sunspots, in the other, they seem to me to provide the true explanation for these erratic appearances, showing how one simple motion, combined with others which are also simple although different from each other, can save so many appearances without introducing any additional difficulties, in fact resolving all the difficulties that attend the alternative theory. So, as I reflect on it, I can only conclude that those who continue to resist this theory either have not heard or have failed to understand these arguments which are so clearly conclusive.

SALVIATI. I shall not label them as either conclusive or inconclusive. As I have said before, my intention has not been to come to any resolution about so great a matter, but only to put forward the arguments in nature and astronomy that I can adduce in favour of both positions, leaving it to others to decide. In the end the conclusion should not be in doubt, since one of the two models must necessarily be true and the other necessarily false, so—remaining within the limits of human reasoning—the arguments in favour of the true position cannot fail to be as conclusive as those supporting the opposite view are vacuous and ineffective.

SAGREDO. It's time, then, for us to hear the opposing arguments in the book of conclusions or investigations that Simplicio has brought.

SIMPLICIO. Here is the book, and here is the place where the author first briefly describes the world system according to the theory of Copernicus: 'The Earth and the Moon therefore with the whole of this elemental world, Copernicus . . .'.

SALVIATI. Stop there a moment, Simplicio. Right from this [384] first opening statement the author seems to me to have a very poor understanding of the position that he is undertaking to refute. If he says that according to Copernicus the Earth and the Moon together complete a revolution of the ecliptic in a year, moving from east to west, this is both false and impossible, and Copernicus never suggested that they did. What he does say is that they move in the contrary direction, that is, from west to east, following the order of the signs of the zodiac, so that it appears to be the annual motion of the Sun, which is fixed and at rest at the centre of the zodiac. What over-confidence this shows, to set out to refute another's theory without understanding its fundamental principles, on which the greater and more important part of the whole edifice rests! This is not a good start for gaining the reader's confidence. But let us proceed.

Arguments put forward ironically in a certain book, in opposition to Copernicus.

SIMPLICIO. Having explained the system of the universe, the author starts to put forward his arguments against this annual motion. The first he advances ironically, mocking Copernicus and his followers, and saying that this fantasy structure of the world leads us to make the most ridiculous statements: that the Sun, Venus, and Mars are below the Earth, and that heavy matter naturally moves upwards and light matter moves downwards; that Christ our Lord and Redeemer ascended into hell, and descended into heaven when he approached the Sun; that when Joshua commanded the Sun to stand still, the Earth came to a stop, or else the Sun moved in the opposite direction to the Earth; that when the Sun is in Cancer the Earth is moving through Capricorn, and that the winter star signs bring summer and the summer signs bring winter. It means that, instead of the stars rising and setting to the Earth, the Earth rises and sets to the stars, and that the east begins in the west and the west in the east; in short, that almost the whole course of the world is turned upside down.

SALVIATI. All these points are fine, except for the way he mixes references to sacred Scripture, which is always to be treated with reverence and awe, with these scurrilous inanities. He shouldn't use sacred matters to wound those who are indulging in philosophy light-heartedly and playfully, neither affirming nor denying but making suppositions or hypotheses [385] as the basis for a discussion among friends.

SIMPLICIO. To tell the truth I was quite shocked as well, especially when he goes on to add that even if the Copernicans respond, albeit in great indignation, to these and similar arguments, they will not be able to give a satisfactory response to the points that follow.

SALVIATI. This is the worst of all, because he is claiming to have points that are more effective and conclusive than the authority of holy writ; so please let us revere Scripture and move on to natural and human reasoning. Mind you, if the arguments he produces from natural reason are no more coherent than those he has come up with so far, we might as well abandon the whole enterprise, because I'm certainly not going to waste words responding to such silly nonsense. And he is quite wrong to say that the Copernicans have responded to these points, because it's not credible that anyone should waste their time on such a pointless exercise.

Supposing the annual motion to belong to the Earth, a fixed star must be bigger than the whole of the great sphere.

SIMPLICIO. I concur in this judgement; so let us hear the other arguments that he brings forward as being much more weighty. Here, as you see, he concludes by means of very accurate calculations that if the great sphere in which the Earth, according to Copernicus, makes its annual revolution around the Sun, is so small as to be imperceptible compared to the immensity of the sphere of the fixed stars, within which Copernicus says it must be placed, then we must necessarily say that the fixed stars are unimaginably distant from us, and that the smallest of them are larger than the great sphere itself, while others are much greater than the whole of the sphere of Saturn. These dimensions are so vast as to be unbelievable and incomprehensible.

SALVIATI. I have already seen a similar point that Tycho made against Copernicus, and this is not the first time that I have exposed the fallacy, or rather fallacies, of this line of argument, constructed as it is on the basis of completely false hypotheses and on something that Copernicus said which his opponents have taken in the most narrowly literal way. They are like litigants in a court case who, when they are in the wrong on the substantive merits of the case, seize on a single word that the opposite party has spoken in passing and never stop making a great fuss about it. To explain this more clearly, when Copernicus set out the wonderful consequences of the Earth's annual motion for the other planets—especially the forward and retrograde motions of the three outer planets—he went on to add that these apparent changes, which were more evident in Mars than in Jupiter because Jupiter is more distant [386] from us, and even less evident in Saturn because it is more distant than Jupiter, were imperceptible in the fixed stars because of their immense distance from us in comparison with the distance of Jupiter or Saturn. The opponents of this view jump on this and take it that when Copernicus said their motion was imperceptible this meant that it was really and absolutely non-existent. They go on to say that even the smallest of the fixed stars is still perceptible, since it is apparent to our sense of sight; and then they make a series of calculations, introducing other false assumptions, concluding that Copernicus's theory must allow that a fixed star is far larger than the orbit of the Earth. Now, to expose the vacuity of this whole line of reasoning, I shall show that if we take a fixed star of the sixth magnitude to be no bigger than the Sun, we can demonstrate conclusively that the distance of these fixed stars from us is such that the annual motion of the Earth would not be perceptible in them, even though it causes such great and

An argument of Tycho founded on false hypotheses.

Litigants who are in the wrong seize on a word spoken in passing by the opposition.

The apparent differences in the motions of the planets are not seen in the fixed stars.

Supposing that a fixed star of the sixth magnitude is no larger than the Sun, the great changes among the planets remain imperceptible in the fixed stars.

The distance of the Sun contains 1208 radii of the Earth.

The diameter of the Sun is half a degree.

The diameter of a fixed star of the first magnitude and of the sixth magnitude.

The amount by which the apparent diameter of the Sun exceeds that of a fixed star.

The distance of a fixed star of the sixth magnitude, assuming it to be equal in size to the Sun.

The variation in the appearance of the fixed stars caused by the Earth's orbit is scarcely more than that caused in the Sun by the size of the Earth.

The sixth-magnitude star posited by Tycho and the author of the booklet is ten million times greater than it need be.

observable variations in the planets. I shall also show separately the great fallacies in the assumptions made by the opponents of Copernicus.

I begin with the supposition, which I share with both Copernicus and his opponents, that the radius of the Earth's orbit, that is the distance from the Earth to the Sun, contains 1208 of the Earth's radii.* Secondly, again in agreement with the same views and with the truth, I posit that the apparent diameter of the Sun at its average distance is approximately half a degree, or 30 prime minutes, that is 1800 seconds, which are 108,000 third-order divisions. Now the apparent diameter of a fixed star of the first magnitude is no more than five seconds, that is 300 thirds, and that of a star of the sixth magnitude is 50 thirds (this is the greatest error of Copernicus's opponents). So the diameter of the Sun contains the diameter of a sixth-magnitude fixed star 2160 times. If therefore we supposed that a sixth-magnitude fixed star was really the same size as the Sun, and no bigger, this would be the same as saying that if the Sun were to move away from the Earth so that its diameter appeared to be one 2160th of its apparent size now, its distance would be 2160 times greater than its actual distance now. [387] This is the same as saying that the distance of a fixed star of the sixth magnitude is 2160 radii of the Earth's orbit. And since by common consent the distance of the Sun from the Earth contains 1208 radii of the Earth, and the distance of the fixed stars is, as we have said, 2160 radii of its orbit, it follows that the radius of the Earth is much greater—almost twice as much, in fact—than the radius of its orbit in relation to the distance of the stellar sphere. Therefore the difference in aspect of the fixed stars caused by the diameter of the Earth's orbit will be barely more than the difference we observe in the Sun due to the radius of the Earth.

SAGREDO. That's a big drop for a first step.

SALVIATI. It is indeed. According to this author's calculations a fixed star of the sixth magnitude would have to be as large as the Earth's orbit to be compatible with what Copernicus says. Yet assuming it to be only as large as the Sun, which is less than one ten-millionth part of that orbit, is enough to nullify this objection to Copernicus.

SAGREDO. Be so kind as to make this calculation for me.

SALVIATI. The calculation is easy and very short. Both sides agree that the diameter of the Sun is eleven radii of the Earth, and the diameter of the Earth's orbit contains 2416 of these radii; so the diameter of the Earth's orbit contains approximately 220 times that of the Sun. Now since spheres are to each other as the cube of their diameters, the cube of 220, i.e. 10,648,000, will show that the orbit is ten million, six hundred and forty-eight thousand times larger than the Sun. According to this author a star of the sixth magnitude would have to be equal to this orbit.

Calculation of the size of a fixed star in comparison with the Earth's orbit.

SAGREDO. Their error, then, consists in hugely mistaking the apparent diameter of the fixed stars.

SALVIATI. It does, but that is not their only error. Indeed, I am greatly astonished that so many astronomers, including such famous names as Alfraganus, Albategnius, Thabit,* and in modern times the likes of Tycho and Clavius*—in short, all the predecessors of our Academician—have been so greatly mistaken in determining the sizes of all the stars, fixed and moving stars alike, apart from the Sun and Moon. They have failed to take account of the surrounding radiation which deceptively makes the stars appear a hundred or more times larger than when they are seen shorn of their rays. There is no excuse for this oversight of theirs, because it was in their power to see the stars without their surrounding rays whenever they wished. They had only to observe them when they first appear in the evening or just before they vanish at dawn. If nothing else, Venus should have made them aware of their mistake, since it can often be seen at midday so small that one has to focus one's eyes sharply, and then the following night it appears like a great torch. I can't believe that they thought Venus's true size was as it appeared in the pitch dark rather than when it was seen in daylight. Our own lights that appear large when seen at night from a distance, but then prove to be just a small circumscribed flame when we see them close to, should have been enough to make them cautious. Actually, to give my honest opinion I don't believe that any of them ever took the trouble to observe and measure the apparent diameter of any stars apart from the Sun and the Moon, not even Tycho,

[388]

The common error of all astronomers concerning the sizes of the stars.

Venus leaves astronomers no excuse for their error in determining the sizes of the stars.

who spared no expense in constructing such large and accurate astronomical instruments and took such care in using them. I think, rather, that one of the ancient astronomers made an arbitrary statement based on guesswork that it was so, and then those who followed never checked but simply repeated what had been said before. If any of them had applied themselves to make a fresh observation they would undoubtedly have realized their mistake.

SAGREDO. They didn't have a telescope, and you have said that it was this instrument that enabled our friend to come to an understanding of the truth, so perhaps they should be excused rather than accused of negligence.

A method for measuring the apparent diameter of a star.

SALVIATI. That would be true if it were not possible to achieve this result without a telescope. It's true that this makes the procedure much easier, because it shows the body of the star unadorned and magnified a hundred or a thousand times. But it's also possible to achieve the same result without a telescope, albeit not so accurately. I have done so a number of times, and the method I adopted* is as follows. I suspended [389] a light rope in the direction of a star—I used Vega,* which rises between the north and north-east—and then by moving towards and away from the rope, I found the point from which the thickness of the rope exactly hid the star from me. Then I measured the distance from my eye to the rope, which is one side of the angle formed at my eye and subtended by the thickness of the rope; this is the same as the angle in the sphere of the fixed stars subtended by the diameter of the star. From the ratio of the thickness of the rope to the distance between the rope and my eye, the table of arcs and chords immediately gave me the size of this angle. I did however exercise the usual caution in measuring such acute angles, not to put the intersection of the visual rays at the centre of my eye, where they are only refracted, but at the point beyond the eye where the width of the pupil makes them truly converge.

SAGREDO. I understand this caution, although I do have some doubts about it. What troubles me more, though, is that if this procedure is followed at night when it is dark, it would seem that we are measuring the diameter of the star with its surrounding radiation, and not of the true bare star itself.

SALVIATI. No, because when the rope blocks the bare body of the star it takes away its halo, which exists in our eye and not in the star itself; so if the star's disc itself is hidden the halo disappears as well. If you make the observation you will be surprised at how thin a rope blocks a great shining torch that you would have thought would be hidden only by a much larger obstacle. To make the measurement as accurately as possible and find how many thicknesses of the rope fit into the distance between the rope and the eye, I don't just measure the diameter of the rope once, but I lay several pieces of the rope side by side on a table so that they are touching, and then use a compass to measure the whole space taken up by fifteen or twenty of them. Then I measure this against the distance from the rope to the meeting point of the visual rays, which I have already established using another finer thread. This very precise procedure enables me to establish that the apparent diameter of a fixed star of the first magnitude, commonly estimated as two prime minutes (and as much as three prime minutes by Tycho in his *Astronomical Letters*, page 167), is no more than five seconds, which is one twenty-fourth or one thirty-sixth of what has been believed. Such is the gravity of the errors on which their theories are based.

The diameter of a first-magnitude fixed star is not more than five seconds.

SAGREDO. I understand this very well. Before we move on, however, I would like to put forward the doubt that I have about placing the meeting point of the visual rays beyond the eye when one observes an object contained within a very acute angle. My difficulty arises because it seems to me that this meeting point can be more or less distant, not so much because of the larger or smaller size of the object itself, but rather because of another factor even when observing objects of the same size.

[390]

SALVIATI. I can see where Sagredo's perspicacity is leading, diligent observer of nature that he is. I would be willing to bet that of a thousand people who have observed the very marked contraction and dilatation in a cat's eyes, there are not two or even one who have noted a similar effect in the human eye, depending on the degree of brightness in the surroundings. In broad daylight the aperture of the pupil contracts markedly, so that when looking at the Sun it becomes smaller than a grain of

The opening of the pupil of the eye dilates and contracts.

millet, whereas when looking in darker surroundings at an object that is not bright in itself the pupil expands to the size of a lentil or even larger. In short, this dilatation and contraction varies tenfold or more; and hence it is clear that when the pupil is widely dilated, the angle at which the rays come together must be more distant from the eye. This is what happens when we look at objects in poor illumination. This is a new principle that Sagredo has shown me, and it alerts us to the fact that we should establish this point of convergence in the course of any observations that require a high degree of accuracy and that have major consequences. But for our present purpose of exposing the error of astronomers, such a degree of accuracy is not necessary, because even if we made the assumption most favourable to their case and supposed that the point of convergence was on the pupil itself it would make little difference, since their error is so great. I don't know, Sagredo, whether this was what you had in mind?

SAGREDO. It was exactly this point, and I'm glad to be reassured that it was not an unreasonable one, since you agree with me. But I would be glad to take this opportunity to hear how one can establish the point at which visual rays converge.

[391]

How to find the distance of the point at which visual rays converge from the pupil of the eye.

SALVIATI. The method is very easy, and it is this. I take two sheets of paper, one black and one white, making the black sheet half the width of the white one. I attach the white sheet to a wall, and place the other, on a stick or some other support, some fifteen or twenty *braccia* from it. Then I move away an equal distance from the second sheet in the same direction. It is clear that at this distance straight lines leaving the edges of the white sheet would just touch in passing the edges of the second sheet placed at the mid-point; whence it follows that if the eye is placed at this point of convergence, the black sheet would exactly conceal the white sheet behind it—assuming the observer was looking from a single point. If however we find that the edge of the white sheet is still visible, this must mean that the visual rays are not coming from a single point, and the eye must be brought closer in order for the white sheet to be hidden by the black one. Bringing the eye closer until the black sheet completely obscures the white one, we note the distance by which we needed to come closer, and this will give a definite figure for the distance from the eye of the point at which the visual rays truly converge. Moreover, it will also give us the diameter of the pupil, or the aperture from which the visual rays proceed.* This diameter will be in the same proportion to the width of the black sheet of paper as the distance from the meeting point of the visual rays, when the white sheet was first obscured by the black one, is to the distance between the two sheets.

If, therefore, we wanted to measure exactly the apparent diameter of a star using the method described above, we would have to compare the diameter of the rope with the diameter of the pupil. So if we found, say, that the diameter of the rope was four times that of the pupil, and the distance from the eye to the rope was, for example, 30 *braccia*, then we can say that the true point of convergence of the lines running from the edges of the diameter of the star to the edges of the diameter of the rope will be at a distance of 40 *braccia* from the rope. This will observe the correct proportions between the distance from the rope to the convergence of the rays and the distance from this convergence to the location of the eye, as this must be the same as the proportion between the diameter of the rope and the diameter of the pupil.

SAGREDO. I understand this fully. So now let us hear what Simplicio has to say in defence of the opponents of Copernicus.

SIMPLICIO. It's true that Salviati's argument has considerably [392] modified the great and quite incredible difficulty that these opponents of Copernicus have raised, but I don't think he has removed it altogether, and it still has enough force to defeat Copernicus's view. If I have understood his final conclusion correctly, then assuming the sixth-magnitude stars to be as large as the Sun—an assumption that I find hard to believe—it would still remain the case that the Earth's orbit causes changes and variations in the stellar sphere comparable to that which the Earth's radius produces in the Sun, which is clearly observable. And since no such change, or even a small one, is visible in the fixed stars, it seems to me that this completely destroys the annual motion of the Earth.

SALVIATI. Your conclusion would be sound, Simplicio, if there was nothing else to be said in support of Copernicus, but in fact there are many more points to consider. As for your objection, there is nothing to prevent us from supposing that the distance of the fixed stars is much greater that we have said. You yourself, and anyone else who wants to adhere to the propositions accepted by the followers of Ptolemy, must admit that it would fit your theory very well to locate the sphere of the fixed stars at a much greater distance than we said it should be just now.

Astronomers agree that the slower orbiting of some spheres is the result of their larger size.

All astronomers agree that the slower orbiting of some planets is caused by the larger size of their spheres; hence Saturn orbits more slowly than Jupiter, and Jupiter more slowly than the Sun, because in each case the slower planet has to describe a larger circle than the one below it. So Saturn, for example, whose sphere is nine times further away than that of the Sun, has an orbital period 30 times longer than that of the Sun. Now in Ptolemy's system the stellar sphere completes a revolution in 36,000 years, that of Saturn in 30 years, and that of the Sun in one year. Applying the principle of similar proportions, we can say: if the sphere of Saturn is nine times larger than that of the Sun and has an orbital period 30 times greater, what will be the size of a sphere that has a period 36,000 times greater? We shall find that the distance of the stellar sphere [393] must be 10,800 radii of the Earth's orbit, which is exactly five times greater than we calculated just now if a sixth-magnitude star was the same size as the Sun. So you can see the extent to which, in this case, the variation in the fixed stars caused by the annual motion of the Earth would be even less. And if we were to use a similar proportion to deduce the distance of the fixed stars from Jupiter and Mars, this would give us 15,000 radii of the Earth's orbit for the former, and 27,000 for the latter, i.e. seven and twelve times, respectively, more than the distance we found when we took a fixed star to be the same size as the Sun.

Using another supposition taken from astronomers, the distance of the fixed stars is calculated to be 10,800 radii of the great sphere.

The proportions of Jupiter and Mars show the stellar sphere to be much more distant still.

SIMPLICIO. In reply to this I think observers since Ptolemy's time have found that the motion of the fixed stars is not as slow as he thought; in fact I think I've heard it was Copernicus himself who observed this.

SALVIATI. You're quite right, but saying this does nothing to support the cause of the Ptolemaics. They have never denied the figure of 36,000 years for the motion of the stellar sphere on the grounds that this would make it too vast; but if such vastness is considered to be impossible in nature, they should have rejected such a slow orbital motion long ago because it could only be in proportion to an impossibly large sphere.

SAGREDO. Please, Salviati, let's not waste any more time discussing proportions with these people who are ready to accept completely disproportionate ideas. It's quite impossible to make any progress with them by going down that road. On one hand they insist that the most appropriate way of arranging the celestial spheres is in the order of their orbital periods, placing the slower spheres above the faster ones with the fixed stars as the highest, as the one that has a longer period than all the others. But then they add another sphere above this, and therefore larger than it, which they say completes its revolution in 24 hours, while the one below it takes 36,000 years. What could possibly be more disproportionate than that? But we said enough about these absurdities yesterday.

SALVIATI. Simplicio, I'd like you to set aside for a moment your loyalty to those who share your opinion and tell me frankly: do you think they really comprehend in their own mind the immensity that they say is too great to be attributed to the [394] universe? Speaking for myself, I don't believe so; I think that just as the imagination boggles and is unable to grasp numbers once you start going beyond thousands of millions, the same thing happens when we try to grasp immense sizes and distances. I think our reasoning has the same experience as our senses: when I look at the stars on a clear night, my senses lead me to judge that they are only a few miles distant, and that the fixed stars are no further away than Jupiter or Saturn or even the Moon. You have only to think of the controversies that arose between astronomers and Peripatetic philosophers about the distance of the new stars in Cassiopeia and Sagittarius.* The astronomers placed them among the fixed stars, and the philosophers thought they were nearer than the Moon: such is the inability of our senses to distinguish between distances that are large and those that are immense, even though the latter are many thousands of times greater than the former. Finally I ask you, O foolish man: are you able to comprehend with your

Immense sizes and distances are impossible for our intellect to comprehend.

imagination the immensity of the universe, that you judge it to be too vast? If you comprehend it, do you think your understanding extends further than the power of God, that you are able to imagine things greater than God can effect? And if you don't comprehend it, why do you claim to pass judgement on things that you don't understand?

SIMPLICIO. All this is true. No one denies that the size of the heavens can exceed our imagination, or that God could have created them a thousand times greater than they are. But must we not admit that nothing in the universe was created in vain or without a purpose? And when we see the planets perfectly ordered and placed around the Earth at distances exactly proportioned to produce effects on Earth that are beneficial to us, what would be the point of inserting a vast space with no stars at all between the outermost planet Saturn and the sphere of the fixed stars? It would be superfluous; what purpose would it serve, and who would benefit from it?

Nature and God are concerned with the care of humankind as if they had no other concerns.

An example of God's care for the human race, taken from the Sun.

SALVIATI. I think, Simplicio, that we claim too much for ourselves if we think that everything the divine wisdom and power does and ordains is solely to care for us, and that nothing else is done for any other purpose. I would not want us to limit God's hand to that extent; rather, I think we should be content with the certainty that God and nature so concern themselves with the governing of human affairs that they could not do more if the care of the human race was their only concern. [395] I think we can illustrate this with a very apt and worthy example taken from the effect of sunlight. When the Sun draws up vapours or warms a plant, it does so as if it had nothing else to do. More than that, when it ripens a bunch of grapes or even a single grape, it is applied to that process so effectively that it could not be more so if the whole purpose of the Sun's activity was applied solely to the ripening of that one grape. Now if that grape receives from the Sun all that it is capable of receiving, and it is deprived of nothing if the Sun simultaneously produces a thousand other effects, it would be very foolish and jealous of that grape to think or demand that the action of the Sun's rays should be used for its benefit alone. I am certain that divine Providence omits nothing that is needed for the governing of human affairs; but that there can be no other affairs in the universe that also depend on his infinite wisdom is something that my reason will not let me believe.

I would not resist the reasoning of a higher intelligence if the facts should prove to be otherwise; but in the meantime, if anyone says to me that it would be pointless and vain to insert an immense space devoid of stars between the planetary spheres and the sphere of the fixed stars, or that the immensity of the sphere of the fixed stars itself, which is beyond our comprehension, is superfluous, I reply that it is presumptuous for us to set up our feeble understanding to judge the works of God, or to call superfluous and vain anything in the universe that does not serve a purpose for us.

It is the height of presumption to call superfluous anything in the universe that we do not recognize as being made for our benefit.

SAGREDO. It would be better if you were to say 'that we do not know serves a purpose for us'. I consider it one of the greatest acts of arrogance—indeed of madness—we could commit, to say 'because I don't know what good Jupiter or Saturn are to me they are superfluous, and have no place in nature'. O foolish man! No more do I know what good my arteries or cartilage or spleen or gall bladder are to me. In fact I wouldn't even know that I had a gall bladder or spleen or kidneys if they hadn't been shown to me in numerous dissected bodies, and I would only know what effect the spleen has on me if I had it removed. To know the effect this or that heavenly body has on me—since you insist that [396] everything they do is directed at us—the body in question would have to be removed for a while, and then whatever effect I felt to be lacking would be attributed to that star. And in any case, who is to say that the space between Saturn and the fixed stars which they say is too vast and serves no purpose, is in fact devoid of any other bodies? If we assume this because we can't see them, are we to say that the four satellites of Jupiter and the companions of Saturn only came into the sky when we first saw them, and that they weren't there before? Or the innumerable other fixed stars—were they not there until men could see them? And were the nebulae just small patches of white in the sky, until with the aid of the telescope we made them become clusters of many beautiful shining stars? How presumptuous, not to say arrogant, men are in their ignorance!

Removing a star from the heavens could enable us to discover the effect it has on us.

There can be many things in the heavens that are invisible to us.

SALVIATI. I don't think we need waste any more time on these fruitless questions, Sagredo. Let's continue with our project, which is to examine the validity of the arguments put forward by each side without coming to any conclusions, leaving this to the judgement of those who know more about it than us. So, to come back to our natural human reasoning, words like 'large', 'small', 'immense', 'tiny', etc., are not absolute terms but relative. The same thing can sometimes be called immense and sometimes imperceptible, let alone small, depending on what it is compared to. This being so, I ask those who say that the stellar sphere in the Copernican system is too vast, in comparison to what? It can only be so defined, in my view, in relation to some other entity of the same kind; so let us take the smallest entity of the same kind, namely the sphere of the Moon. Now if the stellar sphere is deemed to be too vast in comparison with that of the Moon, anything else that is larger in the same degree or more in comparison with some other entity of its own kind must also be deemed to be too vast, and for that reason to be impossible. If that were the case then elephants and whales would undoubtedly be figments of the poetic imagination, because elephants are too vast in comparison with ants, which are land creatures, and whales in comparison with minnows, which are fish and clearly part of the order of nature. The elephant and the whale definitely exceed the ant and the minnow by a far greater proportion than the stellar sphere exceeds that of the Moon, assuming it is large enough to have its place in the Copernican system.

'Large', 'small', 'immense', etc., are relative terms.

Vanity of the argument that the stellar sphere is too vast in the Copernican system.

How large, in any case, is the sphere of Jupiter or of Saturn, each of which is allotted as the location of a single star, and a very small one in comparison with a fixed star? Certainly if every fixed star had to be given the same share of space in the universe as these, the sphere containing such an innumerable multitude of stars would have to be many thousand times larger than suffices for the system of Copernicus. What's more, do you not call the fixed stars very small, even those that are most clearly visible, never mind those that elude our sight? This is because they are very small in comparison to the space around them. Now, if the whole stellar sphere were one single resplendent body, surely it is clear that it could be located at such a distance

[397]

The space allotted to a fixed star is far less than that of a planet.

A star is considered small in relation to the size of the space around it.

The whole stellar sphere could appear as small as a single star if seen from a great distance.

in infinite space that this shining sphere would appear as small to us, or even smaller, as a single fixed star now appears to us from the Earth; and in that case we would judge to be small what now, from here, we call immeasurably large.

SAGREDO. It seems to me a very great folly to think that God created the universe in proportion to the meagre capacity of our understanding rather than to His immense, indeed infinite, power.

SIMPLICIO. All this is fine; but your opponents' objections are based on having to concede that a fixed star should be not just equal in size to the Sun, but so much larger than it, when they are both individual bodies within the stellar sphere. On this the questions posed by this author seem to me to be very pertinent: 'To what end, and for whose benefit, do such vast structures exist? Are they for the Earth, such an insignificant little dot as it is? And why are they at such a great distance, so that they appear so small and are quite unable to produce any effect on Earth? What is the purpose of such a disproportionately huge gulf between them and Saturn? Anything not supported by probable arguments exists in vain.'

Objections by the author of the booklet, expressed as questions.

SALVIATI. From the questions this man raises I think we can gather that, assuming the heavens and the stars remain as they are, with the distances, number, and size he has always believed that they have (not that he has ever given any firm or comprehensible estimate of their size), he fully understands and appreciates the beneficial effects that they bring to the Earth. On this view the Earth is no longer tiny and insignificant, and the stars no longer so remote that they appear so small, but rather large enough to act upon the Earth. The distance between the stars and Saturn is perfectly proportioned, and he has very probable arguments in support of all these things, although I would have been glad to hear some of them.

Replies to the questions asked by the author of the booklet.

The booklet's author is confused and contradicts himself in his questions.

[398]

Seeing, however, how he gets confused and contradicts himself just in these few words, I think he must be short of these probable arguments, and that what he calls arguments are in fact fallacies, or rather vainly imagined shadows. So my question to him now is, do these heavenly bodies really produce effects on Earth, and are they placed as they are, with the sizes and distances that they have, in order to produce these effects,

Questions to the booklet's author, showing the futility of the questions he asked.

or do they have nothing to do with what happens on Earth? If they have nothing to do with the Earth, it is a great folly for us earthlings to want to be judges of their size or arbiters of their location, since we are entirely ignorant of their actions or purposes. If on the other hand he says that they do have effects on Earth and are directed towards that goal, then he is affirming what he denies elsewhere and celebrating what he condemned just now, when he said that the heavenly bodies are so far distant from the Earth that they appear insignificant and are unable to produce any effect on it. But my good man, the stellar sphere, established at its actual distance which you consider appropriate for its influences to be felt in earthly affairs, has a huge number of stars that appear to be insignificant, and a hundred times more that are entirely invisible to us—that is, whose appearance is even less than insignificant. Therefore you must either contradict yourself by denying that they have any effect on Earth, or concede that their insignificant size does not detract from their effectiveness, again contradicting what you said earlier; or else you must concede and acknowledge—and this would be the most genuine and modest concession—that our judgement of the size and distance of the stars is meaningless, not to say presumptuous and arrogant.

SIMPLICIO. To tell the truth, when I read this passage I immediately saw this evident contradiction, where he says that the stars according to Copernicus appear so small that they could not act upon Earth, not realizing he had already conceded that there is an effect upon the Earth from the stars of Ptolemy, which are also his own, and which appear not just very small but are for the most part invisible.

Demonstration that distant objects appear small because of a defect in our eyesight.

SALVIATI. Consider another point. On what basis do we say that the stars appear to be so small? Surely it's because that is how we see them, which depends on the instrument we use to look at them, namely our eyes. The truth of this is clear because by changing the instrument we can enlarge them at will. Who [399] knows, perhaps to the Earth, which looks at them without eyes, they may appear much larger, and as they really are?

But it's time for us to move on from these trifles and come to more weighty matters. I have demonstrated two things. The first is the distance at which the firmament should be placed in order that the diameter of the Earth's orbit should make no more difference to its appearance than that which the diameter of the terrestrial globe makes to the appearance of the Sun. Then, I have shown that for a star in the firmament to appear to us with the size that it does, there is no need to suppose that it is larger than the Sun. I would now like to know whether Tycho or any of his followers has ever tried to establish whether there is any discernible appearance in the stellar sphere that would allow us more definitively to either assert or deny the annual motion of the Earth.

SAGREDO. If I were to reply on their behalf I would say no, they haven't, and they don't need to because Copernicus himself said that no such difference existed. Then arguing *ad hominem*,* they accept this, and on this assumption show the improbability of the consequence that follows from it, namely that the sphere would have to be so immense that for a fixed star to appear to us with the size that it does, it would have to be so enormous in bulk that it would be larger than the Earth's orbit—something which, as they say, is completely incredible.

Tycho and his followers have not attempted to see whether there is any appearance in the firmament against or in favour of the annual motion.

SALVIATI. I agree, and I think they argue against him *ad hominem* more in defence of another man than out of any desire to come to a knowledge of the truth. I don't believe that any of them has made any effort to undertake such an observation. What's more, I'm not sure that any of them know what differences should appear in the fixed stars because of the motion of the Earth, if the stellar sphere were not so distant that any such difference would be too small to be seen. Giving up any such investigation and relying simply on what Copernicus says may well be enough to refute the man, but it doesn't help to clarify the fact. It may be that such a difference exists but Copernicus did not look for it, or that he failed to perceive it because it was so small or he did not have accurate instruments.* This wouldn't be the first thing that he did not know, because of his lack of instruments or of some other shortcoming; and yet, basing himself on other well-founded conjectures, he affirmed [400] what appeared to be contradicted by the things that he could not comprehend. So, as we have said, without a telescope it is not possible to know that Mars is sixty times larger, or Venus forty times larger, in one position than in another; in fact their

Astronomers have perhaps not realized what appearances would follow from the annual motion of the Earth.

Some things that were unknown to Copernicus because of a lack of instruments.

differences appeared to be much smaller than they really are. Yet it has since been established beyond doubt that these changes are exactly what the Copernican system required.

It would be a very good thing to investigate, with the greatest degree of accuracy possible, whether the change that should be discernible in the fixed stars given the annual motion of the Earth can in fact be observed. I am quite certain that no one has done this hitherto, and indeed, as I've said, I doubt whether many even understand what they should be looking for. And I don't say this by chance: I have seen an unpublished text by one of these anti-Copernicans* arguing that, if Copernicus's opinion is correct, it must necessarily follow that the elevation of the pole would be continuously rising and falling every six months, as the Earth moves in this time alternately towards the north and towards the south over a distance equal to the diameter of its orbit. So he considered it reasonable, indeed necessary, that as we move with the Earth the elevation of the pole should be greater when we are in the northern phase than in the southern one. Another mathematician*—a very intelligent one, and what's more, a follower of Copernicus—fell into the same error when he said he had observed changes in the polar elevation and that it varied between summer and winter. Tycho reports this in his *Progymnasmata*, page 684, and denies the validity of the claim but does not question the procedure; in other words he denies that there is any observable variation in the polar elevation, but he does not dismiss the enquiry as inappropriate for establishing the point at issue. This amounts to an acknowledgement that he too considers the variation or non-variation of the polar elevation in a six-monthly cycle as a valid proof for excluding or allowing the annual motion of the Earth.

Tycho and others argue against the annual motion because of the unchanging elevation of the pole.

SIMPLICIO. Indeed, Salviati, this seems logical to me as well. I don't think you will deny that if we travel just sixty miles towards the north the polar elevation will be a degree higher, and then if we continue another sixty miles northwards it will increase by another degree again, and so on. Now if moving just sixty miles towards or away from the pole can make such a perceptible difference in its elevation, what must be the effect of the Earth moving not sixty but sixty thousand miles northwards, carrying us with it?

SALVIATI. If we are to observe the same proportions it [401] would mean the pole's elevation would increase by a thousand degrees. So you see, Simplicio, how powerful a deep-seated impression can be. You have had it so firmly fixed in your mind for so many years that it is the heavens that revolve in twenty-four hours and not the Earth, and therefore that the poles of this revolution are in the heavens and not in the terrestrial globe, that you can't for a moment set aside this habit and presume the opposite. Imagine that it is the Earth that is in motion, just for long enough to grasp what would follow if this deception were true. If it is the Earth that rotates around itself every twenty-four hours, then the poles, the axis, the equator—that is, the great circle described by the point equidistant from the poles—are all in the Earth, as are the infinite other parallels, of greater or lesser degrees, described by the points on its surface more or less distant from the poles. These are all in the Earth and not in the stellar sphere, which being immobile has none of them. We only locate them there in our imagination, extending the Earth's axis to terminate at two points above our poles, and extending the plane of the equator so that it appears to be a circle corresponding to it in the sky. Now if the true axis, the true poles, and the true terrestrial equator do not change as long as you remain at the same place on the Earth, then the Earth can move wherever you like and your position relative to the poles, the circles, and anything else on the Earth will still not change. This is because any such motion would be common to you and everything else on Earth, and where motion is in common it is as if it did not exist. So, just as your position relative to the terrestrial poles will not change (relative, that is, to their greater or lesser elevation), so also it will not change relative to the imaginary poles in the sky, as long as we understand the celestial poles to be, as we have defined them already, those two points which would be marked by the terrestrial axis if it were extended to the stellar sphere. It is true that these points in the heavens can change, when the Earth moves in such a way that its axis points to other parts of the immobile celestial sphere. But our position relative to them, which would make one more elevated than the other, still does not change. If you want the points in the firmament

Motion in common is as if it did not exist.

corresponding to the terrestrial poles to change so that one is higher and the other lower, you will have to travel on the Earth's surface towards one and away from the other. Moving the Earth and us along with it, as I have said, has no effect whatsoever.

An example to show that the elevation of the pole should not change because of the annual motion of the Earth.

SAGREDO. Salviati, be so good as to let me clarify this point [402] with an illustration which may be crude but is, I think, apt for this purpose. So imagine, Simplicio, that you are on a ship, and that you stand at the stern and point a quadrant or other astronomical instrument at the top of the foremast, to take its height. Suppose that this is, say, 40 degrees. It's clear that if you walk along the deck 25 or 30 paces towards the mast and then point the same instrument at the top of the mast again, you will find that its elevation has increased by, say, 10 degrees. But if instead of walking those 25 or 30 paces towards the mast you had remained where you were in the stern, but made the whole ship move in the same direction, do you think that the elevation of the mast would increase by 10 degrees because of the 25 or 30 paces that the ship had travelled?

SIMPLICIO. I understand that it wouldn't increase even by an inch if the ship travelled a thousand or a hundred thousand miles, never mind 30 paces. But I do believe that, if you looked along the sights of the quadrant towards the top of the mast and saw a fixed star aligned with it, and the ship sailed sixty miles in the same direction, then keeping the quadrant fixed it would still point towards the top of the mast as before, but it would no longer be pointing at the star which would be one degree higher.

SAGREDO. But you don't doubt, do you, that the sights on the quadrant would be directed at the point in the stellar sphere corresponding to the direction of the top of the mast?

SIMPLICIO. I don't doubt that, no, but that point would have changed, and it would be lower than the star that was originally observed.

SAGREDO. Exactly so. But in this illustration the elevation of the top of the mast corresponds not to the star but to the point in the firmament that is aligned with the eye and the top of the mast. In the same way, in the case we are examining the point in the firmament corresponding to the pole of the Earth is not a star or some other fixed point, but that point in which the Earth's axis would terminate if it were extended that far.

This point is not fixed, but follows the changes made by the Earth's pole. Therefore Tycho or others who raised this objection should have said that if the Earth's motion was real it should be [403] observable in changes in the elevation or depression of some fixed stars near the place corresponding to our pole, not of the pole itself.

The annual motion of the Earth can lead to some changes in the fixed stars, but not in the pole.

SIMPLICIO. I fully understand this misunderstanding on their part. But I don't see that this diminishes the force of the objection, which I find very strong, if it refers to changes in the fixed stars rather than of the pole. If the ship moving just sixty miles raises the elevation of a fixed star by one degree, surely there would be a similar change, and even a greater one, if the ship were to move in the direction of the same star by a distance equivalent to the diameter of the Earth's orbit, which you say is twice the distance from the Earth to the Sun?

SAGREDO. There's another misunderstanding here, Simplicio, which you understand although you don't realize it; so I'll try to remind you of it. Tell me, therefore, if you direct the quadrant at a fixed star and find its elevation to be, say, forty degrees, and then without moving yourself you tilted the side of the quadrant so that it appeared to be higher, would you say that the star's elevation had thereby increased?

Resolving the misunderstanding of those who think that the annual motion should cause large variations in the elevation of a fixed star.

SIMPLICIO. Of course not, because the change would be in the instrument rather than in the observer moving towards the star.

SAGREDO. And if you sail or walk on the surface of the Earth, would you say that, provided you didn't tilt the quadrant but kept it at its original angle, there was no change in its elevation in relation to the heavens?

SIMPLICIO. Let me think about that for a moment. No, I wouldn't say that it remained unchanged, because my journey is not on the level but on the circumference of the terrestrial globe, which changes its inclination towards the heavens with every step that I take. Therefore the inclination of the instrument, which I am keeping constant, must also change.

SAGREDO. You're quite right. And you will also understand that the larger the circle on whose circumference you were moving, the more miles you would have to travel for the elevation of the star to rise by a single degree; and if your

[404] movement towards the star was in a straight line, you would have to travel even further than on the circumference of any circle, however large.

A straight line and the circumference of an infinite circle are the same.

SALVIATI. Yes, because in the end the circumference of an infinite circle and a straight line are the same thing.

SAGREDO. Now that I don't understand, and I don't suppose Simplicio understands it either. It must imply some hidden mystery, because we know that Salviati never speaks randomly or puts forward some paradox unless it leads to some novel idea. So I'll remind you of this at the right time and place, when you can explain how a straight line can be the same as the circumference of an infinite circle, but I don't want us to interrupt the discussion that we have in hand. So, to come back to the point, I invite Simplicio to consider the motion of the Earth towards and away from some fixed star near to the pole as if it is in a straight line, this being the diameter of the Earth's orbit. So it shows a great lack of understanding to try to compare the rising and falling of the pole star due to the motion along this diameter with that due to motion in the tiny circle of the Earth.

SIMPLICIO. This still leaves us with the same difficulty, since there is no evidence even of the small variation that ought to be there. If no such variation exists, then we must admit that the annual motion attributed to the Earth around its orbit also does not exist.

SAGREDO. I won't hold Salviati up any longer. I don't think it was for nothing that he discounted the rising and falling of the pole star or any other fixed star as insignificant, even though no one understands it, since Copernicus himself considered it to be either non-existent or impossible to observe because it is so small.

How to establish what changes, affecting which stars, we should observe as a result of the Earth's annual motion.

SALVIATI. I said earlier that I don't think anyone has undertaken to observe whether there is any change in the fixed stars at different times of year that could be the consequence of the Earth's annual motion. I added, too, that I doubt whether anyone has fully understood what these changes might be and among which stars they should appear; so it is desirable that we should investigate this point carefully. I have seen it argued, but only in general terms, that the annual motion of the Earth should not be accepted because it is implausible that it should produce no visible changes in the fixed stars. I have not heard anyone say specifically what these changes should be, or in which stars we should observe them. This gives me strong [405] grounds for suspecting that those who confine themselves to this generic pronouncement have not understood, and perhaps have not even tried to understand, how these changes might come about or indeed what it is they say we should see. I say this because I know that if the annual motion attributed to the Earth by Copernicus is to be perceived in the stellar sphere, any change will not affect all the stars equally, but will be greater in some, less in others, and in others again completely non-existent, however great the circle of this annual motion is assumed to be. The changes that should be observable, then, are of two kinds: one is in the stars' apparent size, and the other is in their varying elevations at the meridian. Other changes, such as their places of rising and setting, their distance from the zenith, etc., follow as consequences of this.

The fact that astronomers have not specified what changes might derive from the Earth's annual motion shows that they have not fully understood it.

Changes in the fixed stars should be greater in some, less in others, and in others again non-existent.

SAGREDO. I can see these complications adding up to a tangled knot that I don't know if I'll ever manage to unravel. I've thought about it many times and have never been able to find the thread. It's not so much this question of what happens among the fixed stars, but another even more intractable problem that you've reminded me of by mentioning meridian altitudes, latitudes of rising,* distances from the zenith, and so on. I'll tell you now what it is that produces this perplexity in my mind. Copernicus assumes the stellar sphere to be at rest, with the Sun at its centre, also at rest. Therefore any change that we see either in the Sun or in the fixed stars must belong to the Earth, in other words to us. Yet the rising and setting of the Sun in our meridian spans an enormous arc of almost 47 degrees, with even greater variations along the oblique horizons. How then can the Earth's inclination in relation to the Sun vary so strikingly, and yet not at all in relation to the fixed stars, or so little as to be imperceptible? This is the Gordian knot that has defied all my attempts to unravel it, and if you can cut it for me I will consider you more than an Alexander.*

The greatest objection to Copernicus from what is seen in the Sun and the fixed stars.

SALVIATI. These difficulties do credit to Sagredo's ingenuity. The objection is such that even Copernicus himself almost gave up trying to explain it in an intelligible way. This is [406] apparent both because he himself admitted finding it obscure, and because he had to set out to explain it twice in two different ways. I freely confess that I did not understand his explanation until I had made it intelligible by a different method, which is quite plain and clear, but only after I had thought long and hard about it.

Aristotle's argument against the ancients who claimed that the Earth was a planet.

SIMPLICIO. Aristotle saw the same difficulty, and used it as an argument against some ancient thinkers who claimed that the Earth was a planet. He argued that in that case the Earth would have to have more than one motion, in the same way as the other planets, to produce these variations in the rising and setting of the fixed stars as well as in their meridian altitudes. Since he raised this difficulty and did not resolve it, it must necessarily be, if not impossible, at least difficult to unravel.

SALVIATI. The complexity of the knot makes it all the more pleasing and impressive when it is unravelled; but I can't promise to show you this today, and I beg your indulgence until tomorrow. For the moment, let us proceed with our examination of the changes and variations that should be observable in the fixed stars as a result of the annual motion, as we were saying just now. Some points will arise in the course of our explanation that will prepare the ground for resolving the major difficulty.

The Earth's annual motion along the ecliptic and its diurnal motion around its own centre.

To recapitulate, then, two motions are attributed to the Earth (I say two because the third is not a motion at all, as I shall explain in due course), the annual and the diurnal. The first is to be understood as the motion of the centre of the Earth around the circumference of its orbit, that is, of a great circle described in the plane of the ecliptic, which is fixed and unchanging. The second motion, the diurnal, is made by the terrestrial globe around its own centre and axis, which is not vertical to the plane of the ecliptic but inclined to it at an angle of about 23 and a half degrees. This inclination remains constant throughout the year, and—this is the most important point—always towards the same part of the heavens, so that the axis of diurnal motion is permanently at the same angle. So, if we imagine this axis extended as far as the fixed stars, then as the centre of the Earth completes a revolution around the ecliptic in the course

The inclination of the Earth's axis remains constant; it describes a cylindrical surface inclined to its orbit.

of a year, the axis will describe the surface of an oblique cylinder, one base of which is this annual circle, and the other a similar imaginary circle described by its extremity or pole among the fixed stars. This cylinder is oblique to the plane of the ecliptic according to the inclination of the axis that describes it, [407] which as we have said is 23 and a half degrees. This angle is permanently the same, apart from a very small change in thousands of years which is irrelevant to our present purpose; so the inclination of the terrestrial globe neither increases nor decreases but remains constant. It follows that any change that might be observed in the fixed stars solely as a result of the annual motion would look the same at any point on the Earth's surface as it would at its centre; so for the purpose of this explanation we shall make use of the centre as if it were any point on the surface.

The terrestrial globe does not vary in its inclination but remains constantly the same.

A diagram will make it easier to understand all this. First we draw this circle, ANBO, in the plane of the ecliptic, taking A and B as its northern and southern extremities, in other words the beginning of Cancer and Capricorn respectively. We extend the diameter AB without limit through D and C towards the fixed stars. Now the first thing to note is that none of the fixed stars placed in the plane of the ecliptic will vary in elevation because of any change in the Earth's position on that plane, but will always appear in the same plane. They will, however, move closer or further away from the Earth, by a distance equal to the diameter of the Earth's orbit. The diagram makes this clear: star C will always appear along the same line ABC, regardless of whether the Earth is at point A or point B, although the distance BC is less than CA by the length of the diameter BA. Therefore the most that can be

The fixed stars placed in the plane of the ecliptic never rise or fall because of the Earth's annual motion, but they do move closer and further away.

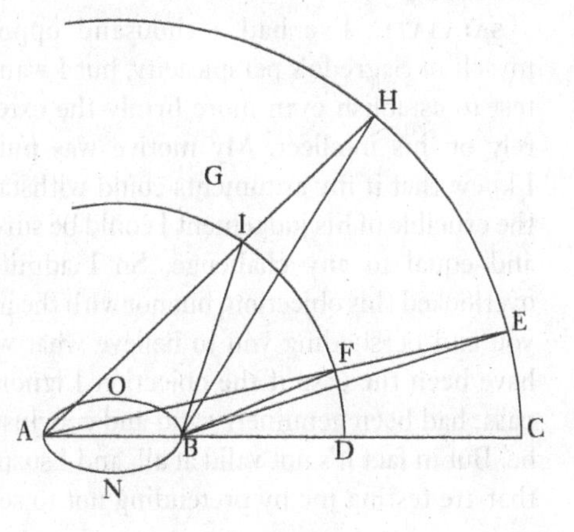

observed in star C or in any other star placed in the ecliptic will be an increase or decrease in its size as it is closer or further away from the Earth.

Objection to the Earth's annual motion, based on the fixed stars placed in the ecliptic.

SAGREDO. Stop there a moment, if you will, because I have a doubt that is troubling me. It's this: I understand perfectly well that star C is seen along the same line ABC regardless of whether the Earth is at point A or point B; and I also understand that the same would be true of any point along the line AB, assuming that the Earth moved from A to B along this line. But since we suppose it to move along the arc ANB, it's clear that [408] when the Earth is at point N, or at any other point apart from A and B, the star is seen not along the line AB but along one of many other lines; and if being seen along different lines produces differences in its appearance, then there must be some alteration in how it is seen. In fact, I will take advantage of the liberty to speak freely that should exist among friends discussing philosophy together, and say that I think you are contradicting yourself and denying what you told us only today was a great truth. I mean what you told us, to our surprise, about the planets, in particular the three outer planets: that as they are located continuously in the ecliptic or very close to it, not only do we see them now close to us and now at a great distance, but they are so different in their regular motions that we see them sometimes stationary and sometimes many degrees retrograde, and all for no other reason than because of the annual motion of the Earth.

SALVIATI. I've had a thousand opportunities to assure myself of Sagredo's perspicacity, but I wanted to use this latest test to establish even more firmly the extent to which I could rely on his intellect. My motive was purely selfish, because I knew that if my arguments could withstand the hammer and the crucible of his judgement I could be sure that they are sound and equal to any challenge. So I admit that I deliberately overlooked this objection, but not with the intention of deceiving you and persuading you to believe what was false; that would have been the case if the objection I ignored, and that you let pass, had been genuinely valid and conclusive, as it appeared to be. But in fact it's not valid at all, and I suspect that it's now you that are testing me by pretending not to see its ineffectiveness.

So I'm going to outwit you and force you to say what you have craftily tried to conceal. Tell me then, on what basis do you know the stations and retrogradations of the planets resulting from the annual motion, and how do you know that these are large enough for at least some trace of them to be visible in the stars in the ecliptic?

SAGREDO. There are two parts to your question that I must reply to. The first is your accusation that I was speaking deceptively, and the second concerns the appearances in the stars, etc. As to the first point, allow me to say that it's quite untrue that I was pretending not to see the ineffectiveness of this objection. To prove it, I can say that I now understand its ineffectiveness very well.

SALVIATI. I don't see how you were not speaking deceptively when you said you didn't understand its invalidity, if you now admit that you understand it very well. [409]

SAGREDO. The fact that I admit understanding it now should assure you that I wasn't pretending when I said I didn't understand it before. If I'd wanted to deceive you and still wanted to do so, what would stop me from continuing the same pretence now and still insisting that I couldn't see why it was invalid? So I repeat that I didn't understand it then, but it is now clear to me, thanks to your awakening my intellect by so firmly declaring that the objection has no substance, and by asking me more generally about how I knew of the stations and retrogradations of the planets.

The stations, direct and retrograde motions of the planets are known in relation to the fixed stars.

We understand these by comparing the planets with the fixed stars, and we perceive their motions towards the west or towards the east or their remaining as if at rest, all in relation to the fixed stars. But beyond the fixed stars, there is no even more distant sphere visible to us with which we could compare the stars that we see, and therefore there is no apparent change in the fixed stars comparable to what is apparent to us in the planets. I think this is what you wanted to force me to say.

An indication in the fixed stars, similar to what is seen in the planets, in support of the annual motion of the Earth.

SALVIATI. Yes indeed, with the extra benefit of your penetrating sharp wit. If a verbal provocation on my part opened your mind, for your part you remind me it's not entirely impossible that at some point we might discover an observable phenomenon in the fixed stars that would allow us to see where the annual revolution resides, and then the fixed stars would join the planets and the Sun itself in testifying that this motion belongs to the Earth. I don't believe that the stars are distributed on the surface of a sphere, all equidistant from the centre; I think their distance from us varies so much that some of them can be two or three times more remote than others. So if the telescope revealed a very small star in close proximity to one of the larger stars, but at a very great distance away, there could be some variation in the relation between the two, comparable to the variations we see in the superior planets.

Fixed stars that are not in the ecliptic rise and fall to a greater or lesser degree, depending on their distance from the ecliptic.

Let this suffice for now on the particular case of the stars in the ecliptic, and move on to the fixed stars that are away from the ecliptic. Suppose we have a great circle that is vertical to the [410] plane of the ecliptic, as for example a circle in the stellar sphere corresponding to the solstitial colure.* We mark this as CEH, which will also be a meridian, and take a star that is outside the ecliptic, marked here at E. Its elevation will vary with the motion of the Earth, because when the Earth is at point A it will be seen along the ray AE and its elevation will be the angle EAC, whereas when the Earth is at point B it will be seen along the ray BE with the elevation of the angle EBC. This is greater than

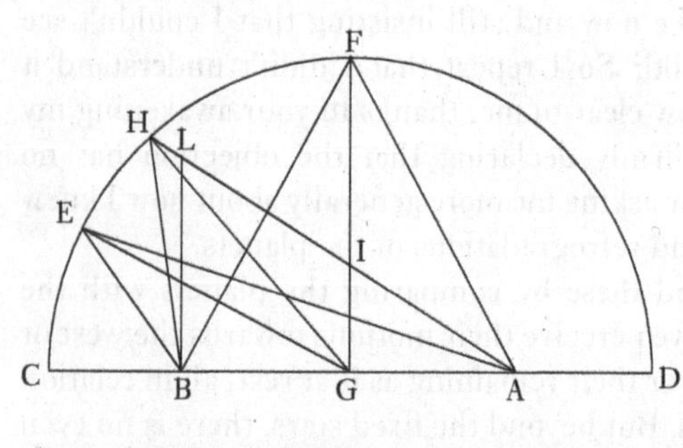

the angle EAC, as EBC is external to the triangle EAB and EAC is internal and opposite to it. So the distance of the star E from the ecliptic will be seen to have changed, and its elevation in the meridian when the Earth is at point B compared to when it is at point A will also have increased, by the amount that the angle EBC is greater than the angle AEC, i.e. by the size of the angle AEB. This is because if the side AB of the triangle EAB is extended to C, the external angle EBC will be equal to the two internal and opposite angles E and A, and therefore it will exceed A by the size of the angle E. And if we take another star in the same meridian that is further away from the ecliptic, as for example star H, the difference in its appearance when seen from points A and B will also be greater, by the amount that the angle AHB exceeds the angle E. Moreover, this angle will continue to increase as the star observed is further distant from the ecliptic, until the apparent difference would be at its greatest in a star placed at the pole of the ecliptic. This can be demonstrated fully as follows.

Let AB be the diameter of the Earth's orbit with G at its centre, and let AB be extended to the stellar sphere at points D and C. From the centre G let the axis of the ecliptic GF also be extended to the stellar sphere, and suppose that a meridian DFC, vertical to the plane of the ecliptic, is described in this sphere. Take points H and E as the location of fixed stars anywhere on the arc FC, and join the lines FA, FB, AH, HG, HB, AE, GE, and BE, so that the angle of difference, or parallax, of the star located at the pole F is AFB, that of the star located at H is the angle AHB, and that of the star located at E is the angle AEB. Now the greatest angle of difference is that of the pole star F, and the others are greater the nearer they are to the pole; so angle F is greater than angle H, which in turn is greater than angle E.

[411] Now suppose that a circle is described about the triangle FAB. The angle F is acute, because its base AB is less than the diameter DC of the semicircle DFC, so it will be located in the larger part of the circumscribed circle intersected by the base AB; and since AB is divided in half and at right angles by the line FG, the centre of the circumscribed circle will be at a point on the line FG. Let us call this point I. Now the longest of the lines drawn from point G, which is not at its centre, to points on the circumference of this circle, will be the one passing through its centre; so GF will be longer than any other line drawn from point G to this circle's circumference. This circumference will intersect the line GH, which is equal to GF, and it will also intersect the line AH. Let the point of intersection be L, and draw the line LB. Now the two angles AFB and ALB will be equal, because they are in the same part of the circumscribed circle; but ALB, being external, is

greater than the internal angle H; therefore the angle F is greater than the angle H.

The same procedure will show that angle H is greater than angle E. Given that the circle described about the triangle AHB has its centre on the perpendicular line GF, and the line GH is closer to this perpendicular than the line GE, the circumference of this circle will intersect both GE and AE; hence the truth of the proposition is clear. Therefore we can conclude that the apparent difference—or to use the technical term, the parallax—of the fixed stars is greater or less depending on whether they are closer to or further away from the pole of the ecliptic; and that for the stars that are in the ecliptic itself this difference disappears.

The distance of the Earth from the fixed stars on the ecliptic varies as the diameter of its orbit.

As for the Earth's moving towards or away from the stars as a result of this motion, it approaches or retreats from the stars in the ecliptic, as we have seen, by a distance equal to the entire diameter of the Earth's orbit. For the stars around the pole of the ecliptic this motion towards or away from them is virtually non-existent, and for the other stars the difference is greater the closer they are to the ecliptic.

There is a greater difference in the stars that are closer to the Earth than in those that are further away.

In the third place, we can see how this difference in appearance is greater or less depending on whether the star in question is closer to us or further away. If we draw another meridian less distant from the Earth, as for example DFI here, then a star located at F when the Earth is at A will be seen along the ray AFE. When however it is observed from the Earth at B it is seen along the ray BF, and the angle of difference BFA, being external to the triangle BFE, will be greater than the first angle AEB.

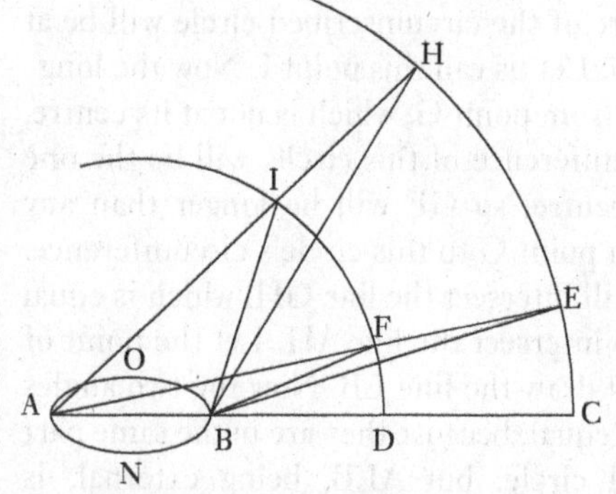

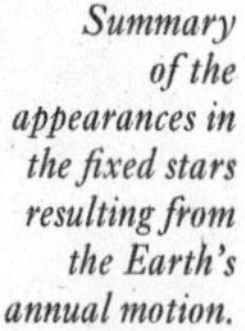
Summary of the appearances in the fixed stars resulting from the Earth's annual motion.

SAGREDO. I have listened to your exposition with great enjoyment and also with profit, and to make sure I have [412] understood it correctly I will try briefly to summarize your conclusions. You have, I think, explained that we can observe two kinds of variations in the appearance of the fixed stars resulting from the Earth's annual motion. One is that their apparent size varies according as we on Earth approach or retreat from them; the other—which also depends on this same motion towards or away from them—is that they appear with greater or lesser degrees of elevation on the same meridian. You also tell us, and I understand this very well, that these variations do not appear equally in all the stars but to a greater extent in some, less in others, and in still others not at all. The motion towards or away from the Earth by which the same star ought to appear larger at one time and smaller at another is imperceptible to us, and practically non-existent, for the stars that are close to the pole of the ecliptic, but is at its greatest in the stars located in the ecliptic, and moderate in those that are between the two. The opposite is true of the other variation; that is, there is no apparent change in the elevation of the stars located in the ecliptic, but the change is at its greatest in those close to the pole of the ecliptic, and moderate in those in between. Moreover, both these variations are more apparent in the stars that are closer to us, less so in those that are further away, and finally they disappear altogether in those that are extremely remote. I say this much speaking for myself; it remains now to convince Simplicio, who I don't think will be easily persuaded that such variations can be imperceptible when they derive from movements of the Earth so vast that they transport it between locations twice as far distant from each other as our distance from the Sun.

SIMPLICIO. I must say frankly that I am very reluctant to concede that the distance of the fixed stars is so great that the variations you say occur in them should be completely imperceptible.

SALVIATI. Don't give in entirely to despair, Simplicio; perhaps there is a remedy for your difficulty. First of all, you [413] shouldn't find it improbable that we see no perceptible change in the appearance of the stars, since men's estimates in such matters are clearly highly susceptible to error, especially in

A small movement towards or away from a very remote shining object is imperceptible.

looking at brightly shining objects. If you yourself were looking at a burning torch at a distance of, say, 200 paces, and you were to move three or four *braccia* towards it, do you think that you would notice that it appeared to be larger? I'm sure I would not, even if I came twenty or thirty *braccia* closer. In fact, I have sometimes had the experience of seeing such a light at such a distance and being unable to decide whether it was approaching or moving away from me, when in fact it was coming closer. But in any case, if the approach and retreat of Saturn, which varies by twice the distance from us to the Sun, is imperceptible, and that of Jupiter is barely visible, how do you think it will be with the fixed stars, which I think you would not hesitate to place twice as far away from us as Saturn? As for Mars, when it comes closer to us...

SIMPLICIO. There's no need to labour this point any further; I understand that it can easily be as you say as regards the apparent size of the fixed stars. But what of the other objection, that we see no variation in their aspect?

SALVIATI. I may be able to put your mind at rest on this point as well. To come straight to the point: would you be satisfied if it were possible to see those changes in the stars that you consider necessary if the annual motion belongs to the Earth?

SIMPLICIO. On this particular point, yes, certainly.

If some annual motion were observed in the fixed stars, there could be no contradicting the motion of the Earth.

SALVIATI. I would like you to say that if such changes could be observed there would no longer be any reason to doubt the mobility of the Earth, given that there could be no other explanation for such appearances. But even if they could not be observed, this would not rule out the Earth's mobility or make it necessary to conclude that it is at rest. It may be that, as Copernicus affirms, the immense distance of the stellar sphere makes such minimal changes imperceptible. They may not even have been investigated hitherto, or if there has been any [414] investigation it has not been carried out with the accuracy and attention to detail that it requires. Such accuracy is difficult to achieve, both because of the imperfections of our astronomical instruments which are subject to much variation, and through the shortcomings of those who do not use them with the appropriate care. For a conclusive proof of the unreliability of such observations, we need look no further than the differences between astronomers in the locations they give for the fixed stars themselves, even for the altitudes of the pole which in most cases differ from one another by many minutes, to say nothing of the location of new stars and comets. The truth is that no one using a quadrant or sextant, which at most will have sides only three or four *braccia* in length, can be sure of not erring in setting a perpendicular or aligning an alidade* by two or three minutes, which on such a circumference will be no more than the thickness of a grain of millet. Besides which, it is almost impossible that the instrument will have been constructed and maintained with absolute accuracy. Ptolemy was reluctant to trust an armillary sphere* constructed by Archimedes himself to measure the Sun's entry into the equinox.

Proof of the unreliability of astronomical instruments in detailed observations.

Ptolemy did not trust an instrument made by Archimedes.

SIMPLICIO. If the instruments are so unreliable and observations are so uncertain, how will we ever be able to know anything for certain and free ourselves from error? I have heard great things spoken about Tycho's instruments, that they were made at enormous expense and that he is exceptionally thorough in his observations.

Tycho's instruments were made at great expense.

SALVIATI. I grant you all this, but neither Tycho's instruments nor his careful observations will suffice to give us certainty in such an important question as this. I would like us to use instruments that are a great deal larger than Tycho's, are highly accurate and can be made with very little expense. Their sides are four, six, twenty, thirty, and fifty miles, so that the width of one degree is a mile, a minute is fifty *braccia*, and a second a little less than one *braccio*. What's more, we can have these instruments as large as we please and it won't cost us a penny. When I was at a villa of mine near Florence I was able clearly to observe the Sun's arrival at summer solstice and its subsequent departure. As it was setting one evening it sank behind a cliff in the mountains of Pietrapana, about sixty miles away, leaving just a thin edge visible towards the north, its width less than a hundredth of the Sun's diameter. The following evening when it set in the same place a similar part of its surface was visible, but noticeably thinner than the night before—clear evidence that the Sun had begun to move away from the tropic, even though the

What instruments are suitable for very precise observations.

Exact observation of the Sun's entrance into and departure from the summer solstice.

A suitable place for observing the fixed stars as they are affected by the Earth's annual motion.

[415] difference in its position between the first and second observations cannot have been even a second on the horizon. Observing it later with an exquisite telescope that magnifies the Sun's disc more than a thousand times was both easy and delightful.

I would like us to use such an instrument to observe the fixed stars, taking one of those whose changes should be most noticeable—that is, as we've seen, those that are furthest from the ecliptic. One of these, Vega, a very bright star close to the pole of the ecliptic, would be very suitable to observe in northern latitudes, following the method that I shall describe, although with a different star. I have already located a suitable place for my own observations: it is an open plain with a very high mountain rising above it to the north, with a small chapel built on its summit. The chapel is oriented from west to east, so that the ridge of its roof can intersect at right angles with the meridian of a house situated in the plain. I intend to fix a beam parallel to this ridge, about a *braccio* above it. Then I shall find a place on the plain from which one of the stars in the Plough disappears behind this beam as it passes through the meridian, or, if the beam is not wide enough to hide the star, a place from which the star is bisected by it; this effect can be seen very accurately with an accurate telescope. If the place is marked by a house, so much the better, but if not I will plant a stake firmly in the ground with a sign marking the point from which the viewer should look again whenever the observation is to be repeated. The first such observation should be about the summer solstice, then continued month by month, or at a time of my choosing, until the winter solstice. This observation will make it possible to measure the variation in the star's elevation, even if it is very small. Imagine the gain it would represent for astronomy if this procedure succeeded in establishing some variation: not only would it provide confirmation of the annual motion, but it would also allow us to discover the size and distance of the star itself.

SAGREDO. The whole procedure is perfectly clear to me. It [416] would be reasonable to think that Copernicus himself or some other astronomer had already put it into practice, since it seems so easy and apt for the purpose.

SALVIATI. On the contrary, if anyone had put it into practice it's hard to believe that they would not have published the result if it favoured one side of the argument or the other. Besides, there is no record of anyone having used this method of observation either for this purpose or for anything else; and in any case it would be difficult to carry it out effectively without an accurate telescope.

SAGREDO. What you say resolves my doubts on this point. But while we still have plenty of time before nightfall, be so kind as to resolve for us those problems that you asked just now to put off until tomorrow; otherwise I will not be able to rest tonight. So please don't insist on the concession we gave you, but put all other arguments aside and explain these points to us. Given the motions that Copernicus attributes to the Earth and assuming the Sun and the fixed stars to be at rest, how can it follow that the variations in the elevation of the Sun, the changes in the seasons, the disparities in the length of day and night, and so on, all occur in exactly the way that is so easily explained in the Ptolemaic system?

The Copernican system is hard to understand but straightforward in its application.

SALVIATI. I can't and I mustn't refuse a request from Sagredo. My only reason for asking to defer this until tomorrow was to give myself time to recollect the principles underlying a clear, comprehensive account of the phenomena you have mentioned, and how they follow as well in the Copernican system as in the Ptolemaic one. In fact, they follow much more naturally and simply in the former system than in the latter; and it will become clear that the Copernican hypothesis is as straightforward in its application in nature as it is hard to comprehend in theory. Nonetheless I hope that by explaining it in different terms from those used by Copernicus I can make it much less difficult to grasp. To do so I put forward the following propositions that are well known and understood.

Propositions necessary for a right understanding of the consequences of the Earth's motion.

First. Assuming that the Earth, a spherical body, rotates around its own axis and poles, any point marked on its surface describes the circumference of a circle, which is larger or smaller depending on its distance from the poles. The largest such circle is that described by a point equidistant from the poles. All these circles are parallel to each other; we shall refer to them as *parallels*.

[417] Second. The Earth, being a sphere made of an opaque material, is continuously illuminated by the Sun on half its surface, while the other half remains in darkness. The line separating the part that is illuminated from the part that is in darkness is a great circle, which we shall call *the circular boundary of the light*.

Third. If the circular boundary of the light passed through the Earth's poles, being a great circle, it would cut all the parallels into equal parts. As it does not pass through the poles it will cut them all into unequal parts, with the sole exception of the middle circle which, being itself a great circle, is cut into equal parts.

Fourth. As the Earth rotates around its poles, the length of the days and nights is determined by the arc of the parallels intersected by the circular boundary of the light. The arc falling in the illuminated hemisphere defines the length of the day, and the remainder defines the length of the night.

A very simple figure illustrating the Copernican model and its consequences.

Given these propositions, let us draw a figure to show more clearly the points that remain to be explained. First we shall draw the circumference of a circle to represent the Earth's orbit, described in the plane of the ecliptic. Then we shall divide this circle into four equal parts by drawing two diameters, Capricorn–Cancer and Libra–Aries, which will also represent the four cardinal points, i.e. the two solstices and the two equinoxes. We mark the Sun, O, fixed and immobile at the centre. Now we draw four equal circles, each with one of the four cardinal points Capricorn, Cancer, Libra, and Aries at its centre. These represent the Earth as it is placed at different times, as in the space of a year it moves around the circumference Capricorn–Aries–Cancer–Libra in the

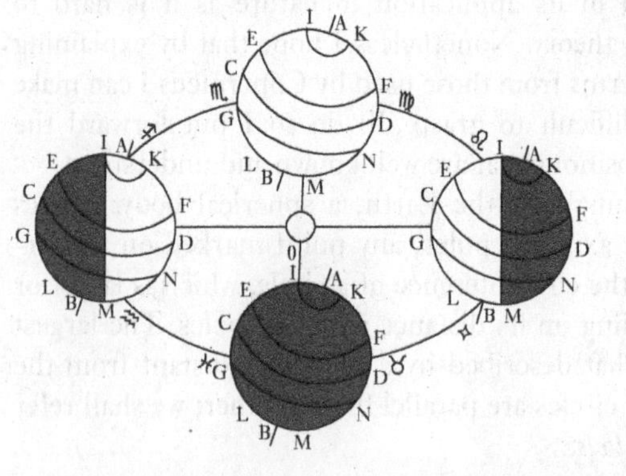

The annual motion of the Sun as it follows from the Copernican model.

order of the zodiac signs, i.e. from west to east. It is clear from this that when the Earth is in Capricorn the Sun will appear in Cancer, and as the Earth moves through the arc Capricorn–Aries the Sun will appear to move through the arc Cancer–Libra; in other words, it will appear to move through the signs of the zodiac in the space of a year. This first argument, then, accounts beyond dispute for the apparent annual motion of the Sun along the ecliptic.

We come now to the other motion, namely the Earth's diurnal motion around its own axis. For this we must establish its axis and its poles, and the point to note is that its axis is not perpendicular to the plane of the ecliptic—not, in other words, parallel to the axis of the Earth's orbit—but inclined away from the perpendicular at an angle of about 23½ degrees, with its north pole towards the orbital axis when the centre of the Earth is at the solstitial point in Capricorn. So, taking the terrestrial globe to have its centre at that point, we shall mark the poles and its axis AB, inclined at 23½ degrees from the perpendicular to the Capricorn–Cancer diameter, so that the angle A–Capricorn–Cancer is the complement of a quarter of a circle, or 66½ degrees. We must understand this inclination to be immutable, and we shall take the upper angle A to be the north and the other angle B to be the south. [418]

Now imagine the Earth rotating around the axis AB in twenty-four hours, again from west to east; all the points marked on its surface will describe parallel circles. With the Earth in this first position we shall mark the great circle, CD; the two that are at 23½ degrees from it, EF above it and GN below; and the two furthest from it, IK and LM, at a similar interval from the poles A and B. With these five parallels marked we can understand there to be any number of others, all parallel to them, described by any number of points on the globe's surface.

Now let us assume that the Earth moves with its annual motion to the other points noted above, always following this rule: that its axis AB never changes in its inclination to the plane of the ecliptic, and also that the direction of its inclination never changes. So, remaining always parallel to itself, it looks continuously towards the same part of the universe, or rather of the firmament. If we imagine the Earth's axis to be extended, its upper extremity would describe a circle parallel and equal to the great sphere Libra–Capricorn–Aries–Cancer, and this would be the upper base of a cylinder described in its annual [419] motion on a lower base Libra–Capricorn–Aries–Cancer. Hence, given this constancy in its inclination, we can mark it in these other three positions centred on Aries, Cancer, and Libra, similar in every respect to the first position we marked centred on Capricorn.

Consider now the figure of the Earth in this first position. The axis AB is inclined away from the perpendicular to the diameter Capricorn–Cancer by 23½ degrees towards the Sun, O. Since the arc AI is also 23½ degrees, the Sun's light will illuminate the hemisphere of the globe that is exposed to it (of which only half is visible here), divided from the part that is in darkness by the boundary of the light IM. This boundary will divide the parallel CD, which is a great circle, into equal parts, but it will divide all the other parallels into unequal parts, because it does not pass through their poles A and B. The parallel IK and all the other parallels between it and the pole A will be entirely in the illuminated area, and conversely all the parallels between LM and the pole B will be in darkness.

Moreover, the two arcs IKF and AFD will be equal, since the arc AI is equal to the arc FD and the arc AF is common to them both; and they will both be equal to a quarter of a circumference. The entire arc IFM being a semicircle, the arc MF will be a quadrant, and will be equal to the other arc FKI; therefore, when the Earth is in this position, the Sun O will be directly overhead for someone at point F. But the diurnal rotation around the fixed axis AB means that every point on the parallel EF passes through this same point F; so the Sun at noon on this day will be directly overhead all those living on the parallel EF, and they will see the Sun in its apparent motion describe the circle that we call the Tropic of Cancer. For the inhabitants of all the parallels above EF, towards the north pole A, the Sun declines from its zenith towards the south. Conversely for the inhabitants of the parallels below EF, towards the equator CD and the south pole B, the midday Sun is elevated above their zenith towards the north pole A.

It can be seen, too, that the longest parallel CD is the only one that is divided into equal parts by the boundary of the light IM; the others, both above and below this longest parallel, are all divided into unequal parts. Of the parallels that are above the line, the arcs that are in the area of the Earth's surface illuminated by the Sun are greater than those that are in darkness. The reverse is true of the remaining parallels below the longest CD, towards pole B; in these the arcs that are illuminated are less than those that are in darkness. It is also clear that the difference between these arcs increases the [420] nearer the parallels are to the poles, to the point where the parallel IK is entirely in the illuminated area, so that its inhabitants have a day of twenty-four hours with no night, whereas the parallel LM is entirely in darkness, so that it has a night of twenty-four hours with no day.

Let us come now to the figure of the Earth in its third position, centred on Cancer, from which point the Sun appears to be at the first point of Capricorn. It is immediately clear that the aspect and situation of the Earth is exactly the same as it was in the first figure, since the inclination of its axis AB has not changed but remains parallel to itself; the only difference is that the hemisphere that was illuminated by the Sun is now in darkness, while the one that was dark is now illuminated. Hence, the differences and variations in length between the day and the night in the first position are now reversed. It can be seen first of all that where the circle IK was entirely in the light in the first figure it is now entirely in the dark, and the opposite circle LM that was previously all in darkness is now all in the light. Then, in the parallels between the equator CD and the pole A the arcs that are in daylight are shorter than those that are in the dark, where previously the reverse was true; and in the same way the parallels towards the pole B now have the arcs that are illuminated longer than those that are in the dark, the opposite of what happened when the Earth was in its other position. The Sun can now be seen to be overhead for the inhabitants of the tropic GN, and for those on the parallel EF it declines towards the south through the whole of the arc ECG, that is, through 47 degrees. In other words, it has passed from one tropic to the other, crossing the equator and rising and falling at the meridian by 47 degrees. And all these changes derive not from any declination or elevation of the Earth, but on the contrary from the fact that it never has any declination or elevation at all, but rather remains constant in the same position relative to the universe. The only change is its circulation around the Sun, which is located at the centre of its revolution and in the same plane, in its annual motion.

A marvellous phenomenon arising from the fact that the inclination of the Earth's axis does not change.

Here we should note a marvellous phenomenon arising from this. The fact that the Earth always maintains the same direction towards the universe—that is, towards the farthest sphere of the fixed stars—means that the Sun appears to us to vary in its elevation by 47 degrees, whereas we see no variation at all in [421] the elevation of the fixed stars. If on the other hand the Earth's axis always maintained the same inclination towards the Sun—that is, towards the axis of the zodiac—then we would see no variation in the elevation of the Sun, which would mean that the inhabitants of a given place would always have the same season and the same difference between day and night: for some it would be always winter, for some always summer, for some spring, and so on. Conversely, we would see a great variation in the elevation and inclination of the fixed stars, again ranging through 47 degrees.

To clarify this, let us look again at the Earth as it appears in the first figure, where the axis AB has its upper pole A inclined towards the Sun. Contrast this with the third figure, where because this axis has maintained the same inclination towards the stellar sphere, remaining parallel to itself, its upper pole A is no longer inclined towards the Sun but rather has declined from its original position by 47 degrees and is now inclined towards the opposite side. So to restore the original inclination of pole A towards the Sun the terrestrial globe would have to move, in the course of its motion around the circumference ACBD, by the same 47 degrees towards E; and any fixed star observed at the meridian would appear to have risen or fallen by the same amount.

For the points that remain to be clarified, let us consider the Earth as it is placed in the fourth figure, with its centre at the first point of Libra, so that the Sun appears at the beginning of Aries. We recall that the Earth's axis in the first figure was inclined above the Capricorn–Cancer diameter, and therefore was in the plane intersecting that of the Earth's orbit in the Capricorn–Cancer line and perpendicular to it. Transferring this to the fourth figure, since—as we have repeatedly said—it remains parallel to itself, it will again be in a plane perpendicular to the surface of the Earth's orbit and intersecting that surface at right angles in the Capricorn–Cancer diameter. Therefore the line that runs from the centre of the Sun O to the centre of the Earth in Libra will be perpendicular to the axis BA. But this same line from the centre of the Sun to the centre of the Earth is always also perpendicular to the circular boundary of the light; so in the fourth figure this circle will pass through the poles A and B, and the axis AB will be in the same [422] plane. But when the great circle passes through the poles of the latitudes, it divides them all into equal parts; so the arcs IK, EF, CD, GN, and LM will all be semicircles, the hemisphere looking towards us and towards the Sun will be illuminated, and the circular boundary of the light will be the same as the circle ACBD. So when the Earth is in this position, all its inhabitants will experience the equinox. The same applies to the second figure, where the illuminated hemisphere of the Earth is facing the Sun but the dark hemisphere is visible to us, and its arcs that are in darkness are again all semicircles; so in this position too it is the equinox.

Finally, since the line produced from the centre of the Sun to the centre of the Earth is perpendicular to the axis AB, which is also perpendicular to the greatest parallel CD, this line from O to Libra must necessarily pass through the plane of CD, intersecting its circumference at the midpoint of its diurnal arc CD. Therefore the Sun will be directly overhead for anyone who is at that point of intersection. But all the inhabitants of the parallel CD will pass through this point in the course of the Earth's diurnal rotation; so they will all have the midday Sun directly overhead, and all the Earth's inhabitants will see the Sun appear to describe the greatest parallel, called the equinoctial circle.

A further point should be clear from this. We have seen that when the Earth is at the two solstitial points, one of the polar circles IK and LM is entirely in the light while the other is in darkness; and that when the Earth is at the equinoctial points, half of each of these polar circles is in the light and the other half is in darkness. It should not be difficult to understand, therefore, that as the Earth moves from, say, Cancer (where the whole of the parallel IK is in darkness) to Leo, part of the parallel IK towards point I begins to enter the light, and the boundary of the light IM begins to withdraw towards the poles A and B, so that its intersection with the circle ACBD is no longer at points I and M but at two other intermediate points on the arcs IA and MB. So the inhabitants of the circle IK will begin to enjoy the light, while those on the circle LM begin to experience darkness.

You can see, then, how by attributing to the terrestrial globe two simple motions, made in periods well suited to their size and moving not against each other but from west to east in common with all the other moving bodies in the universe, it is possible to provide a satisfactory account of all these observed phenomena. But if you wanted to save all these appearances while also maintaining that the Earth is at rest, you would have [423] to abandon the observable symmetry between the size of a moving body and its velocity, and attribute an unimaginable speed to a sphere which is vaster than all the others, while the other smaller spheres were all moving very slowly, and what's more in the opposite direction. To heighten the improbability even more, you would have to suppose that the outermost sphere was dragging all the lower spheres in a direction contrary to their own intrinsic motion. Now I leave it to you to judge which of these accounts seems more likely to be true.

SAGREDO. As far as I can see there is a great difference between the simplicity and ease with which these effects are produced by the means envisaged in this new system, and the multiplicity, confusion, and difficulties needed under the old, commonly accepted view. If the universe was indeed ordered according to this multiplicity of causes, we would have to abolish many of the axioms that are commonly accepted by all philosophers—that nature does not multiply entities unnecessarily, that she uses the simplest and easiest means to produce her effects, that she does nothing in vain, and so on. I confess that this is the most remarkable thing I have ever heard, and I can't imagine that the human mind has ever been able to speculate more subtly. I wonder what Simplicio thinks?

Axioms commonly accepted by all philosophers.

SIMPLICIO. To speak frankly, these seem to me to be the kind of geometrical subtleties that Aristotle criticized in Plato, when he told him that his excessive study of geometry was taking him away from sound philosophy. I have known and heard very great Peripatetic philosophers discourage their pupils from studying mathematics, because it produces minds that quibble and are incapable of good philosophy. This is the exact opposite of what Plato teaches, since he did not admit anyone to the study of philosophy who had not first mastered geometry.

Aristotle criticizes Plato for his excessive study of geometry.

SALVIATI. These Peripatetics of yours are quite right to dissuade their pupils from studying geometry, since there is no art better qualified to expose the fallacies in their arguments. But they are quite different from mathematical philosophers, who would much rather deal with those who are familiar with the principles of Peripatetic philosophy than with those who lack any such knowledge, and who are therefore incapable of comparing one doctrine with another. But leaving this to one side, please tell me which far-fetched or over-subtle conclusions make you find the Copernican system unconvincing.

Peripatetic philosophers condemn the study of geometry.

SIMPLICIO. To tell the truth I haven't entirely understood it, perhaps because I'm not conversant with the reasons that Ptolemy gives for these same effects—the stations, retrogradations, approaches and retreats of the planets, the lengthening and shortening of days, the changing seasons, and so on. But leaving aside the consequences that follow from Copernicus's basic assumptions, the assumptions themselves seem to me to present significant difficulties; and if these assumptions are shown to be without foundation, the whole edifice collapses. Now the whole Copernican system seems to me to rest on unsound foundations, presupposing as it does the mobility of the Earth. Take that away, and there is no need to go on to debate any other questions. And to refute the mobility of the Earth I think that Aristotle's axiom is more than sufficient, when he says that only one simple motion can be natural to a simple body; whereas in the Copernican view the Earth, a simple body, is assumed to have three if not four motions, all quite different from each other. There is its direct motion as a heavy body towards the centre, which is undeniable. Then in

[424]

Four different motions attributed to the Earth.

addition to that it is deemed to have a circular motion in a great circle around the Sun in the course of a year, and a whirling around on itself in twenty-four hours. Finally there is another whirling on its axis, in the opposite direction to its diurnal motion, which it completes in a year. This is the most improbable of all, which is perhaps why you failed to mention it. My mind has the greatest difficulty in accepting this.

Downward motion belongs not to the terrestrial globe, but to its parts.

SALVIATI. As regards the downward motion towards the centre, we have already proved that this does not belong to the Earth, which never has and never will have such a motion; it belongs rather—if it exists at all—to its parts which move to be united with the whole. As for the annual and diurnal motions, these are perfectly compatible because they are in the same direction, in the same way as a ball would spontaneously spin around when it was running down a sloping surface. And as for the third motion that Copernicus attributes to the Earth, that of turning on its axis in the course of a year so as to maintain the inclination of its axis and its direction towards the same part of the firmament, I will tell you a remarkable fact. Far from there being any resistance or difficulty in such a motion, even though it is in the opposite direction from the other annual motion, it is entirely natural in any body that is suspended in equilibrium, without the need for any motive force at all. Such a body, if it is moved around the circumference of a circle, will immediately and of its own accord acquire a rotating motion around its own [425] centre, in a contrary direction to the motion around the circle, at a velocity such that both motions will complete their revolution at exactly the same time.

The Earth's annual and diurnal motions are compatible.

Any body suspended in equilibrium and moving around the circumference of a circle spontaneously acquires a motion around its own axis in the opposite direction to its motion around the circle.

You can see this striking effect in an experiment that illustrates our point. Take a basin of water and put a ball to float in it. If you turn on the soles of your feet while holding the basin in your hands, you will see that the ball immediately starts to revolve in the opposite direction to the motion of the basin, and it will stop when you stop rotating the bowl. This is exactly what the Earth is: a globe suspended and balanced in thin, yielding air, carried around the circumference of a great circle in the course of a year, so that it spontaneously acquires a rotating motion around its own centre in the course of a year, but in the opposite direction to its other annual motion. This is what you

An experiment to demonstrate that two contrary motions can naturally occur in the same moving body.

will see, but if you consider it more carefully you will perceive that it is not a real motion but simply an appearance. When it seems to be rotating around its centre it is in fact maintaining its position in relation to its surroundings, which apart from you and the basin remain unchanged. If you make a mark on the surface of the ball and note where it points in relation to the walls of the room you are in, or to the landscape or the sky, you will see that as you and the bowl rotate the mark on the ball's surface always remains facing the same point. It is only when you compare it to your own motion and that of the bowl that it will appear to keep changing direction and to move in the opposite direction to you and the bowl, as it continually seeks out the points through which the bowl is turning. So it is more accurate to say that you and the bowl are rotating around the ball which remains immobile, rather than that the ball is rotating in the bowl. In the same way the Earth, suspended and balanced on the circumference of the great sphere, and placed in such a way that a fixed point on its surface—for example, the north pole—points towards a given star or a particular place in the firmament, constantly maintains the same direction, despite being carried around the circumference of its orbit in the annual motion.

The third motion attributed to the Earth is rather that it remains unchanging.

Marvellous internal force of the terrestrial globe, that it constantly faces towards the same part of the heavens.

This point alone should suffice to overcome your astonishment, removing any objection and persuading you that there is no lack of a contributory cause for this phenomenon. But what would you say, Simplicio, if we were to add to this a marvellous force intrinsic to the globe itself, that keeps each of its parts facing a determined part of the firmament? I am referring to magnetic force, which is constant in any piece of lodestone. If [426] even the smallest fragment of this stone has this force within it, how much greater will it be in the whole of this terrestrial globe, which so abounds in lodestone that it may well constitute the primary substance of the globe itself?

The terrestrial globe is made of lodestone.

SIMPLICIO. Are you then one of those who subscribe to the magnetic philosophy of William Gilbert?*

The magnetic philosophy of William Gilbert.

SALVIATI. I am indeed, and I think everyone who has read his book attentively and checked his experiments agrees with me. I haven't given up hope that what happened to me may also happen to you, provided you have a curiosity like mine and you recognize that there are infinite things in nature that remain unknown to our human understanding. I hope this might liberate you from slavish adherence to this or that author on matters of natural science, giving free rein to your reason and softening your resistance to the evidence of your senses, so that you do not always refuse to listen to voices that are no longer heard.* But the cowardice, if I may say so, of the majority of thinkers has reached such a point that they blindly give their assent, or rather their submission, to whatever they find written by the authors who were commended by their teachers when they first began their studies. Not only that, but they refuse to listen, let alone to consider, any new question or proposition even though their authors have never examined it, much less refuted it. One such question is the true, primary, intrinsic nature of the material of which our terrestrial globe is made. It never occurred to Aristotle or to anyone else that this material might be lodestone until Gilbert proposed it, and I have encountered many who have shied away like a frightened horse at the first mention of such a suggestion, even though neither Aristotle nor anyone else has refuted it. They have drawn back and refused to discuss the idea, dismissing it as a vain delusion or even as plain madness. Gilbert's book might never have come into my hands had it not been for a very well-known Peripatetic philosopher* who gave it to me, I think because he wanted to keep his library free of such a contagion.

Cowardice of the majority of thinkers.

SIMPLICIO. I freely admit that I am one of the majority of thinkers, and it is only in these last few days, as I have been allowed to take part in your discussions, that I realize I have [427] moved away somewhat from the most common well-trodden paths. But I don't yet feel sufficiently detached from them to dismiss the difficulties surrounding this fanciful new opinion, which seem to me very arduous and hard to overcome.

SALVIATI. If what Gilbert writes is true, it's not an opinion but a matter of scientific knowledge. It's not new, but is as ancient as the Earth itself; and if it's true it can't be hard or difficult, but is simple and straightforward. If you will allow me I will show you that your apprehension is of your own making, and that you are recoiling from something which is not frightening at all, like a boy who is afraid of a bogeyman which is nothing but an empty name.

SIMPLICIO. I should be glad to be enlightened and rescued from error.

SALVIATI. In that case, answer the questions I will put to you. First, tell me whether you think this globe that we inhabit and that we call Earth is made of a single simple substance or a mixture of diverse materials.

SIMPLICIO. I see it as being made up of diverse substances and bodies; first of all, I see its major parts as consisting of water and earth, which are very different from each other.

The terrestrial globe is made up of diverse materials.

SALVIATI. Let's leave the seas and other waters to one side for now, and consider the solid parts. Do they seem to you to be all the same thing, or are they different things?

SIMPLICIO. In appearance I see them as being different. There are large areas of arid sand and others of fertile and fruitful soil. We see an infinity of barren, rocky mountains, full of hard rocks and all kinds of stones—porphyry, alabaster, jasper, and countless other kinds of marble. There are vast mines of every kind of metal, and in short there is such a diverse range of materials that I could go on listing them all day and still not mention them all.

SALVIATI. Now, of all these diverse materials, do you think they make up the globe in equal parts, or is there one part that greatly exceeds all the others and is, as it were, the main material and substance of the whole vast mass?

SIMPLICIO. I think that the stones, marbles, metals, gems, and other diverse materials are as it were external jewels and [428] ornaments on the surface of the globe itself, which in its mass far exceeds all these other things.

SALVIATI. What then do you think is the material of which this vast mass is made, of which all the things you have mentioned are like external ornaments?

SIMPLICIO. I think that it is the simple, or less impure, element of earth.

SALVIATI. What, though, do you mean by *earth*? Is it what is spread across the landscape, that we break up with hoes and ploughs, where we sow seed and plant fruit trees, and where woods grow spontaneously in great profusion—which in short is the habitat of all animals and the womb from which all plant life grows?

SIMPLICIO. I would say that this is the primary substance of which our globe is made up.

SALVIATI. I think you are wrong about that. This earth that we break up and sow and that bears fruit is just a part of the surface of the globe, and a very thin part, as it is quite shallow compared to the distance to the centre. We know from experience that you need to dig only a little way to reach materials that are quite different from this outer crust, being more solid and not at all suitable for growing plants. What's more, it is reasonable to suppose that the parts deeper down are as compressed and hard as the hardest rock, since there is such a great weight pressing down on them. Add to this that there would have been no point in making these deeper materials fertile, since they were never going to produce fruit, but will remain eternally buried in the deepest, darkest abyss of the Earth.

SIMPLICIO. Who is to say that the lower parts closer to the centre are infertile? Perhaps they too produce things that are unknown to us.

SALVIATI. You more than anyone ought to be certain about this. If the integral bodies of the universe exist solely for the benefit of the human race, then surely this globe above all must be intended purely for the convenience of us who are its inhabitants. What benefit can we derive from materials that are so inaccessible and remote from us that we will never be able to make them usable for us? The internal substance of this globe of ours cannot, therefore, be such a fragile and inconsistent [429] material as this surface matter that we call earth; it must be an extremely dense and solid body—in short, a very hard stone. This being the case, what reason is there to be more reluctant to believe that it is lodestone than porphyry, jasper, or any other hard marble? Would you have thought it any less extraordinary if Gilbert had written that the interior of the globe is made of sandstone or chalcedony?

The inner parts of the terrestrial globe must be extremely hard.

SIMPLICIO. I grant you that the inner parts of this globe are more compressed, and therefore more dense and solid, the more so the deeper they are; Aristotle too concedes this. But I see no reason to concede that they are different in nature, or that they are anything other than earth of the same kind that is here on the surface.

SALVIATI. It was never my purpose in broaching this subject to prove to you conclusively that the primary substance of this globe of ours is lodestone. I wanted only to show you that there is no more reason to resist conceding that it is lodestone than that it is any other material. In fact, if you think about it, you will see it is quite likely that it was just this arbitrary choice of name that persuaded people to believe it was made of earth, since we have always commonly used this word *earth* both for the material that we plough and sow and as the name of this globe of ours. Its name could equally well have been taken from stone, and in that case no one would have resisted or contradicted saying that its primary substance was stone. In fact this seems all the more likely because I am certain that if one could peel away the globe's crust, removing a layer of a thousand or two thousand *braccia*, and then separate the earth from the stones, there would be a much bigger pile of stones than of fertile earth.

Our globe would be called 'stone' rather than 'earth' if it had been given this name from the beginning.

I haven't given you any of the reasons that prove conclusively that our globe is indeed formed of lodestone, and this is not the time to do so, especially as you can read Gilbert's book at your leisure. To encourage you to read it, I will simply give you an illustration to show the method that he follows in his philosophy. I know you are well aware how much our understanding of the essence and substance of something is helped by our knowledge of its accidental properties. [430] So I would like you to inform yourself as thoroughly as you can of the accidents and properties that belong exclusively to the lodestone, and not to any other stone or material body. For example, it has the property of attracting iron, and of conferring the same power on the iron simply by being close to it; it can transmit to the iron the property of pointing towards the poles, while retaining the same property itself. More than this, see if you can show experimentally how it has the power not only to make a magnetized needle turn horizontally under a meridian to point towards the poles—a property that has long been recognized—but also something else, just recently observed. If the needle is balanced on a small sphere of lodestone, it will dip towards the surface of the sphere, to varying degrees depending on its proximity or distance from its poles; so that when it is at the pole it stands up vertically, and when it is in the area between the poles it remains

Procedure followed by Gilbert in his philosophy.

The many properties of lodestone.

horizontal. Try to show, too, how the power to attract iron is much stronger near the poles than in the intermediate parts, and also how it is notably stronger at one pole than at the other. This is true of all pieces of lodestone, the stronger pole being always the south-facing one. Note too that when a small lodestone is placed close to a much larger one, the power of its stronger south pole will be weaker and less able to attract the iron than the larger piece's north pole. In short, try to confirm experimentally these and many other properties that Gilbert describes, all of which belong solely to the lodestone, since none of them is to be found in any other material.

Conclusive proof that the terrestrial globe is made of lodestone.

Now, Simplicio, suppose that you were presented with a thousand pieces of different kinds of material, each one covered and hidden under a cloth, and you were asked to guess from external indicators, without uncovering them, what each of them was made of. Suppose you came upon one that clearly demonstrated all the properties that you knew resided only in lodestone and not in any other material. What would you judge to be the essence of that piece? Would you say that it might be a piece of ebony, or alabaster, or tin?

SIMPLICIO. I would say that it was without doubt a piece of lodestone.

[431] SALVIATI. In that case, you should confidently declare that under this cover or crust of earth, stones, metals, water, and so on, is concealed a great lodestone, since anyone who cares to observe them will recognize that it is accompanied by all the effects that belong to a genuine, clearly revealed sphere of lodestone. If nothing else, the fact that a magnetized needle inclines more and more as it is taken towards the north pole, and less as it approaches the equator, until it reaches equilibrium at the equator itself, should suffice to convince even the most sceptical judge. This is to say nothing of the other marvellous effect that is plain to see in any piece of lodestone: that for us in the northern hemisphere, the south pole of the lodestone is stronger than the other, and that the difference between them increases as one moves away from the equator; that at the equator itself the two poles are equal, but both notably weaker; but in southern regions away from the equator its nature changes, and the pole that we found to be weaker becomes stronger than the

other. All this can be compared with what we observe when a small piece of lodestone is in proximity to a larger one: the strength of the large piece prevails over the smaller one and makes it subservient to it, with the same variations as it is moved either side of the midpoint of the large piece as are found in every lodestone when it is moved either side of the Earth's equator.

SAGREDO. I was convinced as soon as I read Gilbert's book. I obtained an excellent piece of lodestone and made many observations with it over a long period of time. They all inspired great wonder, but what I found most astonishing of all was the extent to which its power to hold iron was increased by arming it in the way that this author describes. I found that thus armed the power of my lodestone was increased eightfold, so that whereas without an armature* it could hold barely nine ounces of iron, with the armature it sustained more than six pounds. You may have seen this very piece, holding up two small iron anchors, in the Gallery of your Grand Duke, to whom I ceded it.

A lodestone with an armature holds iron much more powerfully than one that is not armed.

SALVIATI. I have seen it many times, and have greatly marvelled at it. But then I was even more astonished by a very small piece belonging to our Academician, which weighed no more than six ounces, and without an armature could sustain barely two ounces of iron; yet when armed it could hold 160 ounces, in other words eighty times more than when it was not armed, and 26 times more than its own weight. This is a far greater wonder than Gilbert was able to observe, since he writes that he had not been able to find a lodestone that could sustain four times its own weight.

[432]

SAGREDO. This stone seems to me to offer great scope for investigation. I have often wondered how it can be that it confers on its iron armature a strength so much greater than its own, and I have failed to find any satisfactory answer. I have not found much in Gilbert's writing that is helpful on this point; I wonder whether your experience has been the same?

SALVIATI. I have the greatest admiration and indeed envy for this author, for having had such a stupendous insight about a subject that many of the greatest intellects have studied but none had perceived before. I think he deserves the highest praise, too, for the many novel and accurate observations he has made, to the shame of many vain and mendacious authors who write not just what they know to be true, but whatever they pick up from the ignorant multitude, without bothering to check it experimentally. Perhaps they do it just to fill out their books. What I do wish is that Gilbert had been a bit more of a mathematician, in particular that he had a solid foundation in geometry, because that would have made him less ready to accept as conclusive proofs the explanations that he offers for the truths he correctly observes. If the truth be told, these explanations do not have the compelling force to make them necessary and eternal, as is undoubtedly needed for proofs in natural science. But I have no doubt that this new science will be perfected with the passing of time, as new observations are made and even more as new conclusive proofs are added.

Those who are the first to observe and discover new things deserve to be admired.

This is to take nothing away from the recognition that is due to a first observer. My admiration for the original inventor of the lyre is no less, in fact it is greatly increased, by the fact that it was presumably a very crude instrument and was even more crudely played, and that many subsequent artists down the centuries have greatly refined its practice. I find it entirely reasonable that the ancients numbered the inventors of the noble arts among the gods, since the common run of humanity shows so little curiosity and such indifference to rare and refined arts that hearing and seeing them practised by professionals does [433] not inspire them to learn them for themselves. You can't expect minds like these to be inspired to understand the construction of the lyre or the invention of music by hearing the vibrations of the dried sinews of a tortoise or the striking of four hammers.* The ability to discover great inventions starting from very small beginnings, and to perceive marvellous arts hidden under apparently childish appearances, are not things to be found in ordinary minds, but are the qualities and insights of superhuman spirits.

The real reason why the power of a lodestone is greatly multiplied by means of an armature.

Now to answer your question, I too have pondered at length to find the reason why there is such a strong and tenacious bond between the iron armature of a lodestone and another piece of iron that bonds with it. I first established that an armature does not produce any increase in the power of the stone itself, because if a very thin piece of paper, even as thin as gold leaf, is placed between the piece of iron and the armature, it neither holds the iron more strongly nor attracts it over a greater distance; in fact in such a case the stone without an armature sustains more iron than when it is armed. So there is no change in the power of the stone, and yet there is something new in its effect. A new effect must have a new reason; therefore we must ask what new element is introduced in sustaining the iron when an armature is added. The only change that can be discerned is the difference in contact, in that iron was originally touching lodestone and now iron is touching iron; so the conclusion must be that the difference in contact is the cause of the difference in effect.

A new effect must have a new reason.

This difference in contact, as far as I can see, can only derive from the fact that the substance of iron is made up of particles that are finer, purer, and denser than lodestone, the particles of which are coarser, less pure, and more sparse. It follow that if the surfaces of two pieces of iron that are to be brought together are thoroughly smoothed, shone, and polished, they bond together so precisely that the infinite points of contact on one surface meet the infinite points on the other, so that there are far more threads (so to speak) binding them together than there would be between iron and lodestone. This is because lodestone is more porous and less pure, which means that not all the points and threads on the surface of the iron find corresponding points with which to bond on the surface of the lodestone.

Showing that iron is made up of finer particles, purer and denser than lodestone.

That the substance of iron is made up of particles that are much denser, finer, and purer than that of lodestone (especially if the iron has been thoroughly refined, as in the case of the finest steel) is clear from the fact that it can be sharpened to a very fine edge, as in the blade of a razor, something which would be quite impossible with a piece of lodestone. The impurity of lodestone, and the fact that it is mixed in with other kinds of stone, is first of all evident to the senses because it contains patches of a lighter colour. This can then be confirmed by bringing a needle close to it, hanging on a thread: the needle will not rest above these fragments of other stone, but will appear to shun them and to jump towards the lodestone around them. Some of these different fragments are large enough to be clearly visible, so we may suppose that they are present throughout the larger mass even though they are too small to be seen.

[434]

The impurity of lodestone is evident to the senses.

An experiment will confirm what I have said, namely that the firm bond between iron and iron is the result of the large number of points of contact between them. If we bring the sharp point of a needle close to the armature of the lodestone, the attraction between them will be no stronger than it would be to the lodestone alone. This must be because the extent of the contact is the same in both cases, since it is just the single point of the needle. But that is not all. Take a needle and place it on the lodestone so that one end of it projects a little, and then bring a nail close to it. The needle will immediately attach itself to the nail, so that if the nail is then withdrawn the needle will be suspended, with one end attached to the lodestone and the other to the iron. If the nail is then withdrawn further, the needle will be detached from the lodestone, provided the eye end of the needle is joined to the nail and its point joined to the lodestone; whereas if the eye end is towards the lodestone it will remain attached to it when the nail is withdrawn. I conclude that this can only be because the needle is thicker at the eye end and therefore has many more points of contact than at its point.

SAGREDO. I find the whole of this argument entirely conclusive, and these experiments with a needle confirm it with scarcely less certainty than a mathematical proof. I freely confess that nothing I have heard or read in all the discussions about magnetism gives an equally convincing account of its many other marvellous effects. I can think of nothing that would satisfy the mind more than to have these explained as clearly as this.

SALVIATI. You need an element of luck when you start investigating the reasons for conclusions that are unknown to us. If you set off on the path of truth, there is a good chance that [435] you will encounter various other conclusions that you know to be true, either from logical arguments or from experience, and your certainty about them will strengthen the evidence that you are on the right track. This happened to me when I was considering the problem we are discussing. I wanted to find some other comparison to confirm that the reason I was pursuing was the right one, namely that the substance of lodestone really was less pure than that of iron or steel. So I had the craftsmen who work in the gallery of my master the Grand Duke smooth one side of that same piece of lodestone that was once yours, and then give it the highest possible polish and shine; and to my satisfaction it provided just the confirmation I was seeking. It revealed numerous patches that were a different colour from the rest, but shining and polished like any very dense hard stone, while the rest of the surface was smooth to the touch but not polished at all, as if it was under a fog. This latter was the substance of the lodestone, and the polished parts were the pieces of other stone mixed in with it. This became clear visually when the smoothed surface was brought close to iron filings, and a great mass of filings ran towards the lodestone, while not a single one rested on the polished parts, even though these were very numerous—some of them a quarter of an inch across, others somewhat smaller, and many very small and barely visible. This confirmed for me that my original idea was correct, when I conjectured that the substance of lodestone was not fixed and dense but rather porous and like a sponge, except that the cavities and cells in a sponge contain air or water whereas in lodestone they are filled with hard, heavy stone, as is shown by the high degree of polish they receive.

Hence, as I said at the outset, when the surface of iron is applied to the surface of lodestone, the tiny particles of iron, even though they are more densely packed than in possibly any other body (as is shown by the fact that iron can be polished more than any other material), do not all come into contact with pure lodestone; in fact only a few do, and because there are few points of contact the attachment is weak. With an armature, on the other hand, not only is it in contact with a large part of the surface of the lodestone, it also absorbs power from the parts that are close to it even if not actually touching it; so when its carefully smoothed face is applied to the equally smooth face of the iron that it is to sustain, the contact is made between innumerable tiny particles on both [436] surfaces, and the resulting attachment is extremely strong. Gilbert did not make this experiment of smoothing the surfaces of the pieces of iron that are to be brought together; he used convex irons, so that the area of contact was small and the attachment between them was correspondingly weak.

SAGREDO. As I said just now, I find the reasoning you have given scarcely less satisfactory than if it had been a pure geometrical proof. Since it is a matter of physics, I imagine that

Simplicio will also be satisfied by it, as he knows that one should not look for geometrical certainty in questions of natural science.

SIMPLICIO. Indeed, I think Salviati's eloquent words have explained the cause of this effect so clearly that anyone of average intelligence could understand it, even if they have no scientific knowledge. We philosophers, however, keeping to the terminology of our profession, ascribe the cause of these and similar effects to what we call 'sympathy', meaning a certain compatibility and mutual attraction arising between things that have similar qualities. By contrast, we use the term 'antipathy' to describe the hatred and enmity that leads other things naturally to shun and recoil from each other.

'Sympathy' and 'antipathy' are terms used by philosophers as an easy explanation of many natural effects.

SAGREDO. And so these two terms come to account for a large number of phenomena and effects that, not without wonder, we observe in nature. This way of philosophizing seems to me to be in sympathy with the approach that a friend of mine had towards painting. He would take a piece of chalk and write on the canvas, 'I want a spring here, with Diana and her nymphs; some hounds here; a hunter with the head of a stag over here; landscape, woods and hills everywhere else'. Then he would leave the rest for the painter to do, and he persuaded himself that he had painted the story of Actaeon himself when all he had done was write in the names.

An amusing example to show the inefficacy of some philosophical discourse.

But where has this long digression taken us, going against the rules that we agreed? I've almost forgotten the subject we were dealing with before we were sidetracked into this discussion about magnetism, although I know there was something I wanted to say about it.

[437] SALVIATI. We were demonstrating that the third motion attributed to the Earth by Copernicus is not a motion at all, but rather a state of rest, immutably keeping certain parts of the globe constantly facing the same parts of the universe; that is, that the axis of its diurnal rotation remains perpetually parallel to itself and facing the same fixed stars. Then we said that this unchanging position occurs naturally in any body that is balanced and suspended in a fluid and yielding medium, where even if it is moved around it does not change its direction in relation to the things around it; it only appears to rotate in relation to the vessel containing it and the one who is

moving it around. Finally, to this simple and natural phenomenon we added magnetic force, which allows the terrestrial globe to maintain this unchanging position even more firmly, and so on.

SAGREDO. I remember it all now. The point that occurred to me and that I wanted to mention was in relation to the objection raised by Simplicio against the mobility of the Earth. He argued that a multiplicity of motions could not be attributed to a simple body, because in Aristotle's teaching a simple body could naturally have only one simple motion. The point I wanted us to consider was, indeed, about the lodestone, which plainly has three natural motions: one, as a heavy object, towards the centre of the Earth; and second, the horizontal circular motion to restore and maintain its axis towards particular parts of the universe. Now Gilbert has discovered a third motion,* that of inclining its axis from a horizontal position towards the surface of the Earth, in varying degrees depending on its distance from the equator, where it remains parallel to the Earth's axis. And indeed it is not unlikely that it has a fourth motion in addition to these three, namely rotating on its axis if it is balanced and suspended in the air or in some other fluid and yielding medium, if any external or contingent obstacles are removed. Gilbert himself seems to favour this view. So you see, Simplicio, how precarious Aristotle's axiom has become.

Lodestone has three different natural motions.

SIMPLICIO. Not only does this not undermine the axiom, it's not even directed at Aristotle, since he is speaking of the motion that can belong naturally to a simple body and you are opposing him with what belongs to a compound body. What you say is nothing new in Aristotle's teaching, because he allows that mixed bodies can have a composite motion, etc.

Aristotle allows that compound bodies have mixed motions.

[438]

SAGREDO. Stop there a moment, Simplicio, and answer the questions I shall put to you. You say that lodestone is not a simple body but a compound one; so tell me now what are the simple bodies that are mixed together to make up lodestone.

SIMPLICIO. I can't specify the ingredients or say what their exact proportions are. Suffice it to say that they are elementary bodies.

SAGREDO. This suffices for me too. What then are the natural motions of these simple elementary bodies?

SIMPLICIO. The two simple rectilinear motions, *sursum et deorsum*, upwards and downwards.

SAGREDO. So tell me now, do you think the motion that will be natural to this mixed body must be one that can result from combining the two simple motions of the simple bodies of which it is made up, or could it also be a motion that is impossible to form from a combination of the other two?

The motion of mixed bodies must be such as can result from combining the motions of their constituent simple bodies.

SIMPLICIO. I think it will move with the motion resulting from combining the motions of its constituent simple bodies; and that it cannot possibly move with a motion that is impossible to form from a combination of these.

Two rectilinear motions cannot be combined into circular motions.

SAGREDO. But you will never succeed in combining two rectilinear motions to form a circular motion, such as the two or three circular motions belonging to the lodestone. So you see the difficulties that come from following unsound principles, or rather from drawing unsound consequences from valid principles. Now, if you want to maintain that rectilinear motion belongs only to the elements and circular motion to celestial bodies, you are obliged to say that lodestone is a compound body made up of elemental and celestial substances. Your philosophy would be on firmer ground if you said that the integral bodies of the universe, those that in their nature are mobile, all move in circular motion; and that therefore lodestone, as a part of that true primary and integral substance of our globe, partakes of that same nature. But then you will see the fallacy of saying that lodestone is a mixed body and the terrestrial globe a simple body, since the globe is evidently a hundred times more composite, containing as it does many hundreds of different materials, quite apart from a large quantity of lodestone itself which you say is mixed. This seems to me to be the same as saying that bread is a mixed body but Spanish stew is a simple body, even though it contains a considerable amount of bread and dozens of other ingredients.

Philosophers are obliged to admit that lodestone is made up of celestial and elemental substances.

The fallacy of those who call lodestone a mixed body and the terrestrial globe a simple body.

A Peripatetic argument full of fallacies and contradictions.

[439] One of the things I find extraordinary with the Peripatetics is that they accept that our terrestrial globe is, *de facto*, a composite body made up of infinite different materials—something that they cannot deny—and then concede that composite bodies must move with composite motion. This must be made up of rectilinear and circular motion, since the two rectilinear

motions, being contrary to each other, cannot be combined. They affirm that the pure element of earth is nowhere to be found, and they admit that it has never moved with any local motion. Then they want to posit in nature this body that does not exist, and they attribute to it a motion that it has never exercised and never will. As for the body that does exist and always has, they deny it the motion that they previously agreed must naturally belong to it.

SALVIATI. Please, Sagredo, let us not weary ourselves any longer with these details. As you know, our purpose is not to come to a definite conclusion or to accept this or that opinion as the truth, but simply to put forward, for our own interest, the arguments and counter-arguments that can be adduced on either side. Simplicio replied as he did to defend his Peripatetic colleagues; so let us now suspend our judgement and leave it to those better qualified than us to decide. Over the past three days we have examined the system of the universe at considerable length, so I think it is time now for us to come to the major phenomenon which originally gave rise to our discussions, namely the ebb and flow of the tides, which I think can very probably be attributed to the motions of the Earth. But if you agree, we shall leave this for our meeting tomorrow.

An improbable effect of the lodestone accepted by Gilbert.

Lest I forget, however, let me first tell you of one detail that I wish Gilbert had never considered. He allows that if a small sphere of lodestone could be perfectly balanced it would turn on its own axis; but there is no reason why it should do so. The whole terrestrial globe has a natural motion rotating around its own centre in twenty-four hours, and every part of it shares in [440] this motion along with the whole. The parts already have this motion, simply by being on the surface of the Earth and being carried along with it. So to attribute to them a motion around their own centre would be to give them a second motion quite different from the first; they would have two motions, one around the centre of the globe in twenty-four hours and the other around their own centre. But this second motion is arbitrary and there is no reason to introduce it. If a piece of lodestone when it was separated from the larger mass ceased to follow it as it did when they were joined together, so that it no longer participated in the terrestrial globe's rotation around its centre, then it might be plausible to believe that it was about to adopt a new rotation around its own individual centre. But if it continues in its original, natural, and eternal motion when it is separated just as it did when it was joined, what reason is there to burden it with a new one?

SAGREDO. I understand this very well. It reminds me of a similarly false argument which was put forward by some writers on spherical astronomy, including Sacrobosco,* if I remember rightly. He wanted to show that the surface of the element of water, like the Earth, was a sphere, since our globe is made up of these two elements. He cited as conclusive proof of this the fact that tiny particles of water form into a round shape, like dewdrops and the drops that we see every day on the leaves of many plants. Citing the well-known axiom that the same reasoning applies to the whole as to the parts, he argued that as the parts took this shape the same must be true of the element as a whole. I think it is shameful that these writers do not realize the evident shallowness of this argument, or realize that if it was valid then not only small drops but any quantity of water at all, separated from the element as a whole, would form itself into a ball, which is clearly not the case. It is plain to see and easy for the intellect to grasp that as water seeks to take a spherical form around the common centre of gravity to which all heavy bodies are drawn—that is, the centre of the terrestrial globe—it is indeed followed by every part, as the axiom states. The surfaces of every sea, lake, pond, and any body of water in a container do spread out in the form of a sphere, namely the sphere whose centre is the centre of the terrestrial globe, not individual spheres on their own.

A vain argument put forward by some to prove that the surface of the element of water is spherical.

[441]

SALVIATI. This is a truly rudimentary error, which I would excuse if it was only Sacrobosco who made it, but I blush for the reputation of his commentators and other great thinkers, including Ptolemy himself, who also fell into it. But it's late and time we took our leave, to meet as usual tomorrow to bring all our discussions so far to a final conclusion.

FOURTH DAY

[442]

SAGREDO. I don't know whether you really are later than usual in returning to continue our discussion, or whether it is just my eagerness to hear Salviati's views on such a fascinating subject that has made it seem so. I've been waiting at the window for a good hour, expecting any minute to see the gondola appear which I sent to fetch you.

SALVIATI. I think it's your imagination, rather than our lateness, that has made the time seem longer; and so as not to prolong it any further, let us come to the point without more ado. I shall show that the motions we have already attributed to the Earth for every reason except as an explanation of the ebb and flow of the tides, fit perfectly with this phenomenon as well, and that conversely the ebb and flow of the tides confirms the mobility of the Earth. Nature seems to have allowed this, either because the reality is in fact as it appears, or because she wants to play a trick on us and to mock our foolish fancies. So far we have based our arguments for the mobility of the Earth on appearances in the heavens, since nothing that occurs on Earth seemed to be conclusive one way or the other. In our discussions we have shown at length how all the terrestrial effects which are commonly cited as evidence for the stability of the Earth and the mobility of the Sun and the firmament, would appear exactly the same to us if the Earth was moving and the Sun and the firmament were at rest. Only the element of water, being vast, fluid, and not attached or joined to the terrestrial globe as [443] all its other solid parts are, has a degree of freedom and is almost a law unto itself; so it alone in the sublunary world is able to give us some trace and indication of the mobility or otherwise of the Earth. Having studied the effects and variations in the motion of the tides at great length, drawing partly on what I have seen and partly on what I have heard from others, and having read and heard the vain explanations which many people have produced to explain these variations, I have been drawn to reach two firm conclusions (subject to the necessary assumptions). First, if the Earth is immobile, the ebb and flow

Does nature play a trick on us by making the ebb and flow of the tides endorse the mobility of the Earth?

The ebb and flow of the tides and the mobility of the Earth confirm each other.

All terrestrial events, except for the motion of the tides, could equally well confirm that the Earth is mobile or that it is at rest.

First general conclusion: that the ebb and flow of the tides would not be possible if the terrestrial globe were at rest.

of the tides cannot occur naturally; and second, if we grant that the motions of the Earth are as we have suggested, then the sea must necessarily be subject to the ebb and flow of the tides exactly as observation shows them to be.

SAGREDO. This is a tremendously important proposition, both in itself and for the consequences which follow from it, so I shall listen with the greatest attention to your exposition and proof of it.

Recognition of effects leads us to investigate their causes.

SALVIATI. In questions of natural science, such as this matter which we are discussing, it is our knowledge of the effects which leads us to investigate and find their causes. Without this knowledge we are travelling blind—indeed worse than blind, because we don't even know what it is we are looking for, and a blind man at least knows where he wants to go. So we need first of all to have a clear understanding of the effects whose causes we are trying to establish. And in this case you, Sagredo, are much more fully and reliably informed than me, because, besides having been born and having lived for a long time in Venice, where the tides are notable for their size, you have also sailed to Syria and, as someone with an alert and curious mind, you must have made many observations in the course of your travels. I, on the other hand, have only been able to observe for a short time what happens here at this end of the Adriatic, and also down on the shores of our own Tyrrhenian sea;* so I have to rely for much of my information on the reports of other people, which often differ among themselves and so are very unreliable and bring more confusion than confirmation to our investigations. Still, from the features which we do know [444] for certain and which are also the most important, I think I have been able to discover their true primary causes, although I am not so bold as to claim that I can give correct and adequate explanations for those effects which are new to me and which I have not therefore been able to think about carefully. What I am about to say I put forward simply as a key opening the way to a path which no one has trodden before, in the firm hope that more acute minds than mine will be able to explore more widely and to penetrate more deeply than I have been able to with this discovery. And if it transpires that in other distant seas there are variations which do not occur in our Mediterranean, that will not necessarily invalidate the reason and cause which I shall give, provided it is proved to be a correct and complete explanation for what happens in our sea—for there must in the end be just one true primary cause for effects which are of the same kind. So I shall give an account of the effects which I know to be true, and will explain the reason for them which I believe to be correct; and if you, gentlemen, will cite other effects known to you in addition to mine, we shall test whether the cause which I put forward accounts for them as well.

Three periods of the tides: diurnal, monthly and annual.

There are, then, three periods which can be observed in the ebb and flow of the tides.* The first and most important is the biggest and best known, the diurnal motion in accordance with which the waters rise and fall at intervals of a few hours; in the Mediterranean these are of roughly six hours each, that is, the water rises for six hours and falls for six hours. The second interval is monthly, and appears to derive from the motion of the Moon: not that the Moon introduces any new motion, but it affects the range of the motion already described, which is significantly different depending on whether there is a full, new, or half Moon. The third interval is annual, and appears to derive from the Sun; this also affects just the diurnal motion, with differences between the range of the tides at the solstices and those at the equinoxes.

Variations which occur in the diurnal period of the tides.

Let us first consider the diurnal period, as this is the main one, on which the Moon and the Sun seem to exert their secondary influences with their monthly and annual alterations. These hour-by-hour changes can be seen to be of three kinds: in some places the water rises and falls without making any forward motion; in others it moves now towards the east and then back towards the west, without rising or falling; while in still other places their motion varies both in height and in their [445] course, as is the case here in Venice, where the water rises as it comes in and falls as it goes out. This happens at the extremity of a gulf which extends from west to east and ends on a beach which allows the water to spread out as it comes in; if its course was intercepted by mountains or very steep banks, it would rise and fall without any forward motion. Elsewhere the water flows back and forth without any change in level, as happens very strikingly in the Strait of Messina, between Scylla and Charybdis,* where the currents are very swift because of the narrowness of the channel. In the open sea, and around islands surrounded by open sea such as the Balearics, Corsica, Sardinia, Elba, the coast of Sicily facing Africa, Malta, Crete, etc., changes in level are very small but there are substantial currents, especially where there is a narrow stretch of sea between islands or between an island and the mainland.

Now it seems to me that these effects on their own, confirmed and indisputable as they are, are enough to convince anyone who seeks to stay within the terms of natural science that the mobility of the Earth is, at least, highly probable. The alternative—that the Mediterranean basin remains motionless and the water it contains behaves in the way it does—is beyond my imagination and, I would guess, that of anyone who delves beneath the surface in considering this matter.

The cause of the ebb and flow of the tides according to a certain modern philosopher.

SIMPLICIO. These phenomena are not new, Salviati; they have been studied by countless scholars for centuries, and many have exercised their minds to provide one or another explanation for them. A great Peripatetic philosopher* not many miles from here has put forward a cause which he has brought to the surface from a text of Aristotle, whose significance other interpreters had overlooked. He finds that, according to this text, the true cause of the motion of the tides is simply the varying depth of the sea: deeper water, being greater in volume and therefore heavier, displaces the shallow water which then tries to flow back, and it is this constant struggle which causes the ebb and flow of the tides. Then there are many who relate the tides to the Moon, saying that the Moon has a particular domination over water. There is a prelate* who has recently published a treatise arguing that as the Moon moves across the sky, it attracts a body of water which continually follows it, so that the sea is always higher where it lies under the Moon. As for the fact that the water still rises even when the Moon is below the horizon, he [446] says that the only way to account for this is that the Moon, besides having this natural capacity itself, is able also to transmit the same power to the opposite sign of the zodiac. Others, as I expect you know, say that the Moon with its temperate heat has the power to rarefy water so that it rises when it is rarefied.* Then there are also those who...

The cause of the ebb and flow of the tides attributed to the Moon by a certain prelate.

Girolamo Borro and other Peripatetics explain the tides by reference to the temperate heat of the Moon.

SAGREDO. Please, Simplicio, spare us any more: I don't think it is worth wasting time reproducing these theories, let alone wasting words on refuting them. You would be doing an injustice to your intelligence if you were to go along with these or any other such nonsensical ideas, when you have just rid yourself of so many others.

SALVIATI. I am not quite so impatient as you, Sagredo, so I don't mind putting in a few words to answer Simplicio, if he thinks there is anything probable in what he has been telling us. So, Simplicio, water whose outer surface is higher displaces water which is below it, but water which is deeper has no such effect; and once the higher water has displaced the lower it quickly becomes calm and finds its level again. This Peripatetic philosopher of yours must believe that all the lakes in the world which have no tide, and all the seas where the motion of the tides is imperceptible, have a completely level bed—and I was so naïve as to think that, in the absence of evidence from soundings, islands projecting above the surface of the water were a clear indication of the unevenness of the seabed. As for the prelate, you could point out to him that the Moon moves over the whole surface of the Mediterranean every day, but the water only rises at its eastern extremity and for us here in Venice.* To those who say that the Moon's temperate heat is enough to swell the water, tell them to put a saucepan of water on the fire and put their right hand into it until the heat makes the water rise just an inch, and then take their hand out and write about the swelling of the sea; or at least ask them to explain why the Moon rarefies the water in some places and not others, as for example here in Venice and not in Ancona, Naples, or Genoa. It must be said that there are two kinds of poetic inspiration: there are those who have the gift and skill to invent fables, and there are those who are inclined and ready to believe them.

Response to the vain theories adduced as causes of the ebb and flow of the tides.

Islands are an indication of the unevenness of the seabed.

Two kinds of poetic inspiration.

SIMPLICIO. I don't think anyone believes fables once they understand that that is what they are. As for the many opinions about the causes of the tides, I know that any given effect can have only one true primary cause, and therefore I understand very well that only one explanation can be true and all the others are false, like fables. Indeed, it may well be that none of the explanations produced so far is the true one. Actually,

[447]

I believe that that is the case, for it would be very strange if the truth were so obscure that it completely failed to stand out against the darkness of so many errors. But I will permit myself to say, since we have licence to speak freely among ourselves, that introducing the motion of the Earth as an explanation of the ebb and flow of the tides seems to me every bit as much a fable as any of the others I have heard. And until I am given reasons which are more in keeping with natural phenomena, I will have no hesitation in believing that the tides are a supernatural effect, and as such are a miracle inscrutable to human understanding—as indeed are many others which derive directly from the omnipotent hand of God.

The truth is not so obscure that it fails to stand out against the darkness of errors.

SALVIATI. That's a very prudent line of argument, and one which follows Aristotle's teaching, for as you know, at the beginning of his *Mechanics* he defines as miraculous those things whose causes are hidden. But that the true cause of the tides is one of these incomprehensibles, I think the strongest indication is that not one of the explanations which has hitherto been put forward as the true cause can be made, by whatever artificial means we may try, to produce a similar effect. For neither moonlight or sunlight, temperate heat or differences of depth, can artificially make water in a motionless container flow back and forth, or rise and fall, in one place and not in another. But if by simply moving the container, without any artificial aid, I can replicate for you exactly all these changes that we observe in the water of the sea, why should you reject this explanation and fall back on the miraculous?

Aristotle defines as miraculous those things whose causes are unknown.

SIMPLICIO. I will fall back on the miraculous unless you are able to persuade me with other natural causes than moving the basins which contain the water of the sea, because I know that these basins do not move, because the whole of the terrestrial globe is by its nature immobile.

SALVIATI. But you believe, don't you, that the terrestrial sphere could be made mobile supernaturally, that is, by the absolute power of God?

SIMPLICIO. Of course; who could doubt it?

[448] SALVIATI. Well then, Simplicio, since to explain the ebb and flow of the tides we need to introduce a miracle, let's say that the miracle consists in making the Earth move, and that

the motion of the sea follows naturally from that. This will be much simpler—I might say, more natural—as miracles go, just as it is easier to set a sphere moving in a circle, as we see so many other spheres move, than to make an immense body of water move to and fro more quickly in some places than others, to rise and fall more in some places than others and in some places not at all, and all these different motions in the same container. This would be a whole series of different miracles, whereas making the Earth move is just one. What's more, the miracle of moving the water would require another miracle as a consequence, namely making the Earth stand firm against the impetus of the water, which would be powerful enough to move the Earth first one way and then the other if it were not miraculously held in place.

SAGREDO. Please, Simplicio, let's suspend our judgement for a while before we condemn as vain this new opinion which Salviati is trying to explain to us, and let's not be too ready to lump it together with the ridiculous older theories. As for miracles, let's not have recourse to them until we have heard what he has to say within the bounds of nature—although to my way of thinking, all the works of God and nature appear miraculous.

SALVIATI. I think so too; and saying that the motion of the Earth is the natural cause of the ebb and flow of the tides doesn't make it any less of a miracle. So to return to our discussion, I say again that no one has yet explained how it is that the waters in our Mediterranean basin can move as they do, while the basin itself which contains the water remains at rest. The reasons for this difficulty, which so defy explanation, lie in the daily observable facts which I shall now describe; so listen carefully.

Showing that it is impossible for the ebb and flow of the tides to occur naturally if the Earth is at rest.

We are here in Venice; the tide is low, the sea is calm, and the air is still. The water begins to rise, and in the space of five or six hours rises by three feet or more. This is not because the original water has become rarefied; it is water which has newly arrived from elsewhere, of exactly the same kind as the water that was here before, with the same salinity, the same density, the same weight. Ships float in it exactly as before, not lying a hair's breadth lower in the water; a barrel of this new water

[449]

doesn't weigh an ounce more than an equal quantity of the other; it has the same temperature, not a degree different. In short, it is water which we can see has just come in to the lagoon through the gaps and channels of the Lido. Now please tell me where this water has come from, and how it got here. Do you think there are abysses or channels in the bottom of the sea, through which the Earth draws in water and expels it again, like the breath of an immeasurably enormous whale?* In that case why does the water not rise by the same amount in Ancona, Ragusa, or Corfu,* where the variation in the tides is so small as to be barely perceptible? How can anyone possibly pour more water into a motionless container in such a way that the level rises in one defined part of it and not in another?

Perhaps you will suggest that this extra water comes from the wider Ocean, flowing in through the Strait of Gibraltar; but this creates even bigger difficulties, without solving the ones we've mentioned already. To start with, what must be the speed with which this water flows, if it comes in through the Strait of Gibraltar and reaches the furthest shores of the Mediterranean in just six hours? This is a distance of two or three thousand miles—and it would have to flow the same distance in the same time when it runs out. What would become of the ships scattered across the sea? What about those in the Strait itself, on a continuous torrent of an immense volume of water, enough to spread across an area hundreds of miles wide and thousands of miles long in just six hours, all through a channel not more than eight miles wide? What tiger or falcon ever ran or flew at such a speed—enough to travel four hundred miles or more in an hour? No one denies that there are currents running the length of the Mediterranean, but they are slow enough for a vessel equipped with oars to overcome them, albeit with some loss of headway. And there is a further difficulty if the water comes through the Strait of Gibraltar: how, in this case, can it produce such a rise in the sea in places so far away from the Strait, without first causing a similar or greater rise in places which are closer to it? In short, I don't think that the most stubborn obstinacy or the most subtle ingenuity, if we stay within the terms of natural science, can ever find a way round these difficulties while still maintaining that the Earth is at rest.

SAGREDO. I'm entirely convinced by this, and I'm eager to hear how these marvels can follow unimpeded from the motions we have already assigned to the Earth.

SALVIATI. For these effects to follow as a consequence of the motions that belong naturally to the Earth, they must not only occur unimpeded and unopposed; they must follow easily, indeed necessarily, in such a way that they could not possibly be otherwise. For this is what characterizes true natural effects. So, having established that it is impossible to explain the motions of the sea while at the same time maintaining the immobility of the basin which contains it, let us go on to see whether we can explain the observed effects by positing the mobility of the container.

[450]

True natural effects follow easily from their causes.

There are two kinds of motion which can be imparted to a container so that the water it contains acquires the property of flowing towards one end of the container and then the other, and of rising and falling there. The first is when one or the other end of the container is lowered, for then the water flowing towards the lower end rises and sinks alternately at either end. But this rising or falling is simply motion away from or towards the centre of the Earth, and such motion cannot be attributed to the concavities in the Earth which contain the waters of the sea. No motion we might attribute to the terrestrial globe can cause any part of the Earth's surface to move towards or away from its centre.

Two kinds of motion of the containing vessel can make the water contained in it rise and fall.

Concavities in the Earth's surface cannot move towards or away from its centre.

The second kind of motion is when the container, without being tilted at all, moves forward, not uniformly but at varying speeds, sometimes accelerating and sometimes slowing down. The water in the container is not fixed to it like any of its solid parts; rather, being liquid, it is as it were free and separated from the container, and not constrained to participate in every movement which the container makes. So the result of this uneven motion is that when the container slows down, the water retains some of the impetus which it has already acquired, and flows towards the front of the container, so that its level necessarily rises there. Conversely, when the container gains speed, the water retains some of the earlier slowness and falls behind, flowing towards the back of the container and rising there, before it adapts to the new impetus. We can see these effects very clearly with our

Uneven forward motion can make the water in a container move about.

own eyes in the example of one of the barges which continually come from Fusina filled with fresh water to be used in the city. [451] Picture one of these barges moving steadily across the lagoon, with the water it contains perfectly calm. Then its progress is significantly impeded, either because it runs aground or because it strikes some other obstacle. The water which it contains will not lose its forward impetus to the same degree as the barge, but will retain some of it, flowing forwards towards the bow, where it will rise perceptibly, and falling at the stern. If on the other hand the same barge, in the midst of its steady progress, acquires a new burst of speed, the water it contains will not adapt straight away but will retain some of its slowness and will stay back flowing towards the stern, where it will consequently rise, sinking at the bow.

There are three things we should note about this effect, which is plain to see and can be tested at any time. The first is that, for water to rise at one end of the container, there is no need for any additional water, nor does the water have to flow all the way from the other end. The second point is that the water in the middle of the barge does not rise or fall appreciably—unless, that is, the barge was moving very rapidly and the impact or some other obstacle which impeded it was very powerful and sudden; in this case not only might all the water rush forwards, but much of it might splash out of the barge. The same would happen if the barge was moving slowly, and suddenly received a violent impulse propelling it forwards. But if the barge was moving steadily and was moderately retarded or accelerated, the rise or fall of the water in the middle would, as I have said, be imperceptible. In the other parts of the barge it would rise less the nearer it was to the middle, and more the further it was away. The third point to note is that, whereas the water in the middle rises and falls very little in comparison with the water at either end, it runs much further forwards and backwards in comparison with the water at the extremities.

The parts of the terrestrial globe accelerate and slow down in their motion.

Now, gentlemen, the effect of the barge on the water it contains and of the water on the barge which contains it, is exactly the same as the effect of the Mediterranean basin on the waters it contains, and of the waters on the Mediterranean basin which contains them. So we shall now proceed to show how it is that the Mediterranean basin and every other bay or inlet, in short every part of the Earth, moves with a significantly uneven motion, even though we assign nothing but regular and uniform motion to the globe itself.

SIMPLICIO. I'm not a mathematician or an astronomer, but [452] this strikes me at first sight as highly paradoxical. If it's true that the motion of the whole is regular and yet the motion of the parts, even if they remain joined to the whole, can be irregular, this paradox would seem to defy the axiom which states *eandem esse rationem totius et partium*: what is said of the whole also holds for the parts.

SALVIATI. I'll prove my paradox and leave it to you, Simplicio, either to defend the axiom against it or to make the two agree. My proof will be brief and very simple, and will draw on matters which we have dealt with at length in our discussions so far; and I won't say a word in favour of the ebb and flow of the tides.

Showing how the parts of the terrestrial globe are accelerated or retarded.

As we have said, two motions are attributed to the terrestrial globe. The first is the annual motion of the Earth's centre on the circumference of its orbit around the ecliptic, in the order of the signs of the zodiac, that is from west to east. The second is the same globe's rotation on its own axis in twenty-four hours, also from west to east, although its axis is somewhat inclined and is not parallel to that of its annual revolution. Now the composition of these two motions, each uniform in itself, results in a motion which varies in different parts of the Earth. This will be easier to understand if I explain this by means of a drawing.

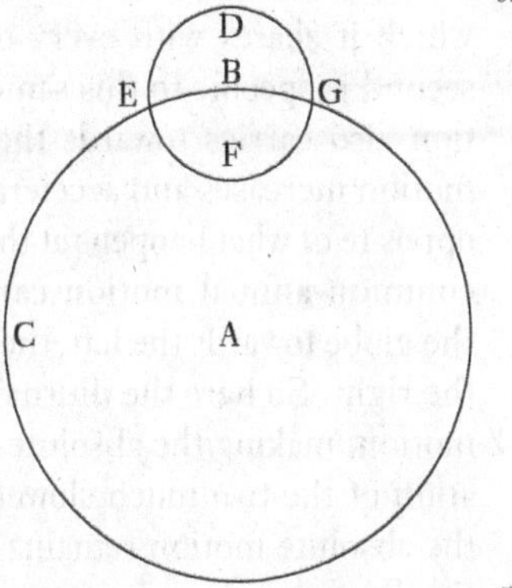

The parts of a circle moving uniformly around its centre move with contrary motions at different times.

First, I will draw the circumference of the Earth's orbit, BC, around its centre, A. Then, taking any point B on this circumference, let us draw this smaller circle, DEFG, with B as its centre, to represent the terrestrial globe. We know that this globe, with its centre B, moves around the circumference of its orbit from west to east, i.e. from B in the direction of C. We also know that the terrestrial globe rotates around its centre B, also from west to east, i.e. in the sequence D, E, F, G, in the space of twenty-four hours. But here we must be careful to note that as the globe rotates on its axis, any given part of it will move at different times with contrary motions. This will be clear if we consider that when the parts of its circumference around point D are moving towards the left, in other words towards E, the opposite parts which are around point F are moving towards the right, in [453] other words towards G. So when the parts around D reach point F their motion will be in the contrary direction to what it was when they were at D; and when the parts at point E are as it were descending towards F, those at point G are ascending towards D.

The mixture of the annual and daily motions causes the unevenness of motion in different parts of the terrestrial globe.

Given these contrary motions of the different parts of the surface of the globe as it rotates on its axis, it follows that when this daily motion is combined with the other annual motion, the result must be an absolute motion for the parts of the Earth's surface which is significantly accelerated at one time and equally slowed down at another. This will be clear if we look at the part around point D. Its absolute motion will be very rapid, being the product of two motions made in the same direction, namely towards the left. The first is part of the annual motion which it shares with every other part of the globe, while the second is specific to this same point D, which the diurnal rotation also carries towards the left; so in this case the diurnal motion increases and accelerates the annual motion. This is the opposite of what happens at the opposite point F. Here, while the common annual motion carries it together with the rest of the globe towards the left, the diurnal rotation carries it towards the right. So here the diurnal motion works against the annual motion, making the absolute motion produced by the combination of the two much slower. Finally, around points E and G the absolute motion remains equal to the annual motion, since the diurnal motion, being upwards and downwards rather than towards the left or right, has little or no effect on these points. We must conclude, therefore, that while the motion of the globe as a whole and of each of its parts would be regular and uniform if they had only a single motion—either just the annual motion or only the diurnal one—it follows that when the two motions are combined they produce unequal motions for the different parts of the globe, sometimes accelerated, sometimes held back, because the daily rotation is either added to or subtracted from the annual revolution. So if it is true (as experience clearly shows that it is) that acceleration and slowing down of the motion of a container makes water flow back and forth along its length, and rise and fall at each end, surely it must be conceded that this effect can, indeed must, affect the waters of the sea, when the basins in which they are contained are subject to such variations. And this must be especially true of seas whose length extends from west to east, as this is the direction along which the containers do in fact move.

Primary and most powerful cause of the ebb and flow of the tides.

[454] This, then, must be the primary and most powerful cause of the ebb and flow of the tides, without which they would not exist at all. But there is also a wide variety of individual phenomena which can be observed at different times and places, which must derive from other concomitant causes even if these are all linked to the primary cause. So we need to identify and examine the various factors which could be the reason for these effects.

Various phenomena which occur in the ebb and flow of the tides. First phenomenon: water which rises at one extremity returns to equilibrium of its own accord.

The first such phenomenon is that when a significant acceleration or slowing down of the container causes water to flow towards one or other of its extremities, it rises at one end and falls at the other, but it does not remain in this state when the primary cause has ceased. Rather, its own weight and its natural inclination to find its own level make it flow rapidly back again; and being both heavy and a liquid, it does not simply flow back to a state of equilibrium, but its own impetus carries it further, so that it rises at the end where it had originally fallen. Even then it does not stop but runs back again, and flows back and forth several times, showing its reluctance to pass instantly from the velocity it had acquired to a state of rest; rather, its motion is gradually reduced, a little at a time. It is exactly the same phenomenon as we see with a weight suspended from a cord, which when it is moved from its state of rest—that is, from the perpendicular—returns to the perpendicular and comes to rest of its own accord, but only after going beyond it many times as it swings to and fro.

In a shorter container the oscillations are more frequent.

The second phenomenon to note is that these reciprocal oscillations occur with greater or lesser frequency—that is, with shorter or longer time intervals—according to the varying length of the basin containing the water. In a shorter space the oscillations are more frequent, and in a longer space they are rarer.

Again this is exactly the same as can be observed in pendulums, where the oscillations of those with a shorter cord are more frequent than in those where the cord is longer.

Greater depth makes the oscillation of the water more frequent.

This leads to the third notable fact, which is that it is not only the varying length of the container that affects the frequency of the water's oscillation. The greater or lesser depth of the water has the same effect. If water is contained in receptacles of equal [455] length but different depths, the deeper water will settle with shorter oscillations, while the oscillations will be less frequent in the water which is shallower.

Water rises and falls at the extremities of the container, and flows in the middle parts.

The fourth point which we should note and observe carefully is that water produces two effects as it finds its level. One is rising and falling alternately at each extremity of the container; the other is flowing back and forth horizontally. These two diverse motions occur differently in different parts of the water. The parts at either end are those which rise and fall the most; those in the middle do not rise or fall at all. Of the other parts, those which are nearer to the extremities rise and fall proportionately more than those which are further away. In contrast, the other motion back and forth is more marked in the parts in the middle, and does not occur at all in those parts at the extremities (unless they rise so much that they spill over the edge and overflow out of their original container; where the banks hold them back they simply rise and fall). This does not, however, prevent the water in the middle from flowing back and forth, and this motion also occurs proportionately in the other parts, which flow to a greater or lesser extent according to their distance from the middle.

Phenomena concerning the motions of the Earth which cannot be illustrated in practice.

We should note the fifth particular phenomenon all the more carefully because it is impossible for us to put its effect into practice experimentally. It is this: when containers which we construct, such as the barges we mentioned above, are set in motion at a greater or lesser speed, the acceleration and deceleration are shared in the same way by the container as a whole and by each of its parts. So, for example, if the barge's motion is held back, the forward part is not slowed down more than the part which follows, but every part shares equally in the same deceleration. The same happens with acceleration; if the barge receives a new stimulus to greater speed, the bow and the stern both accelerate in the same way. But in immense basins such as enormously long seabeds, the extraordinary fact is that their extremities do not increase or decrease in speed jointly, equally, or at the same moment in time, even though these basins are simply hollows formed in the solid surface of the terrestrial globe. What happens is that when the motion at one extremity is very much slowed down, because of the combined effect of the diurnal and annual motions, at the other extremity the two motions are still reinforcing each other to produce a very rapid motion. This will be easier to understand if we go back to the drawing which we made just now.*

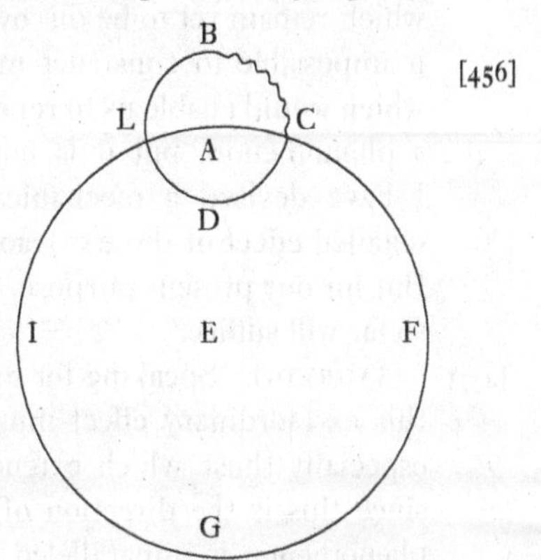

[456]

Let us suppose that a stretch of sea is, say, a quadrant in length, such as the arc BC. We saw above that the parts near point B are in very rapid motion, because of the combined effect of the diurnal and annual motions both moving in the same direction. But at the same time point C is moving very slowly, because the forward motion deriving from the diurnal motion has been cancelled out. So we can see that in a gulf of the sea whose length extends over the arc BC, its extremities are moving at the same time at very unequal rates. The differences would be greatest of all in a stretch of sea extending over a semicircle, such as BCD. Here one extremity, at B, would be in very rapid motion; the other extremity, D, in very slow motion; and the middle parts around C would be in moderate motion. Shorter stretches of sea would be less subject to this strange phenomenon of having their different parts affected by motion at different rates at certain times of the day. Now since in the first case we know from experience that acceleration and deceleration can cause the water in a container to flow back and forth, even when these are shared equally by every part of the container, then what are we to suppose will happen in a container so remarkably situated that the acceleration and deceleration are conferred very unevenly among its various parts?

Surely we can only say that there must be other, even greater and more extraordinary causes of turbulence in the water, which remain yet to be discovered. Many people may consider it impossible to construct machines and artificial containers which would enable us to reproduce and test the effects of such a phenomenon, but it is not completely impossible; in fact I have devised a mechanical model* which illustrates the detailed effect of these extraordinary combinations of motion. But for our present purpose, what you have been able to grasp so far will suffice.

[457] SAGREDO. Speaking for myself, I can well understand how this extraordinary effect must arise in the basins of the sea, especially those which extend a long way from west to east, since this is the direction of the Earth's motions. Since this phenomenon is unparalleled among the motions that we can produce, and gives rise to effects which we cannot recreate artificially, I believe it is impossible for us to duplicate it.

Explanations for the particular phenomena observed in the ebb and flow of the tides.

Secondary causes for the lack of tides in small seas and lakes.

SALVIATI. Now that we have dealt with these points, it is time for us to examine the particular diverse phenomena which can be observed in the ebb and flow of the tides. First, it should not be difficult for us to understand why it is that lakes, ponds, and even small seas have no significant tides. There are two clear reasons for this. The first is that the small size of the basin means that, in acquiring different degrees of velocity at different times of the day, it acquires them with only small differences between all its parts. The parts in front and those behind, i.e. the eastern and western parts, accelerate and decelerate with very little difference between them. Moreover, these changes come about gradually; the motion of the container is not abruptly impeded by a sudden obstruction or suddenly subjected to a rapid acceleration, so that it and all its parts gradually and evenly acquire the same degrees of velocity. This uniformity means that the water in the basin also receives the same velocity with hardly any turbulence or resistance, and so its rising and falling, and its flowing towards one or the other extremity, is barely perceptible. We can see this effect clearly in a small artificial container, where the water will gradually acquire the same degrees of velocity if the container is made to accelerate or decelerate slowly and uniformly. But in basins of the sea which extend a large distance from east to west, the acceleration and deceleration are much more marked and uneven, as one of its extremities will have a very slow motion while the other is moving very fast.

The second reason is the reciprocal oscillation of the water as it settles after the impetus it receives from the motion of its container. As we have seen, this oscillation has a high frequency in a small basin. The reason why the Earth imparts motion to the waters only at twelve-hour intervals is that the maximum [458] acceleration and deceleration in the basin's motion take place only once a day. But the second cause of motion, which derives from the weight of the water as it seeks to return to equilibrium, produces oscillations of only one, two, three hours and so on, depending on how short the basin is. The combining of this motion with the first, which in a small basin is very small in any case, has the effect of making the first motion almost imperceptible. For before the motion deriving from the first cause, which has a period of twelve hours, has fully taken effect, it is overtaken and counteracted by the secondary movement deriving from the weight of the water, which has a frequency of one, two, three, or four hours, etc., depending on the length and depth of the basin. The secondary motion, running counter to the first, disrupts it and cancels it out before it can reach its maximum or even the midpoint of its extent. These conflicting motions have the effect of negating the ebb and flow of the tides, or at least of making them much less apparent. This is quite apart from the continuous change produced by the air, which ruffles the surface of the water so that we would be unable to perceive such very small variations in level, of half an inch or less, as might exist in basins or inlets which are only a degree or two in length.

The reason why the tides are generally made in six-hour periods.

I come, secondly, to the question of why the period of the ebb and flow of the tides commonly appears to be six hours, when the principal cause of the motion of the waters acts at intervals of twelve hours, i.e. once because of the maximum velocity of its motion and once because of its maximum slowness.* The answer is that this could not possibly happen solely as a result of the primary cause, but account must be taken of the secondary causes as well, i.e. those deriving from the greater or lesser length of the containers and the greater or lesser depth of the waters they contain. These have no effect on the motion of the waters; this is the product only of the primary cause, without which there would be no tides at all. They do, however, have a very powerful effect in determining the frequency of their oscillation, so powerful that the primary cause has to submit to them. So a period of six hours is no more inevitable or natural than any other time interval; but it may well be the one that has been most observed, since this is the period that pertains in our Mediterranean, which for many centuries was the only sea that was navigable. As a matter of fact, this period is not observed [459] everywhere even in the Mediterranean. In some of the narrower parts of the sea, such as the Aegean and the Hellespont, the tides have a much shorter period, and even vary a great deal among themselves. Indeed, some say that after Aristotle had observed the tides at length from some cliffs in Euboea, their unpredictability and his inability to find an explanation for it drove him to such desperation that he threw himself into the sea and drowned.*

Why some seas have no tides, despite being very long.

Thirdly, we can readily explain why it is that some seas which are very long, such as the Red Sea, are almost impervious to the effect of tides. The reason for this is that the length of the Red Sea extends not from east to west but from south-east to north-west, and since the Earth's motion is from west to east the impulses moving the water are always across the lines of longitude rather than those of latitude. Therefore seas which extend lengthwise towards the poles, and are narrow in the other direction, have no reason to be subject to tides, apart from any effect they may share with other seas which flow into them and which are subject to large tidal variations.

Why tides are most pronounced at the extremities of a gulf, and most limited in its central parts.

Fourthly, it should be easy for us to see why the rise and fall of the tide is most marked at the extremities of a gulf, and most limited in its central parts. We see this clearly in everyday experience here in Venice, at the extremity of the Adriatic, where the tidal range is five or six feet, whereas in parts of the Mediterranean further from the extremities it is not more than half a foot—on the islands of Corsica and Sardinia, for instance, or on the coasts of Rome or Livorno. We can also see how, in contrast, the movement back and forth is extensive where the rise and fall are limited. It is easy, as I say, to understand the reason for these phenomena, since we have seen them clearly replicated in various kinds of artificial containers, where the same effects follow naturally when we subject the container to uneven motion, i.e. to acceleration and slowing down.

Why waters flow more rapidly through narrow spaces than open ones.

Fifthly, we know that a given quantity of water that moves, albeit slowly, in a wide channel, necessarily flows with greater impetus if it has to pass through a narrow opening. So it is not difficult to understand the reason for the strong currents which [460] occur in the narrow channel separating Calabria and Sicily. All the water that is held back by the extensive coast of Sicily and the Ionian Gulf flows down towards the west, although only slowly because of its vast amount. But when it is channelled into the strait between Scylla and Charybdis* it drops rapidly and becomes very turbulent. A similar but much greater effect must occur between Africa and the large island of Madagascar, where the waters of the Indian and Atlantic* oceans, which the island divides, are channelled into an even narrower strait between it and the coast of Africa. There must also be very strong currents in the Magellan Strait, which links the vast Atlantic and Pacific oceans.

An account of some of the more obscure phenomena which can be observed concerning the tides.

Let us go on now, as our sixth point, to give an account of some of the more obscure and controversial phenomena which can be observed in this matter of the tides; and here we need to consider another important point about their two principal causes, and how they combine and mingle together. As we have already noted more than once, the first and simplest cause is the regular acceleration and deceleration of the different parts of the Earth, which should cause the seas to flow eastwards and westwards at regular intervals in the space of twenty-four hours. The second is the motion deriving from the weight of the water itself which, having once been set in motion by the primary cause, seeks to restore itself to equilibrium by repeated oscillations. These do not have any one fixed frequency, but are as varied as the varying length and depth of the different basins and inlets of the seas, so that this second principle can cause the seas to flow in one direction and back again in an interval of one hour, or of two, four, six, eight, or ten hours, etc. If we now start to combine the primary cause, which has a fixed period of twelve hours, with one of the secondary causes with a period of, say, five hours, it will sometimes happen that the two causes coincide and both impel the water to move in the same direction. This conjunction or, as it were, conspiring together of the two motions will mean that the tides will be very great. At other times it will happen that the primary cause works against the effect of the secondary cause; then, since one of the principles cancels out the effect which the other should have, the motion of the water will be weakened, and the sea will be reduced to [461] a state where it is tranquil and barely moving. There will be other times again when these two principles neither wholly counteract nor wholly reinforce each other, and then they will cause other variations in the growth and diminution of the tides.

It can also happen that the mingling of the two principles of motion can give rise to contrary motions in two large seas linked by a narrow channel, so that as one is rising the other is flowing in the opposite direction; in such a case the channel between them can become exceptionally rough, with opposing currents and very dangerous whirlpools and turbulence. We continually hear reports from those who have experienced such phenomena. These conflicting motions, arising as they do not only from the different orientation and length of the seas but also to a large extent from differences in their depth, will sometimes produce disturbances in the water which are unpredictable and unobservable. The reasons for these have been and remain a cause of great anxiety to sailors, who encounter them without any apparent impetus from the wind or any other major disturbance in the air which might explain them. We should, indeed, recognize that disturbances in the air play a large part in other phenomena, and indeed we should acknowledge them as a third incidental cause which is powerful enough to bring about major changes in the effects we observe from the two primary, more essential causes. It is quite clear, for example, that a strong wind blowing continuously from the east can hold back the water and prevent the tide from going out. This means that when the second phase of the tide comes in, and then the third, their volume is much increased; so that, if the wind holds the water back for several days, the tides rise higher than usual and cause exceptional floods.

We should also note—and this will be our seventh question—another cause of motion, which arises when a large volume of water from rivers flows into a sea which is not very large. In the channels or straits which communicate with such a sea the water will always flow in the same direction, as happens in the Bosphorus at Constantinople, where the water always flows out of the Black Sea towards the Sea of Marmara. The reason for this is that in the Black Sea, since it is not very long, the principal causes of the tides have little effect; but some very large rivers flow into it, and all this volume of water has to pass through the strait, so that there is a notable current in the strait which always flows towards the south. We should note, too, that although this strait or channel is quite narrow, it is not subject to the same perturbations as the Strait of Messina. This is because the Bosphorus has the Black Sea to the north, and the Sea of Marmara, the Aegean, and the Mediterranean to the south, which extend over a long distance; and we have seen that seas which run north–south, however long they are, are not subject to tides. The Strait of Messina, on the other hand, lies between the parts of the Mediterranean which extend for a great distance from east to west, i.e. in the direction of the ebb and flow of the tides, and this is why the turbulence is so great there. It would be even greater between the Pillars of Hercules, if the Strait of Gibraltar were any narrower; and it is said that in the Magellan Strait the water is very rough indeed.

The reason why in some narrow straits the sea's water always flows in the same direction.

[462]

This is all that comes to my mind to say now concerning the causes of this basic diurnal motion of the tides and the phenomena associated with it. If there are any points to be raised we can consider them now, before going on to discuss the other two motions, the monthly and the annual.

SIMPLICIO. I don't think it can be denied that your reasoning is very plausible if you argue *ex suppositione*, as we say—that is, assuming that the Earth moves with the two motions attributed to it by Copernicus. But take these two motions away, and everything else falls to the ground; and your own argument clearly shows that such a hypothesis cannot be maintained. You presuppose the twofold motion of the Earth to explain the ebb and flow of the tides, and then you argue in a circle, because you go on to cite the tides as evidence to confirm these same motions. More specifically, you say that as water is a liquid and so is not firmly fixed to the Earth, it is not obliged to follow the

Against the argument that the hypothesis of the Earth's motion is supported by the ebb and flow of the tides.

Earth's motion in every detail, and from this you deduce the motion of the tides. I shall argue the opposite by following in your footsteps, as follows. The air is much thinner and more fluid than water, and even less attached to the surface of the Earth; water adheres to the Earth if only because its own weight presses down on it, unlike the air which is very light. Therefore the air should be much less obliged to follow the motion of the Earth; and hence if the Earth moved in the way you claim, we who inhabit the Earth would be carried along at the same velocity and would feel a constant unbearably strong wind from the east. Our everyday experience tells us this. If we ride post-haste at only eight or ten miles an hour when the air is still, we [463] feel it against our face like a moderately strong wind; so just think what it would feel like if we were moving at 800 or 1000 miles an hour, in the face of the air which did not share this motion. And yet we feel no such effect.

Response to this objection to the rotation of the Earth.

Water is more apt than air to conserve an impetus imparted to it.

SALVIATI. To this objection, which appears very convincing, my response is as follows. It is true that air is much thinner and lighter than water, and that its lightness means that it adheres less to the Earth than water, which is so much heavier and more bulky. But you draw an erroneous conclusion from this when you say that the air's lightness, thinness, and lesser degree of attachment to the Earth make it less constrained than water to follow the Earth's motions, and that therefore we who share fully in these motions ought to feel the effect of this resistance. In fact, what happens is precisely the opposite. If you remember, I said that the tides were caused by the fact that the water does not follow the uneven motion of its container, but retains the impetus which it had received earlier, and that the impetus does not diminish or increase in the water at the same rate as it does in the container. Since, therefore, conserving and maintaining an impetus received earlier means resisting a new increase or decrease in motion, a body which is more able to conserve an impetus will also be better placed to demonstrate the effect which follows from conserving it. Now we can clearly see how water tends to maintain a disturbance imparted to it even when the cause of the disturbance has ceased, because when the sea is whipped up by strong winds it still remains in motion long after the wind has dropped. The divine Poet described this

Light bodies are more easily moved than heavy ones, but are less apt to conserve their motion.

elegantly when he wrote 'Even as the Aegean Sea', etc.* The continuance of this motion depends on the weight of the water; because as we have said on another occasion,* light bodies are more easily moved than heavy ones, but are correspondingly less apt to maintain the motion imparted to them once the cause of the motion has ceased. So the air, because it is so thin and light, is easily set in motion by the slightest force, but it has very little aptitude for conserving this motion when the moving force ceases. As regards the air surrounding the globe, my view is that it adheres to the Earth and is carried along with it no less than water is. This is especially true of that part of the air which is enclosed in a confined space, such as a plain surrounded by [464] mountains. In fact, I think it is much more reasonable to say that this air is swept around by the Earth's rough surface than it is to claim, as you Peripatetics do, that the higher levels of air are swept around by the motion of the heavens.

It is more reasonable to suppose that the air is swept along by the rough surface of the Earth than by the motion of the heavens.

What I have said so far is, I think, a very adequate response to the objection raised by Simplicio. But I intend to go further in satisfying him, and to confirm the motion of the Earth to Sagredo, by identifying a new objection and providing a new response to it, based on a remarkable experiment. I have said that the air is carried along by the roughness of the Earth's surface, especially those parts of the air which are below the summits of the highest mountains. It would seem to follow from this that if the Earth's surface were not uneven, but smooth and polished, there would no longer be a reason for it to pull the air along with it, or at least it would do so less uniformly. Now the surface of our globe is not all rough and jagged; there are very large areas which are quite smooth, namely the surface of the open seas. As these areas are a long way from any mountain ranges which might surround them, there seems to be no reason why they should be able to carry the air above them along with them; and if they don't carry it along with them, then the consequence of their not doing so should be felt in those places.

The rotation of the Earth confirmed by a new argument derived from the air.

SIMPLICIO. I was going to raise just this objection, which seems to me a very effective one.

SALVIATI. Well said, Simplicio; and so, since we do not feel the effect in the air which would follow as a consequence of our globe's rotation, you argue from this that it does not move. But if this effect, which you consider a necessary consequence of the globe's rotation, were in fact felt and experienced, would you accept this as strong evidence in favour of its motion?

SIMPLICIO. In that case you would have to speak to others and not just to me. If such a thing happened, others might know its cause even though I might not.

SALVIATI. It's impossible to win against you, and it would really be better not to play at all rather than always being on the losing side. But I'll press on, so as not to let down our third companion.

Let me repeat and add some details to what I said earlier. There appears to be no reason why the air, as a thin, fluid body not firmly fixed to the Earth, should be obliged to follow the Earth's motion, with the exception of those parts of the air that [465] are close to the Earth's surface or do not extend far above the top of the highest mountains, and are carried along by the roughness of the Earth's surface. This part of the air ought to be all the less resistant to following the Earth's rotation because it is full of vapours, fumes and exhalations, all of which partake of the qualities of earth, and so naturally follow the same motions. But where the causes of this motion are absent—where there are wide, smooth spaces on the surface of the globe, and where there are fewer vapours from the earth mixed in with the air—then the cause which makes the surrounding air submit entirely to the Earth's rotation should be partly removed. In such places, while the Earth rotates towards the east, one ought to feel a constant wind blowing from east to west; and this wind should be most strongly felt in the places where the Earth's rotation is fastest, i.e. in places which are furthest from the poles and nearest to the great circle of the diurnal rotation.

The vaporous air close to the Earth's surface shares in its motions.

Now experience does indeed strongly endorse this philosophical reasoning. In the open seas furthest from land in the torrid zone—i.e. in the tropics—where also there are no evaporations from the Earth, one does feel a permanent breeze blowing from the east. It is so constant that ships have an easy passage to the West Indies, and the same wind favours ships which sail across the Pacific Ocean from the coast of Mexico to what we call the East Indies, although for them they are in the west. Sailing east from these parts, on the other hand, is difficult and uncertain; ships cannot follow the same routes, but

Perpetual breeze felt in the tropics, blowing towards the west.

Ships sail easily towards the West Indies, but with difficulty when they return.

must stay closer to land so as to find other occasional and unpredictable winds which have other causes, such as we land-dwellers constantly experience. There are many different causes giving rise to these winds, which there is no need to go into now; they are occasional winds which affect every part of the Earth without distinction, causing storms in seas far removed from the equator and which are surrounded by the rough surface of the land. In other words, such seas are subject to those disturbances of the air which work against the primary wind, which would be felt perpetually, especially over the sea, if it were not for these accidental winds blowing against it. So you can see how the effects we see in the water and the air marvellously accord with our observations of the heavens to confirm that our terrestrial globe is in motion.

The seas are disturbed by winds coming from the land.

[466]

SAGREDO. I would like to set the seal on this question by telling you about another circumstance which I think you may not be aware of, which also confirms the same conclusion. Salviati, you have cited the phenomenon encountered by sailors in the tropics, namely the constant wind blowing from the east, which I have heard about from those who have made this voyage several times. I understand, moreover—and this is a significant point—that sailors don't refer to this as a wind, but use some other term which I can't recall, perhaps referring to its steadiness and constancy. Apparently, once they have encountered it they set their sails and then, without having to touch them again, make steady progress even when they are asleep. Now this steady breeze could be recognized as such because it blows continually without interruption; if other winds had disturbed it, it would not have been recognized as a distinctive effect quite different from any other wind. This makes me suspect that our Mediterranean sea may also be subject to the same effect, but that we do not notice it because it is often deflected by other winds which override it. I don't have any firm foundations for saying this, but I put it forward as a likely conjecture, based on what I had occasion to note when I was the Venetian consul in Aleppo. I kept a log and record of the dates of departure and arrival of ships between the ports of Alexandria and Alexandretta* and here in Venice. For curiosity's sake, I compared a large number of these and found that on average the homeward

Another observation derived from the air confirming the motion of the Earth.

Voyages in the Mediterranean from east to west are made in a shorter time than those from west to east.

voyages to Venice, in other words voyages from east to west in the Mediterranean, were completed in about 25 per cent less time than those in the opposite direction. So it appears that in general winds blowing from the east are stronger than those coming from the west.*

SALVIATI. I'm very glad to have this information, which is not insignificant as confirmation of the motion of the Earth. It could be argued that all the water of the Mediterranean flows constantly towards the Strait of Gibraltar because all the rivers which flow into it must be discharged into the Atlantic; but I don't think this is enough on its own to account for such [467] a notable difference. This is clear too from the fact that in the Strait of Messina, water flows no less towards the east than towards the west.

SAGREDO. Unlike Simplicio, I'm not concerned to satisfy anyone other than myself, and I'm convinced by all that has been said in this first part of our discussion. So when you're ready to proceed, Salviati, I am ready to listen to you.

SALVIATI. I'll do as you ask, but first I would like to hear what Simplicio thinks, since his verdict will give me an idea of the reception I can expect my arguments to receive from the Peripatetic schools, if they should reach their ears.

SIMPLICIO. I wouldn't want my opinion to be taken as representing the views of others, or as a basis for speculating on what they might say. As I have said more than once, I have no expertise in these kinds of study; those who have penetrated the innermost secrets of philosophy will be able to answer in ways which would not occur to me, as I have only a nodding acquaintance with the subject. Still, to show that I'm not entirely devoid of ideas, I will reply that the effects which you describe, and especially the last, can equally well be explained by the motion of the heavens, with the Earth being at rest. There is no need to bring in any new idea, but only the opposite of what you yourself have introduced. It has been the received teaching in the Peripatetic schools that the element of fire and also a large part of the air are carried along in the diurnal rotation, from east to west, by their contact with the lunar sphere, within which they are contained. So now, still following in your footsteps, I would like us to determine the amount of air

Reversing the argument, the constant motion of the air from east to west is shown to derive from the motion of the heavens.

which shares in this motion, which comes down almost to the top of the highest mountains, and which would extend down to the surface of the Earth if the mountains themselves did not impede it. This is the counterpart to what you say: you affirm that the air which is confined between mountain ranges is carried along by the rough surface of the Earth as it moves; we say, conversely, that the element of air is all carried along by the motion of the heavens, except for the part confined between mountain ranges, which is prevented from moving by the rough surface of the Earth which is at rest. And whereas you say that if it were not for this roughness the air would not be carried along, we can reply that if it were not for this roughness the air would all continue in motion.

[468] Hence, because the surface of the open sea is smooth and flat, it is exposed to the constant motion of the breeze blowing from the east; and this motion is felt most strongly in the tropics, near the equator, where the motion of the heavens is fastest. And as this celestial movement is powerful enough to carry along with it all the air that is unimpeded, it is reasonable to say that it contributes the same motion to water, which is a liquid and is not bound by the immobility of the Earth. We can be all the more confident in affirming this because, as you have acknowledged yourself, this movement is very small in comparison to its efficient cause. The motion of the heavens, as it goes around the whole terrestrial globe in a natural day, passes over many hundreds of miles in an hour, especially at the equator, whereas the currents in the open sea flow at only a few miles per hour. This means that ships sail quickly and easily westwards not only because of the constant breeze from the east, but also because of the current. This same current may also be the cause of the tides; the water striking the varying alignment of the sea shore turns back in the contrary direction. We can see this effect in the course of rivers, where the water forms eddies and rebounds if it encounters a projection in the river bank or if there is a hollow under the surface. Hence, it seems to me that the same effects which you cite as evidence for, and which you attribute to, the Earth's motion, can be quite satisfactorily explained assuming that the Earth is fixed and that motion belongs to the heavens.

The motion of water derives from the motion of the heavens.

The ebb and flow of the tides may also derive from the diurnal motion of the heavens.

SALVIATI. There's no denying that your argument is ingenious and appears very plausible—but plausible in appearance, not in reality. What you say is in two parts. In the first part you give an explanation for the constant motion of a breeze from the east, and for a similar motion in the water; in the second you suggest that the cause of the tides may also originate from the same source. The first part has, as I have said, some semblance of probability, although much less than my explanation based on the motion of the Earth; the second is not just wholly improbable, but absolutely impossible and false. Coming to the first, you say that the inner surface of the sphere of the Moon sweeps along with it the element of fire and all the air down to the summits of the highest mountains. To this I reply, first, that it is doubtful whether the element of fire exists; and that even if it does, it is even more doubtful whether the lunar sphere, and for that matter all the other spheres, exist as vast solid bodies. Rather, a good number of philosophers are now beginning to believe that beyond the outer limit of the air there is an uninterrupted expanse of a substance much thinner and purer than our air, through which the planets follow their courses. But whichever view is correct, there is no reason to suppose that mere contact with a sphere, which you yourself say is absolutely smooth and polished, should be able to carry the whole element of fire with it in a circular motion alien to its natural inclination. This has been proved and demonstrated at length, citing the evidence of the senses, in *The Assayer*.* This is quite apart from the further improbability that this motion could be transmitted from the very refined element of fire to the much denser element of air, much less from air to water. Whereas it is not only probable but necessary that a body with a rough and mountainous surface, as it rotates, should carry along with it the air which is in contact with it and which strikes its projecting parts. The evidence for this is plainly visible, although even without seeing it I don't believe anyone who thinks about it could doubt it.

The continual motion of air and water can be more plausibly explained by assuming that the Earth moves than by assuming it to be at rest.

[469]

It is not plausible that the element of fire could be carried along by the lunar orb.

As for the second part of your argument, even if the motion of the heavens were imparted to both the air and the water, this motion would not have anything to do with the tides. A single uniform cause can only produce a single uniform effect, and

The tides cannot derive from the motion of the heavens.

therefore the only effect which this motion could produce would be a constant uniform motion from east to west. Moreover, this could only occur in a sea which encircled the whole globe. Such a motion could not happen in an enclosed sea like the Mediterranean, which is blocked at its eastern end, because if its water could be driven westwards by the course of the heavens it would have dried up centuries ago. And in any case, its waters do not just flow towards the west; they also flow back towards the east, and in regular periods. You cite the example of rivers to argue that even if the sea originally flowed only from east to west, the varying alignment of the shore can make some of the water flow back in the other direction, and I grant that this can happen. But, my dear Simplicio, you must realize that if water turns back in this way it does so perpetually, while if it flows forwards, it flows always in the same way, as the example of rivers demonstrates. But to explain the ebb and flow of the tides, you have to find a reason which will explain how they flow now in one [470] direction and now in the other, at the same place; and as these are irregular and contrary effects, they cannot be deduced from a constant and uniform cause. This fact, which destroys the case for saying that the diurnal motion of the heavens contributes to the motion of the sea, also invalidates the argument of those who would attribute it only to the diurnal motion of the Earth, thinking that this alone could explain the motion of the tides. This irregular effect must be the product of an irregular and changeable cause.

SIMPLICIO. I have no reply to make, either on my own account, because of the weakness of my wit, or on behalf of others, because the theory is such a novel one. But I don't doubt that if it becomes generally known in the schools, there will be no lack of philosophers who will be able to challenge it.

SAGREDO. Well, we shall wait for that to happen; in the meantime, Salviati, if you please, let us proceed.

SALVIATI. All that has been said up to now relates to the diurnal motion of the tides. We began with a general demonstration of their primary and universal cause, without which no effect whatsoever would come about. Then, passing on to particular varied and in some ways irregular effects which occur in relation to the tides, we have discussed the secondary and concomitant causes which bring these about. We must go on now to consider the other two periods, the monthly and the annual. These do not introduce any new or different phenomena from those we have already considered in connection with the diurnal period, but they have an effect on the diurnal motions, making them greater or smaller at different times in the lunar month and the solar year. It is almost as if the Moon and the Sun were playing a part in bringing about these effects, an idea which my intellect totally rejects. I can see that the tides are a local, physical movement involving an immense volume of water, and I cannot bring myself to subscribe to such causes as the effect of lights, warm temperatures, the predominance of hidden qualities, or other such fancies. In fact, so far are these from being actual or even possible causes of the tides, that the very opposite is true. The tides gave rise to these ideas, putting them into the heads of those who are more suited to talking and showing off than to investigating and reflecting on the secret works of nature. They would rather hold forth, and sometimes even write, about the most absurd ideas, than pronounce those wise, simple, and modest words, 'I do not know'. Anyone can see that the Moon and the Sun cannot produce any effect on even the smallest [471] container of water by means of their light, their motion, or their great or moderate heat; in fact, to make water rise at all by means of heat it has to be brought almost to boiling point. In short, there is nothing we can do to replicate artificially the motions of the tides, apart from moving the vessel containing the water. Surely this is enough to convince anyone that any other cause that is put forward to explain this effect is a vain fantasy that has nothing whatever to do with the truth?

Changes in effects imply changes in their causes.

I say, therefore, that if it is true that an effect has only one primary cause, and that there is a fixed and constant relationship between cause and effect, it must follow that any fixed and constant change in the effect must come from a fixed and constant change in the cause. The changes which affect the tides at different times of the year and of the month follow a fixed and constant time period; therefore there must be regular changes in the same time intervals affecting the primary cause of the tides. Now the changes which occur in the tides at these times only affect their size—the extent to which the water rises or

Causes of the monthly and annual periods of the tides explained at length.

falls, and the greater or lesser impetus with which it flows; so it follows that the primary cause of the tides must grow or diminish in strength at these determined times. But we have concluded that the primary cause of the tides is the irregularity and unevenness in the motion of the basins containing the water; therefore this irregularity must vary at the corresponding times, sometimes being greater and sometimes less. At this point we must recall that this irregularity—that is, the difference in the velocity of motion of the basins, i.e. the different parts of the surface of the globe—derives from their composite motion, the product of the combined annual and diurnal motions of the globe as a whole. Of these it is the diurnal rotation, sometimes adding to and sometimes subtracting from the annual motion, that produces the irregularity in the composite motion. So the additions to and subtractions from the annual motion as a result of the diurnal rotation are the original cause of the uneven motion of the basins, and hence of the tides. It follows that if these additions and subtractions always affected the annual motion to the same extent, the tides would continue to exist, but they would always be constant in their effect.

Monthly and annual changes in the tides can only derive from changes in the additions to and subtractions from the annual motion as a result of the diurnal motion.

[472] But we need to find the cause which makes the ebb and flow of the tides greater and less at different times, and therefore (assuming we want to maintain the identity of the cause) we must find some change in these additions and subtractions, so that they are more or less powerful in producing the effects which follow from them. And I cannot see how this power or lack of it can derive from anything other than greater or lesser additions and subtractions, so that their composite motion produces greater or lesser degrees of acceleration and deceleration.

SAGREDO. I feel as if I am being led gently by the hand, and although I haven't encountered any obstacles in the way, I feel like a blind man who can't see where his guide is taking him. I can't imagine where this journey is going to end.

SALVIATI. I know there is a great difference between your quick reasoning and the slow pace of my philosophy; but in this matter we are discussing now, I'm not surprised that your perspicacity is still defeated by the dense fog which conceals the goal to which we are travelling. Any possible surprise vanishes when I recall how many hours, days, and even more nights

I have spent speculating over this matter, and how often I despaired of ever getting to the bottom of it. I was like the wretched Orlando,* trying to find consolation by persuading myself that the evidence before my eyes from so many trustworthy witnesses was not true. So don't be surprised if, for this once, you aren't able to anticipate our conclusion; and if you are still surprised, I don't think you will remain so when you see the outcome, unexpected though I think it will be.

SAGREDO. Thank God your despair didn't lead you to the fate we read of poor Orlando, or the perhaps no less fictitious fate which is told of Aristotle.* If it had, I and everyone else would have been deprived of the revelation of something as hidden as it is sought after. So please satisfy my thirst for it as quickly as you can.

The ratio of additions to the annual motion as a result of the diurnal rotation can vary in three ways.

[473]

SALVIATI. I will. What we have to discover is how the additions and subtractions to the annual motion caused by the Earth's diurnal rotation can vary sometimes in a greater and at other times in a lesser proportion, since this variation is the only possible explanation for the monthly and annual changes which we see in the size of the tides. I come now to consider three ways in which the ratio of these additions and subtractions to the annual motion and the diurnal rotation can vary in size.

First, this can happen because of an increase or reduction in the velocity of the annual motion, while the additions and subtractions made by the diurnal rotation remain unchanged. This is because the annual motion is roughly three times greater,* i.e. faster, than the daily motion, even at the equator; so if it is increased even further, the additions or subtractions of the diurnal motion would produce less of a change. Conversely, if the annual motion is slowed down, the changes resulting from the same diurnal motion will be proportionately greater, just as an increase or decrease of four degrees of velocity is proportionately less for something moving with a velocity of twenty degrees than it is for one moving at only ten degrees. The second way would be for the additions and subtractions to be increased and decreased while the velocity of the annual motion remains the same. This is easily understood, since a velocity of, say, twenty degrees is clearly changed more by the addition or subtraction of ten degrees than by the addition or subtraction of four. The third way would be if these two changes were combined, with the annual motion decreasing and the diurnal additions and subtractions increasing. This much, as you can see, is not difficult to follow. What has been much harder for me has been to find out how this can happen in nature; but in the end I have found that nature does indeed make marvellous use of such changes, in ways which could hardly be foreseen.

I say they are marvellous and unforeseeable for us, but not for nature herself, who can bring about things which are infinitely astonishing to us with consummate ease and simplicity; and what is supremely difficult for us to understand is supremely easy for her to achieve. I have shown, then, that the proportions between the additions and subtractions of the diurnal rotation and the annual motion can increase and decrease in two ways; I say two, because the third is simply a combination of these two. Now I will go on to add that nature makes use of both these ways; and I will add further that if she made use of only one of them, then one of the two periodic changes of the tides would have to be eliminated. The monthly periodic change would cease if there were no change in the annual motion, and there would be no annual periodic variation if the additions and subtractions caused by the diurnal rotation continually remained equal.

What is supremely difficult for us to understand is supremely easy for nature to achieve.

If the annual motion did not change, there would be no monthly variation.

If the daily motion did not change, there would be no annual variation.

[474]

SAGREDO. Do you mean to say that the monthly variation in the tides depends on the variations in the Earth's annual motion, and that the annual variation in the tides depends on the additions and subtractions of the diurnal motion? Now I'm more confused than ever, and have even less hope of solving this conundrum, which strikes me as more tangled than the Gordian knot. I envy Simplicio, who I assume from his silence understands everything, and doesn't share any of the confusion which is greatly perplexing me.

SIMPLICIO. Sagredo, I can well believe that you are confused, and I think I know the source of your confusion. I imagine the reason is that you understand some parts of what Salviati has just been saying but not others. And you're right to say that I'm not confused, but not for the reason you think; rather the opposite. It's not that I understand everything; in fact I don't understand anything at all, and confusion is caused by a multiplicity of things, not by nothing at all.

SAGREDO. Look, Salviati, how some of the tugs on the reins that you've given to Simplicio over the last few days have tamed him, and transformed him from a steeplechaser into a docile nag. But please don't keep us both in suspense any longer.

SALVIATI. I shall do my best to explain myself more clearly, and your sharp wits will compensate for any obscurity in my expression. There are, then, two phenomena whose causes we have to discover: the first concerns the variations in the tides which occur in a monthly cycle, and the other relates to the annual cycle. We shall discuss the monthly variation first, and then the annual one; and we shall resolve them both by referring to the principles and suppositions we have already established, without introducing any new concepts, either in astronomy or in general, in order to account for the tides. Rather, we shall show that all the various phenomena that are observable in the tides have their cause in facts which are already known and accepted as true and beyond doubt.

It can be confidently supposed that the revolution of a small circle is completed in a shorter time than that of a larger circle. Two examples show this.

It is, then, naturally and necessarily true that if a movable body is made to rotate by a motive force, it will take a longer time to complete a revolution along a larger circle than along a smaller one. This is a truth which all accept and which is confirmed by every experiment, of which we shall cite some examples.

First example.

In rotary clocks, especially large ones, clockmakers regulate their timekeeping by means of a stick which is free to swing [475] horizontally, with a lead weight attached to each end. If the clock runs slowly, they have only to bring these weights a little nearer to the centre of the stick, and so make it oscillate more frequently; and conversely, to make the clock run more slowly, they move the weights more towards the extremities, making its oscillations more infrequent, and hence lengthening its measurement of the hours. In this example the motive force is the counterweight, and so does not change; the moving bodies are the lead weights, which also do not change; and their oscillations are more frequent when they are nearer the centre, in other words when they are moving in smaller circles.

Second example.

Or again, take equal weights and suspend them from cords of different lengths, and then move them away from the perpendicular and let them go. We shall see that the weights attached to the shorter cords complete their oscillations in a shorter time, since they are moving in smaller circles. More than this, attach one of these weights to a cord which runs over a nail driven into the ceiling, and hold the other end of the cord in your hand. Release the weight and, while it is swinging, pull on the end of the cord in your hand, so that the weight rises. You will see that as the weight rises, the frequency of its oscillations increases, since it is continually moving in smaller circles.

Two particular phenomena which should be noted concerning pendulums and their oscillations.

Here there are two particular points which should be noted. The first is that the oscillations of a pendulum necessarily follow a fixed time interval, which is impossible to change except by lengthening or shortening the cord. You can confirm this by making an experiment straight away. Tie a piece of string to a stone and hold the other end of the string in your hand; then see if there is anything you can do to alter the time it takes to complete its swing, apart from lengthening or shortening the string, and you will find it is quite impossible. The other point which is truly marvellous is that the same pendulum will oscillate with the same frequency, or with only minimal and almost imperceptible variations, whether it is swinging in very large or very small arcs of the same circumference. For whether you move the pendulum from the perpendicular by only one, two, or three degrees, or whether you move it by 70, 80, or even 90 degrees, and then let it go, it will oscillate with the same frequency, even though in the first case it moves through an arc of only four or six degrees, and in the other through an arc of 160 degrees or more.

[476]

Remarkable problems concerning objects descending along a quadrant of a circle, and along any chord of a whole circle.

This can be seen more clearly by taking two equal weights and suspending them from two cords of equal length. If they are moved from the perpendicular, the first by only a small amount and the other by a much larger amount, and then released, they will both swing to and fro in the same time interval, the first in a very small arc and the second in a much larger one. This allows us to solve a most intriguing problem, which is this. Take a quadrant of a circle—call it AB in this sketch which I'll draw here on the ground—and set it up vertically so that it rests on a horizontal plane, touching it at point B. Make an arc with a piece of wood which is smooth and polished on its concave side, following the curve of a circumference ADB, so that a smooth round ball can run freely within it. The frame of

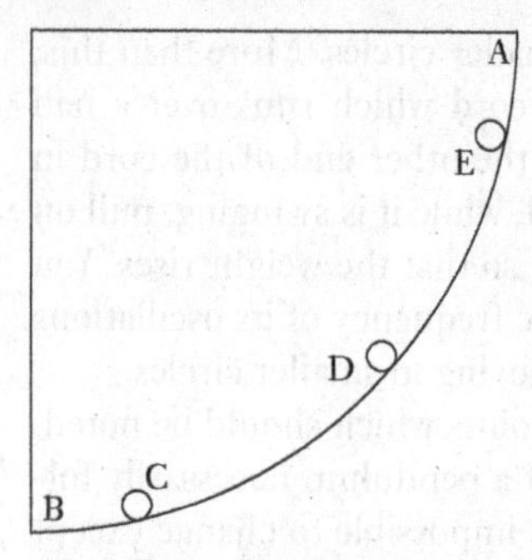

a sieve is very suitable for this purpose. Now wherever you place the ball, as near or far as you like from the lowest point B—placing it at point C, D, or E, for example—when you let it go it will always arrive at B in the same space of time, or with only imperceptible differences. It makes no difference whether it starts from C, D, or E, or any other point—a truly remarkable phenomenon.* And there is another phenomenon which is no less intriguing. If you draw a chord from B to C, D, E, or any other point, not only on the quadrant BA but anywhere on the circumference of the entire circle, a moving body will fall along these chords in exactly the same time. So it would fall along the whole diameter of the circle erected perpendicular to the point B in the same time as it would along the chord BC, even if it subtended only one degree or even less. Add to this a further remarkable fact, which is that a body falls along any arc of the quadrant AB in a shorter time than it would along the chord of that same arc. So the most rapid motion and the shortest time for a body to travel from A to B will not be along the straight line AB—even though that is the shortest distance from A to B—but along the circumference ADB. And if you take any other point on this same arc, such as point D, and draw two chords AD and DB, the body will travel from A to B in less time along the chords AD and DB than it would along AB—but the shortest time of all would be if it travelled along the arc ADB. And the same would apply to any other smaller arc above the lowest point B.

[477] SAGREDO. Please stop—I can't take in any more. You're so overloading me with wonder, and distracting my mind in so many different directions, that I'm afraid I won't have enough mental capacity left to concentrate on the main subject that we're discussing, which is quite obscure and difficult enough in itself. But I do hope you will favour me by staying a few more days after we've finished our investigation of the tides, and will honour this house of mine which is also yours for a little longer. There are so many other problems to discuss which we have left

in suspense, which I think are no less fascinating and intriguing than this question we have been discussing in these past days, and which we must conclude today.

SALVIATI. I shall be glad to; but we shall need more than just one or two sessions if we are to deal, not just with the other questions which we have set aside to discuss separately, but also with the many matters relating to local motion—both natural motion and that of projectiles—which our friend the Lincean Academician has dealt with at length. But to come back to our original subject, we were saying that objects moved in a circle by a motive force which remains constant have a period for their circulation which is fixed and determined, and cannot be made longer or shorter; and we gave examples and cited experiments we could make to confirm this. Now we can also confirm that the same is true of our experience of the celestial motion of the planets, where we can see that the same rule applies; the planets which move in larger orbits take a longer time to complete them. This can be very clearly observed in the Medicean planets, which take only a short time to complete their revolutions around Jupiter. So there is no reason to doubt, in fact we can confidently affirm, that if the Moon, for example, were to be drawn gradually into a smaller orbit while still continuing with its own intrinsic motion, it would acquire a tendency for its orbital period to become shorter. This would be just like the pendulum where we reduced the length of the cord in the course of its oscillations, so shortening the radius of the circumference on which it was moving.

The Earth's annual motion along the ecliptic is unequal, because of the motion of the Moon.

Now what I have just cited as an example with the Moon does in fact occur in reality. You'll remember that we concluded, in agreement with Copernicus, that the Moon cannot [478] be separated from the Earth, and that it moves around the Earth in a month; this is generally agreed. We recall, too, that the Earth, always accompanied by the Moon, revolves in its great orbit around the Sun in a year, in which time the Moon revolves around the Earth almost thirteen times. This revolution of the Moon means that it is sometimes closer to the Sun and sometimes much further away: closer when it is between the Sun and the Earth, and further away when the Earth is between it and the Sun. In short, it is closer to the Sun when it is in conjunction, at the new Moon, and further away when it is in opposition, at full Moon. The difference between its greatest and its shortest distance from the Sun is the same as the diameter of the lunar orbit. Now if it is true that the force moving the Earth and the Moon around the Sun is constant; and if it is true, further, that the same body, moved by the same motive force but in unequal circles, passes over similar arcs of smaller circles in a shorter time; then we must conclude that when the Moon is at a shorter distance from the Sun, i.e. when it is in conjunction, it passes through a larger arc of the Earth's orbit than when it is further away, i.e. when it is in opposition, at full Moon. This inequality in the Moon must also be shared by the Earth. If we imagine a straight line from the centre of the Sun, through the centre of the Earth, and extended as far as the orbit of the Moon, this will be the radius of the orbit which the Earth would follow uniformly if it were unaccompanied. But now place another body, carried along by the Earth, on this radius, locating it sometimes between the Earth and the Sun and at other times beyond the Earth, at a greater distance from the Sun. It must follow that the shared motion of both bodies along the circumference of the orbit will be somewhat slower in the latter case, when the Moon is further away from the Sun, than in the former, when the Moon is between the Earth and the Sun and so closer to the Sun. So what happens is just the same as when the clockmaker adjusts the clock's timekeeping, with the Moon functioning like the lead weight which is placed sometimes further from the centre, to make the stick oscillate more slowly, and sometimes closer, to make its oscillations more frequent. Hence it is clear that the Earth's annual motion in its orbit along the ecliptic is not uniform, and that its irregularities derive from the Moon, varying in a monthly cycle.

[479] Now we concluded earlier that the monthly and annual periodic variations in the tides could only derive from the varying ratios between the additions and subtractions caused by the daily rotation in relation to the annual motion. We concluded, further, that these varying proportions could arise in two ways: either by a change in the annual motion, with the size of the additions remaining constant, or by a change in the latter with the annual motion remaining constant. We have now identified the first of these two reasons, arising from the irregularity of the annual

motion caused by the Moon with its monthly cycle. Hence, for this reason the tides must increase and decrease in a monthly cycle. So now you can see how the cause of the monthly cycle lies in the annual motion, and also how the Moon plays a part in this process even though it has nothing to do with the sea and the water.

SAGREDO. If you were to show a very high tower to someone who had no knowledge of any kind of stairs or ladder, and asked them if they thought they could get to the top of it, I'm quite sure they would say no, as they wouldn't be able to see any way of reaching it other than by flying. But if you then showed them a stone not more than half a *braccio* high and asked them if they thought they could step onto it, they would certainly say yes, and equally they wouldn't deny that they could easily climb onto it not just once but ten, twenty, or a hundred times. So when they were shown a staircase, by which they could by their own admission easily climb to the point which they had just said was impossible to reach, I'm sure they would laugh at themselves and acknowledge their own lack of insight. Well, Salviati, you have led me so gently step by step that, to my astonishment, I have reached a height which I never thought I could get to, and with minimal effort. The staircase, indeed, was so dark that I didn't realize I was approaching the top, or that I had got there until I came out into the daylight and discovered a great expanse of sea and countryside. And just as there is no effort in climbing a single step, so each of your propositions seemed so clear that I thought I was making little or no progress, as there was little or nothing in it that was new to me. So I am all the more amazed at the unexpected outcome of this argument, which has led me to understand something which I thought was inexplicable. I am left with just one difficulty which I would very much like to have resolved for me, and it is this. If the motion of the Earth [480] around the zodiac, together with the Moon, is irregular, surely this irregularity should have been observed and noted by astronomers; and yet as far as I know this has not happened. I know you are informed about such things, so please resolve my difficulty for me, and tell me how the matter stands.

SALVIATI. That's a very reasonable question. My reply is that, although astronomy has made great strides over many centuries in investigating the structure and motions of the heavenly bodies, it is still at a stage where many things remain in doubt, and I dare say there are many more still which are as yet undiscovered. I think it likely that the first people who observed the heavens recognized only the motion common to all the stars, that is, the diurnal motion. It may well have taken them only a few days to realize that the Moon was inconstant in its motion in relation to the other stars, but then it was probably many years before they identified all the planets. In particular, I think that Saturn and Mercury were probably the last to be recognized as planets or wandering stars—Saturn because it moves so slowly, and Mercury because it is seen so rarely. Then I expect many more years passed before they observed the stations and retrogradations of the three outer planets, and likewise their variable distance from the Earth, which makes it necessary to introduce eccentrics and epicycles. These were unknown even to Aristotle, since he makes no mention of them. How long did Mercury and Venus keep astronomers in suspense with their baffling appearances before they were able to establish their location, never mind anything else? So just the order of the celestial bodies and the overall structure of the parts of the universe which are known to us were in doubt until the time of Copernicus. It was he who finally showed us the true structure and system according to which these various parts are ordered, so that now we know for certain that Mercury, Venus, and the other planets revolve around the Sun, and that the Moon revolves around the Earth. But we still have not established beyond doubt how each planet behaves in its individual revolution, and what exactly is the structure of its orbit—the study commonly called planetary theory. For evidence of this you have only to look at Mars, which causes so many problems for modern astronomers;* and even the Moon has had various theories applied to it since Copernicus so greatly altered the Ptolemaic theory.

There may still be many things in astronomy which have not yet been observed.

Saturn and Mercury were among the last planets to be observed, Saturn because it moves so slowly and Mercury because it is seen so rarely.

The individual structures of the planetary orbits are still not definitively resolved.

[481] But let's come to the particular question we are discussing, namely the apparent motions of the Sun and the Moon. A great irregularity has certainly been observed in the motion of the Sun, since there is a marked difference in the time it takes to pass through the two halves of the ecliptic, divided by the equinoctial points:* it takes around nine days longer in passing over one half than the other, which as you can see is a very significant

The Sun moves through one half of the zodiac in nine days less than through the other.

difference. But no one has yet observed—perhaps no one has even investigated—whether its motion through a small arc, such as each of the twelve signs of the zodiac, is entirely regular, or whether it moves more rapidly at some points and more slowly at others, as it must if the annual motion really belongs to the Earth accompanied by the Moon, and only apparently to the Sun. As for the Moon, its cycle has been studied mainly in relation to eclipses, for which it is enough to have an exact knowledge of its motion around the Earth; its progression through particular arcs of the zodiac has not been investigated so closely. So there is no reason to doubt that as the Earth and the Moon travel through the zodiac, they accelerate somewhat at new Moon and slow down at full Moon, simply because this irregularity has not been observed. There are two reasons for this: first, because it has not been looked for, and secondly, because it may not be very great.

The motion of the Moon has been investigated principally in relation to eclipses.

The irregularity doesn't need to be very great to produce the effect which we see in the variation of the size of the tides, because not only these variations but the tides themselves are very small compared to the size of the bodies which they affect, even though they may seem large to us from our small perspective. One degree more or less of speed added to where there are naturally 700 or 1000 degrees can hardly be called a large alteration, either in the force which confers it or in the body which receives it. The water in the Mediterranean travels at about 700 miles per hour as a result of the diurnal rotation, although this motion is imperceptible to us because it is the common motion which it shares with the Earth. The tidal motion which is apparent to us in the currents of the sea is less than one mile an hour (in the open sea, that is, not in narrow straits), and it is this that alters the great primary and natural motion. This is a significant alteration for us and for ships; for a vessel which can be rowed at, say, three miles an hour in still water, the difference between having such a current following it or against it will effectively double its speed—a very significant difference in the motion of the boat, but tiny in relation to the motion of the sea, which is modified by just one seven-hundredth of its speed. The same can be said of a rise and fall of one, two, or three feet, or perhaps four or five feet at the end of a gulf more than two

The ebb and flow of the tides is very small compared to the vastness of the seas and the rapidity of the Earth's motion.

[482]

thousand miles long where the water is hundreds of feet deep. An equivalent variation would be much less than if the water in one of the barges carrying fresh water to Venice were to rise in the prow of the barge by the thickness of a leaf when the boat stops. So we can conclude that tiny variations in the sea, in proportion to its immense size and great speed, are enough to produce great changes in relation to the smallness of ourselves and our affairs.

SAGREDO. I am fully satisfied on this point. It remains for you to explain to us how the additions and subtractions produced by the diurnal rotation can vary in size. You hinted that the annual increase and decrease in the size of the tides depended on these variations.

SALVIATI. I shall make every effort to make myself understood, but I am daunted by the difficulty of the phenomenon, and the capacity for abstract thought which is needed to understand it. The variations in the additions and subtractions which the diurnal rotation produces in the annual motion derive from the inclination of its axis to the plane of the Earth's orbit, namely the ecliptic. This inclination means that the equator intersects the ecliptic and is inclined and oblique to it by the same amount as the inclination of the axis. The additions are equal to the whole diameter of the equator when the centre of the Earth is at the solstitial points, and become less and less as the centre approaches the equinoctial points, which are the points where the additions are the smallest.* This is the whole explanation, but couched in rather obscure terms, as you can see.

Causes for the variations in the additions and subtractions to the annual motion produced by the diurnal rotation.

SAGREDO. Or rather as I can't see, since so far I understand nothing at all.

SALVIATI. I feared as much. Let's see if we can shed some more light by means of a drawing, although it would be better if we could represent it with solid bodies rather than just a drawing; but we'll use perspective and foreshortening to help us out. So I shall draw here, as we did before, the circumference of the Earth's orbit, with point A as one of the solstitial points, and the diameter AP as the section of the solstitial colure* which it has in common with the plane of the Earth's orbit, or ecliptic. If we take the centre of the terrestrial globe to be at point A, its axis CAB, which is inclined to the plane of the

[483]

Earth's orbit, falls in the plane of this colure, which passes through the axes of both the equator and the ecliptic. For simplicity's sake we shall just draw the circle of the equator and mark it with these letters, DGEF. The line DE is the section which this circle has in common with the plane of the Earth's orbit, so that half of the equatorial circle, DFE, is below this plane, and the other half, DGE, is above it. Now let us assume that the equator is turning in the sequence D, G, E, F, and that the centre, A, is moving towards E. Since the centre of the Earth is at A, and its axis CB (which is perpendicular to the equatorial diameter DE) falls, as we have said, in the solstitial colure, of which the diameter PA is the section which it has in common with the Earth's orbit, this line PA must also be perpendicular to DE, because the colure is perpendicular to the Earth's orbit. Therefore DE is tangent to the Earth's orbit at point A. So when the Earth is in this position, the motion of its centre along the arc AE, which amounts to one degree per day, varies hardly at all; in fact it is the same as it would be along the tangent DAE. Finally, since the daily rotation carrying point D through G to E adds to the motion of the centre, which has moved along practically the whole length of the diameter DE, while the motion of the other semicircle EFD falls short of the motion of the centre by the same amount, it follows that at this point, i.e. at the solstice, the additions and subtractions will be measured by the whole diameter DE.

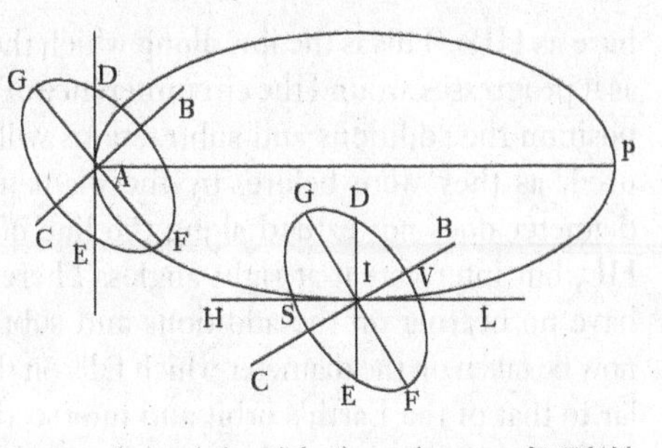

Now let us go on to see whether they would still be of the same magnitude at the equinoxes. We shall move the centre of the Earth to point I, ninety degrees from point A, and we shall keep the same equator, GEFD, its common section DE with the Earth's orbit, and the same axial inclination CB. But the tangent of the Earth's orbit at point I will no longer be DE, but another line intersecting it at a right angle which we can mark

[484] here as HIL. This is the line along which the centre, I, will move as it progresses around the circumference of its orbit. Now in this position the additions and subtractions will no longer be measured, as they were before, by the diameter DE, because this diameter does not extend along the line of the annual motion, HL, but intersects it at right angles. Therefore points D and E have no bearing on the additions and subtractions; these must now be taken on the diameter which falls on the plane perpendicular to that of the Earth's orbit and intersecting the line HL. Let us draw this diameter, GF; now what we might call the additional motion will be made by the point G along the semicircle GEF, and the rest will be the subtracted motion around the other semicircle, FDG. But the additions and subtractions are not determined by the whole length of this diameter GF, because it is not in the same line as the annual motion, HL, but intersects it, as we can see, at point I—the point G being elevated above the plane of the Earth's orbit and the point F depressed below it. So the extent of the additions and subtractions will be taken from that part of the line HL which lies between two perpendiculars drawn to it from points G and F, which are GS and FV. The extent of the additions is therefore the line SV, which is less than either GF or DE, which was the extent of the additions at the solstice, A.

Hence we can establish the extent of the additions and subtractions for wherever the centre of the Earth is placed along the quadrant AI. If a tangent is drawn from such a point, and perpendiculars are dropped to it from each end of the equatorial diameter determined by the plane through this tangent perpendicular to the plane of the Earth's orbit, the part of the tangent between these perpendiculars will give the extent of the additions and subtractions; and this will always be smaller at a point nearer to the equinoxes and greater when it is nearer to the solstices. As for the difference between the maximum and minimum extent of the additions, this is easy to find, because it is the same as the difference between the whole axis or diameter of the sphere and the part of it that lies between the polar circles. This is approximately a twelfth less than the whole diameter, assuming the additions and subtractions are made at the equator. They will be less at other latitudes in proportion to their diminishing diameters.

This is as much as I am able to tell you on this matter, and I dare say it is as much as it is possible for us to know with any certainty, since as you know, certain knowledge can come only from conclusions which are firm and constant. Such are the three general periods of the tides, deriving as they do from causes which are invariable, simple, and eternal. But these pri- [485] mary and universal causes are intermingled with secondary and particular causes which are capable of producing many changes. Some of these, such as the changing winds, are inconstant and unpredictable. Others, such as the length of the sea's inlets, their different geographical orientations and the great variations in the depth of the water, are determined and fixed, but cannot be observed because they are so many and diverse. It would take very prolonged observations and absolutely reliable reports to compile an account of them which could provide a sound basis for theories about how they combine to produce all the appearances, not to say peculiarities and anomalies, which can be found in the motion of the tides. So I shall content myself with pointing out that such accidental effects occur in nature and are capable of producing extensive changes. I shall leave it to those who have practical knowledge of the various seas to observe them in detail.

To conclude our discussion I shall add just one further point which needs to be considered in relation to the exact timing of the tides. It seems to me that these are affected, not only by the varying length and depth of the sea's inlets, but also to a great extent by the confluence of different stretches of sea, all differing in their length and also in their position or orientation. This diversity is apparent here in the Adriatic, a gulf which is much smaller than the rest of the Mediterranean and quite different in its orientation. The Mediterranean is closed at its eastern end by the coast of Syria, whereas the Adriatic is closed at its western end. And since the tidal range is much greater at the extremities—indeed these are the only places where there is a very large rise and fall—it is very likely that the times of high tide in Venice coincide with low tide in the rest of the Mediterranean, which in a sense dominates the Adriatic because it is so much bigger and because it extends directly from west to east. So it would not be surprising if the effects due to the primary causes did not occur at the times and following the periods we would expect in the Adriatic, but rather at those prevailing in the rest of the Mediterranean. But such details require prolonged observations, such as I have not so far been able to undertake, nor do I expect to be able to do so in future.

SAGREDO. I think you have already done a great deal by pointing out the way for us to follow in investigating this profound question. Even if you had only expounded your first [486] general proposition, it seems to me so far superior to the foolish ideas advanced by so many others that just thinking about them again makes me feel ill. I don't see how there can be any possible counter-argument to your demonstration that, if the basins containing the waters of the sea are fixed, it would be impossible in the ordinary course of nature for the motions which we observe in the tides to come about; and that on the contrary, if we assume the motions which Copernicus for other reasons attributes to the terrestrial globe, such alterations in the seas must necessarily follow. I find it quite astonishing that among so many men of profound understanding, not one recognized the incompatibility between the immobility of the vessel containing the water and the motion of the water which it contains; for this incompatibility now seems to me to be self-evident.

SALVIATI. What's more astonishing is that some of them, who did show more understanding than most by attributing the cause of the tides to the motion of the Earth, were not able to follow their insight through to a conclusion. They didn't realize that it was not enough to identify a simple uniform motion such as the daily rotation of the Earth; the explanation required an uneven motion, faster at some times and slower at others, because if the motion of the containers is uniform the waters they contain will adapt to it and will never change. An ancient mathematician* is reported to have said that the tides were caused by the opposition between the motion of the Earth and the motion of the lunar sphere. This is totally unfounded, not only because no explanation is offered for how this might happen, but also because there is no opposition between the rotation of the Earth and the motion of the Moon, since they both move in the same direction. So this is manifestly false. And any other views which have so far been put forward or

A simple motion of the terrestrial globe is not enough to explain the motion of the tides.

The opinion of the mathematician Seleucus criticized.

imagined are, in my view, completely invalid. But of all the great men who have speculated on this marvellous effect of nature, the one who most astonishes me is Kepler. Enlightened and acute thinker as he was, he had grasped the motions attributed to the Earth, and yet he still listened and assented to the notion of the Moon's dominion over the water, and occult properties, and similar childish ideas.*

Kepler respectfully corrected.

SAGREDO. In my view these speculative thinkers were in much the same situation as I now find myself in, of not being able to grasp how these three periods, the annual, monthly, and diurnal motions, interact with each other, or how they appear to depend on the Sun and Moon even though the Sun and the Moon have nothing to do with water. I need more time to apply my mind at length to the whole question before I can understand it fully; at the moment it still eludes me because of its novelty and difficulty. But I don't despair of being able to master it if I spend some time reflecting on it in solitude and silence, to digest what for the moment is simply stored up in my imagination.

[487]

In our discussions over these last four days we have, then, seen strong evidence in favour of the Copernican system. Three arguments in particular seem to be quite conclusive: the first based on the stations and retrogradations of the planets, and their approaches to and recessions from the Earth; the second, on the rotation of the Sun and the observations concerning the sunspots; and the third on the ebb and flow of the tides.

SALVIATI. We could perhaps add a fourth, and maybe even a fifth. The fourth argument is based on the fixed stars, if accurate observation should confirm the minute mutations which Copernicus posits but says are imperceptible.* And a fifth novelty has just arisen which could also provide evidence for the mobility of the terrestrial globe, thanks to the very refined observations which are being made by signor Cesare, of the noble family of Marsili in Bologna, who is also a member of the Lincean Academy.* He has written a most learned account of how he has observed a continual, although very gradual, change in the line of the meridian, which I have recently read with amazement, and which I hope he will make available to all those who study the marvels of nature.

The motion of the meridian observed by Cesare Marsili.

SAGREDO. This is not the first time I have heard mention of this gentleman's great learning, and of his assiduous support of all men of letters; and if this or any other work of his is published, we can be sure that it will be something of great note.

SALVIATI. Now that the time has come for us to bring our discussions to an end, it remains for me to beg your indulgence if, as you reflect at more leisure on the points I have made, you should come across difficulties or doubts which I have not adequately resolved. Please excuse my shortcomings, partly because of the novelty of the ideas, partly because of the weakness of my own intellect, and partly because of the sheer magnitude of the subject matter. Excuse me, finally, because I do not expect and never have expected others to give the assent which I myself do not give to this fantasy, which I could [488] readily admit to be a vain illusion and a tremendous paradox. In our discussions you, Sagredo, have often applauded and shown yourself convinced by the conjectures I have put forward. I believe this is, in part, more because of their novelty than their certainty, but even more because, courteous as you are, you wanted to give me that gratification which we all feel when our efforts are approved and praised. And as I am obliged to you for your kindness, so I have appreciated the ingenuity of Simplicio; his constancy in so strongly and resolutely defending his master's teaching has greatly endeared him to me. So I thank you, Sagredo, for your kindness and consideration, and I beg Simplicio's pardon if I have sometimes offended him with my excessively bold and assertive talk. I assure him that I was not motivated by any ill will; I simply wanted to give him more opportunity to put forward his own conjectures so as to be better informed myself.

SIMPLICIO. There is no need for you to apologize, least of all to me; I am used to academic and public debates, and I have often heard the disputants become heated and angry, and indeed trade insults and sometimes come close to blows. As regards the discussions we have had, and in particular this last one on the causes of the tides, the fact is that I am not entirely convinced; but I confess that, from the rather vague idea of it that I have formed in my mind, your theory strikes me as much more ingenious than the others I have heard, although I don't regard it as conclusively proved. Rather, I always keep in mind a very sound doctrine that I was taught by a most learned and eminent person,* to which one must perforce submit. I know that both of you, if you were asked whether God in his infinite power and wisdom could endow the element of water with its observable oscillating motion by some means other than that of moving the basin in which it is contained, you would reply that he could have done such a thing in many different ways, which might indeed be inconceivable to us. So I conclude that, this being the case, it would be the height of presumption to try to limit or restrict the divine power and wisdom to any one particular fantasy.

SALVIATI. A fine and truly angelic doctrine, to which I will [489] add another, also divine, which is entirely in keeping with it. It is this: that while we are free to debate about the structure of the universe, we shall never discover all God's handiwork,* lest perhaps we cease to exercise our human minds and become idle. So let us then exercise those activities which God allows and ordains for us, that we may recognize and wonder all the more at his greatness, the more we find the profound depths of his infinite wisdom outstrip our ability to penetrate them.

SAGREDO. Let this, then, be the conclusion of our four days of discussions. We must now curb our curiosity and allow Salviati an interval of rest, if he chooses to take it, but on the condition that he returns at a time convenient to him to satisfy our wish—mine especially—to resolve the problems we put to one side. I wrote these down as questions to put to him in one or two further sessions, as we agreed. I am anxious above all to hear the elements of our Academician's new science* concerning local motion, both natural and constrained. Meanwhile let us go, as is our custom, to enjoy the cool of the evening for an hour in the gondola which awaits us.

EXPLANATORY NOTES

For details of the works cited in abbreviated form in the notes below, see Note on the Text and Translation, p. xxxii.

4 *Fr. Niccolò Riccardi, Master of the Holy Apostolic Palace*: Riccardi's name was added to the *imprimatur* without his knowledge or approval. See the Introduction, pp. xxviii–xxix.

5 *Most Serene Grand Duke*: Ferdinando II (1610–70) was the son of Cosimo II (1590–1621), who had appointed Galileo his 'mathematician and philosopher' in 1610.

the great book of nature: the origin of the metaphor 'book of nature' is biblical. Years earlier, Galileo had written down two passages, one from Isaiah 34: 4: 'The sky will be rolled up like a scroll', the other from Revelation 6: 14: 'The sky receded like a scroll, rolling up' (Galileo, *Early Notebooks: The Physical Questions*, translated by William A. Wallace (Notre Dame, IN: Notre Dame University Press, 1977), 94). Galileo's apologist, Thomas Campanella (1568–1639), also speaks of 'the world which is the book of God' (Thomas Campanella, *A Defense of Galileo*, translated by Richard J. Blackwell (Notre Dame, IN: University of Notre Dame Press, 1994), 64, 71). The convergence between Campanella and Galileo is merely apparent, however. What is decisive in this metaphor are the characters in which one assumes that the book of nature is written. For Campanella these characters are the figures and shapes of the constellations while for Galileo they are strictly mathematical relations.

7 *TO THE DISCERNING READER*: this preface was required to be added by the Roman Censor, Niccolò Riccardi, as a condition of his granting an *imprimatur*, i.e. permission to print. In the first edition of the *Dialogue* it was printed in a different typeface from the body of the text, a discrepancy which was noted as one of the complaints against Galileo at his trial; see *Selected Writings*, 363. On the circumstances of the printing of the *Dialogue* see the Introduction, pp. xxvi–xxix.

A salutary edict: Copernicus's *On the Revolutions of the Heavenly Spheres* was placed on the Index of Proscribed Books on 3 March 1616.

the opinion of Pythagoras concerning . . . Earth: Copernicus believed that Pythagoras (sixth century BC) had taught that the Sun is at rest and that the Earth is in motion.

adjudicators . . . inexperienced in astronomical observations: this refers to the panel of eleven theologians who were asked to assess Copernicus's *On the Revolutions of the Heavenly Spheres*. See the Introduction, p. xix, and William R. Shea and Mariano Artigas, *Galileo in Rome* (Oxford: Oxford University Press, 2003), 81.

7 *I was in Rome at the time*: having heard that Copernicanism was under suspicion, Galileo went to Rome at the end of 1615 and remained there for almost six months in the hope of showing that the Earth was not at rest.

anyone working north of the Alps: an allusion to Kepler and other non-Catholic astronomers in northern Europe.

purely as a mathematical hypothesis: on 12 April 1615, Cardinal Roberto Bellarmine had written to an advocate of Copernicanism named Paolo Antonio Foscarini to say that he and Galileo were 'acting prudently in speaking hypothetically and not in absolute terms, as I have always understood Copernicus to have done'. Bellarmine's letter is in *Selected Writings*, 94–6.

Peripatetics only in name: 'Peripatetic' is a name given to the followers of Aristotle. It originally derived from the colonnades (*peripatoi* in Greek) of the school in Athens where Aristotle and his students would walk up and down during their discussions.

8 *cause of the tides*: in 1616 Galileo wrote a *Discourse on the Tides* that became the Fourth Day of the present *Dialogue*.

Giovan Francesco Sagredo: Sagredo (1571–1620) was a Venetian patrician and a talented amateur of science. He had studied with Galileo at Padua and became his close friend. In the *Dialogue*, he speaks for the intelligent layman who is already half-converted to the new astronomy.

Filippo Salviati: Salviati (1538–1614) was a noble Florentine who had often invited Galileo to his villa near Florence. Galileo wished to perpetuate his memory by making him his spokesman in the *Dialogue*.

a Peripatetic philosopher: this philosopher could have been Cesare Cremonini (1550–1631), who was Galileo's colleague at the University of Padua and was famous for his commentaries on Aristotle.

9 *Simplicius*: a sixth-century interpreter of Aristotle. It is generally agreed that the Simplicio of the *Dialogue* is a composite of Galileo's Peripatetic opponents.

FIRST DAY

11 *Aristotelian and Ptolemaic system . . . Copernican system*: Nicolaus Copernicus (1473–1543) was a Polish astronomer who had studied in Italy. His book, *On the Revolutions of the Heavenly Spheres*, was published in the year he died. He defended the belief, which he ascribed to Pythagoras (sixth century BC), that the Earth moves around the Sun, against Aristotle (385–323 BC), who maintained that it was at rest, a position that was developed by Claudius Ptolemy (flourished in Alexandria around AD 150).

12 *splendid proofs . . . 'continuity'*: here is how Aristotle put it: 'A magnitude if divisible in one direction is a line, if in two directions a surface, and if in three directions a body. There is no magnitude not included in these. For, as the Pythagoreans say, the whole world and all things in it are summed

up in the number three (Aristotle, *On the Heavens*, book 1, chapter 1, section 268a 1–268b 10).

13 *necessary demonstrations*: Galileo stresses that his arguments are based on rigorous mathematical reasoning ('necessary demonstrations') as well as careful observation and experimentation ('the experience of our senses').

laughing-stock . . . Senate itself: pestered by his mother about what had been debated at the Senate, Papirius told her that it concerned the question whether it would be better to allow one man two wives, or one woman two men. The result was that a large and noisy delegation of townswomen appeared before the Senate to argue for the latter alternative. The anecdote is told by Macrobius, *Saturnalia*, 1.6, 18–26.

14 *which you know already*: this is the first of a number of allusions to the Platonic theory of recollection.

16 *I will follow Aristotle*: Aristotelians insisted that qualitative properties disclose the nature of things whereas Galileo maintained that quantitative relations are the ones that provide genuine clues to an understanding of reality. For him, mathematics is the grammar of science, and his most famous description of the language of nature occurs in his *The Assayer*: 'Philosophy is written in this great book which is continually open before our eyes—I mean the universe—but before we can understand it we need to learn the language and recognize the characters in which it is written. It is written in the language of mathematics, and its characters are triangles, circles, and other geometrical figures, without which it is humanly impossible to understand a word of what it says. Without these, it is just wandering aimlessly in a baffling maze' (quoted in *Selected Writings*, 115).

17 *simple bodies . . . natural principle of motion*: the four Aristotelian elements naturally moved downwards (earth and water) or upwards (air and fire).

21 *the name 'ordered world' itself*: the reference is to the word *cosmos* which in the original Greek means both 'order' and 'world'. The ancient Greeks considered that the world was perfectly harmonious and impeccably ordered. We now use *cosmos* without the idea of perfect order, but 'cosmetic' retains something of the earlier meaning.

22 *integral bodies*: the planets and the Moon. See Sagredo's summary at the beginning of the Second Day: '. . . that the universe has no such difference between its parts, and that the Earth enjoys the same perfection as its other integral bodies' (p. 132).

23 *worthy of Plato*: Galileo has in mind a passage in Plato's *Timaeus*, 38–9, but he has taken great liberties with the text. He returns to this idea in the Fourth Day of his *Two New Sciences* (see the translation by Stillman Drake (Madison, WI: University of Wisconsin Press, 1974), 232–4).

the Lincean Academician: Galileo is referring to himself. In 1611, as a recognition for his telescopic discoveries he had been made a member of the Lincean Academy by its founder, Prince Federico Cesi.

24 *If a body . . . impact*: this and later passages enclosed in square brackets were added by Galileo in his own copy of the first edition.

25 *two hundred braccia*: the unit of length that Galileo uses is the *braccio* (plural *braccia*) which is 58.4 cm or about an inch less than 2 feet. The numbers used by Sagredo to illustrate free fall, two hundred *braccia* in less than ten pulse beats, are merely illustrative. Assuming a regular pulse beat of 72 per minute, and hence 8.3 seconds for 10 pulse beats, a cannon ball falling from a height of 200 *braccia* would reach the ground in about 7 seconds if the correct value of gravity (980 cm/sec^2) is used. But Galileo could not know this correct value.

the cannon ball continually gains impetus: the word *impetus* that Galileo uses is borrowed from medieval physics and is not a mathematically defined concept. It is a quality possessed by a moving body which can be conserved or communicated to other bodies.

29 *proofs concerning local motion*: see the Second Day, pp. 233–7, and the selection from the Third Day of Galileo's *Two New Sciences* in *Selected Writings*, 382–92. The Academician is Galileo (see note to p. 23).

32 *maintaining its allotted speed*: from the viewpoint of Newtonian physics, which was developed half a century later, the 'Platonic' cosmology produces no gain. Rather it implies two major miracles. First, it involves changing instantaneously the direction of the movement of the falling planets, which is as difficult as conferring instantaneously a determined velocity to a body. Indeed, in the natural order of things it is impossible. Secondly, it implies that the force of attraction of the Sun has doubled at the very moment when the circular motion is substituted for a downward one. But neither of these considerations can be said to hold for Galileo, who considers the operation of conferring motion on a body at rest and that of changing its direction as altogether different. In the first case, something new has to be produced, but in the other the change is merely accidental. As to the doubling of the force of attraction, Galileo has no need for it whatsoever since, for him, circular motion is inertial and does not engender centrifugal forces, so that no force of attraction from the Sun is necessary to make the planets describe the particular orbits and stay in them. Furthermore, Galileo does not work on the assumption that the Sun attracts the planets; they move towards the Sun by virtue of an inclination that has its origin in their bodies. For a fuller discussion, see William R. Shea, *Galileo's Intellectual Revolution*, 2nd edition (New York: Science History Publications, 1977), 121–9.

33 *a hundred braccia in four pulse beats*: this passage was added by Galileo in his copy of the *Dialogue*. One hundred *braccia* in four pulse beats gives 1,040 cm/sec^2 for the value of g and is close to the true value of 980 cm/sec^2. This is a better result than the numbers given previously (see note to p. 25). See also the Second Day, p. 235.

36 *the concave surface of the moon's orbit*: Simplicio uses the word 'orb' or 'globe' to refer not to the path of the Moon as we now consider it but to the crystalline sphere in which the Moon was supposed to be embedded.

the same reasoning . . . to the parts: this axiom is quoted by Aristotle, *On the Heavens*, book 1, chapter 3, section 270a, 11.

37 *'there is no arguing . . . first principles'*: see Aristotle, *Physics*, book 1, chapter 2, section 185a, 3.

even though he proved it . . . contrary motion: see Aristotle, *On the Heavens*, book 4, chapters 4–5, sections 311a, 15–312b, 7.

38 *Logic . . . is the organ which we use*: 'organum' was the word commonly used to designate Aristotle's works on logic.

41 *exempt from contraries*: Galileo has paraphrased a passage in Aristotle, *On the Heavens*, book 1, chapter 3, section 270a, 14–17.

44 *two whole books*: Aristotle's *On Generation and Corruption* consists of two 'books'.

45 *horned arguments known as sorites*: the source of the 'liar's paradox' is the letter of the apostle Paul to Titus where he warns him not to rely on the Cretans for, 'as one of themselves said, "All Cretans are liars"' (Titus 1: 12). This was called a 'horned argument' or a 'forked question' by medieval logicians. It is not a sorites, whose name derives from the Greek word *soros*, meaning 'pile' or 'heap'. The classic example of a sorites is the paradox that arises when one considers a heap of sand from which grains are individually removed. Is it still a heap when only one grain remains? If not, when did it change from a heap to a non-heap? Note that it is the Aristotelian, Simplicio, who makes the mistake of confusing a horned argument with a sorites.

46 *the sphere in which they rotate*: ancient astronomers considered the Earth as a small stationary sphere at the centre of a much larger one that carried the stars and rotated once every 24 hours. Unlike the Earth, which was composed of four elements (earth, air, fire, and water), this larger sphere was made of a special fifth element that was considered unalterable and transparent except where the stars, which were held to be denser parts of this matter, were located.

47 *Cremonini*: see note to p. 8. Simplicio is paraphrasing what Cremonini wrote in his *Disputatio de Coelo* (Venice: Tommaso Baglioni, 1613), 23–4.

51 *a priori . . . a posteriori*: the terms *a priori* and *a posteriori* are used to distinguish two types of knowledge or kinds of arguments. *A priori* knowledge or justification is independent of experience (e.g. mathematical proofs); *a posteriori* knowledge or justification is dependent on experience or experimentation (e.g. the statement 'some students are bilingual').

52 *Abyla and Calpe*: Calpe is the Rock of Gibraltar, and Abyla is a hill on the African side of the Strait. Abyla and Calpe were called the Pillars of Hercules by the ancients.

53 *Cain carrying a bundle of thorns*: Dante, *Inferno* 20.126 and *Paradiso* 2.49–51, mentions the popular belief that the dark spots on the Moon show Cain condemned to carry for all eternity a bundle of thorns.

55 *Pythagoras sacrificed a hundred oxen*: ancient Greek writers say that Pythagoras sacrificed an ox, some even one hundred oxen, but Cicero already doubted the veracity of the tale (Cicero, *On the Nature of the Gods*, 3.88).

55 *two new stars*: a very bright star, now known as a supernova, suddenly appeared in 1572 and was studied by the Danish astronomer Tycho Brahe (1546–1601). A second supernova appeared in 1604 and was the subject of well-attended lectures that Galileo delivered at the University of Padua.

56 *Anti-Tycho*: the title of a book published in 1621 by the Italian astronomer Scipione Chiaramonti (1565–1652), who was a learned astronomer but firmly convinced that the Earth was stationary.

parallaxes: 'parallax' is the technical word for the difference in the position of an object in the sky when it is observed from different positions on Earth.

What does this modern author say . . . about the sunspots?: the modern author is Chiaramonti, who discussed the nature of the novae and the comets but did not consider the sunspots in this work. Galileo takes him to task in the Second and the Third Days. See the Introduction, p. xxiii.

57 *Various opinions on the sunspots*: see the Introduction, pp. xvi–xvii and the excerpts from Galileo's *Letters on the Sunspots* in *Selected Writings*, 33–54.

the Sun's eccentric sphere: in the Ptolemaic system the Sun's centre of revolution is not at the centre of the Earth but is slightly off-centre, hence the word 'eccentric'. For the devices used by Ptolemy to explain the supposedly circular motions of the Sun and the planets see the Introduction, p. xxii.

58 *Demosthenes*: a prominent statesman and orator in ancient Athens, lived 384–322 BC.

59 *Mark Welser*: see the Introduction, p. xvi.

61 *Prytaneum*: the town hall of a Greek city-state, normally housing the chief magistrate and the common altar or hearth of the community.

65 *nature does nothing in vain*: this aphorism was often quoted by the Scholastic philosophers and is found in Aristotle's *On the Heavens*, book 1, chapter 4, section 271a, 33.

66 *that the Moon is inhabited by humans*: in antiquity Plutarch (AD 45–127) had speculated about human life on the Moon in his *On the Face of the Moon*. Galileo's contemporary, Johann Kepler, did likewise in his *Ad Vitellionem Paralipomena* that appeared in 1604 (translated by William H. Donahue (Santa Fe, NM: Green Lion Press, 2000)). Catholic theologians viewed the idea with suspicion.

70 *if the Moon had an epicycle*: an epicycle is a small circle whose centre moves round the circumference of a larger one. Copernicus never abandoned the notion that perfect motion had to be circular and Galileo shared this idea. In the case of the Moon, Copernicus used not only one but two epicycles, the second rotating on the first that was attached, in turn, to the circumference of a third and larger circle called the deferent (Copernicus, *On the Revolutions*, book 4, chapter 2).

Antichthons: the Greek word *antichthon* means Counter-Earth. The Pythagoreans held that the Earth had an exact counterpart that revolved about the 'central fire' but in such a way as never to be seen by us. See Aristotle, *On the Heavens*, book 2, chapter 13, section 293a, 15–30.

71 *the ray from their eye*: on Galileo's theory of optics see note to p. 87.

on the meridian: literally, 'in this or that belly of its Dragon'. The path of the Moon is tilted with respect to the ecliptic by some 5°. Hence its course among the stars appears to undulate and this suggested to the ancients the idea of a dragon. The head and tail of the dragon were placed at what are now called the Moon's nodes, namely the points where its orbit cuts the ecliptic. The most northerly position was called the upper belly of the dragon, and the most southerly the lower belly.

74 *the heavens are impenetrably hard*: actually Aristotle only affirms the existence of orbits and says nothing about their hardness or the impenetrability of the heavens (*On the Heavens*, book 2, chapter 8, sections 289, b1–290, a8).

75 *a distinguished professor in Padua*: Cesare Cremonini (see note to p. 8), who discusses at length whether the heavens are intangible in his *Disputatio de Coelo* (Venice: Tommaso Baglioni, 1613), 113, 231–3.

76 *in the Assayer and his Letters on the Sunspots*: for the first, see Galileo Galilei, *The Assayer*, translated by Stillman Drake in *The Controversy of the Comets* (Philadelphia: University of Pennsylvania Press, 1960), 279–80, and for the second, Galileo Galilei and Christoph Scheiner, *On Sunspots*, translated by Eileen Reeves and Albert Van Helden (Chicago: University of Chicago Press, 2010), 284–5.

83 *'we should not expect . . . luminous body'*: the text that Simplicio is made to quote may be by an author who has not been identified or it could be a literary device that Galileo uses to show how obscure is the material upon which Simplicio bases his arguments.

87 *darkness is the absence of light*: Galileo mocks Simplicio by having him solemnly introduce a trivial and obvious definition of darkness. Aristotle uses the same words but in a less naïve context (*On the Soul*, book 2, chapter 7, section 418b ff., and *On Sensation*, section 439a, 20).

his visual rays: Galileo was both influenced and hampered by the traditional description of how light-rays travel. We are all familiar today with the correct theory, which is that of *intromission*, meaning that vision is caused by rays of light entering the pupil, whereas the rival theory of *extromission*, which accounted for vision by rays streaming from our eyes, was more commonly accepted in Galileo's day. Whether the rays originate from the object or from the eye, the geometrical description of the situation is the same because the direction of the rays does not alter the way they are traced. We still speak of 'eye contact', of hard stares, and of gazing as 'looking outward'. In the *Sidereal Message* Galileo considers the rays as being carried from the eye to the object when lenses are placed between the eye and what is being observed, and in the letter to the Jesuit Christopher Grienberger on 1 September 1611, he writes, 'our visual rays leave our eye as from the vertex [of a triangle] and stretch out spherically until they reach the perimeter of the Moon' (*Opere*, 11, p. 118). In a comment written in 1619 Galileo still speaks of 'visual rays proceeding from the eye' (note on Orazio Grassi's *Discourse on the Comets*, *Opere*, 6, p. 107). In the Third Day of the *Dialogue* Galileo states that the pupil is 'that aperture from which the visual rays proceed' (p. 383).

94 *pillar of cloud . . . bright by night*: the reference is to Exodus 13: 21–2.

96 *forty times more*: Galileo is comparing the surfaces of the Earth and the Moon when he should have compared their volumes, which are as 14 to 1.

98 *booklet of conclusions*: the full title reads: *Mathematical Disquisitions on Astronomical Controversies and Novelties, defended by Johann Georg Locher* (Ingolstadt, 1614). Locher was a student of Christoph Scheiner, who taught at the Jesuit University of Ingolstadt in Bavaria. As was the custom, Locher publicly defended his doctoral thesis, which was a critique of recent views about the motions of the Earth. Scheiner, who had actually written most of the text, used this opportunity to expand the lecture into a booklet of just under 100 pages.

Cleomedes, Vitellio, Macrobius: Cleomedes was a first-century Greek writer known only through his book *On the Circular Motions of the Celestial Bodies*, which is a compendium of Greek sources. Vitellio (or Witelo) was a Polish friar who lived in Italy in the 13th century and wrote an important work on perspective. Macrobius (AD 370–430) was a Roman grammarian and Neoplatonist philosopher. His commentary on Cicero's *Dream of Scipio* contains the idea mentioned here.

some other modern author: the Jesuit François de Aguilon, who published a book on optics in 1613.

101 *a man . . . secret device*: Giambattista della Porta (1535–1615) mentions such a device in the preface to the seventh book of his *Natural Magick* (London, 1658, p. 190; the original edition in 20 books appeared in Naples in 1589), and also in the appendix to his *Taumatologia* (published in Luisa Muraro, *Giambattista della Porta mago e scienziato* (Milan: Feltrinelli, 1978), 189).

102 *ancient spots*: the dark areas that can be seen on the Moon without the aid of the telescope.

108 *Michelangelo's*: Michelangelo Buonarroti (1475–1564), whose reputation as a supremely gifted artist was already established in his lifetime. Galileo was a close friend of his nephew, Michelangelo Buonarroti the younger (1568–1646).

109 *Archytas*: a Pythagorean philosopher (428–350 BC) and a contemporary of Plato. He was born in Taranto in southern Italy and is said to have experimented with toy flying machines, including a wooden dove that flew by 'the secret blowing of air enclosed inside', perhaps a primitive compressed-air mechanism.

110 *generally accepted propositions*: this passage was seized upon by the Inquisition; see the complaints cited against Galileo at his trial in *Selected Writings*, 363, and the Introduction to this volume, pp. xxix–xxxi.

SECOND DAY

112 *a fifth essence*: see note to p. 46.

113 *a highly esteemed doctor in Venice*: probably Santorre Santorio (1561–1636), a medical doctor who practised in Venice but also taught at the University of Padua and was Galileo's colleague. He mentions two such incidents in his *Methodus vitandorum errorum omnium qui in arte medica contingunt* (*How to Avoid All Errors in the Art of Medicine*) (Venice, 1602; pp. 198–9 in the 1632 re-edition).

the Galenist doctors and the Aristotelians: Aristotle maintained that the nerves originated in the heart (*Generation of Animals*, book 5, section 781a, 20), but Galen believed that the brain was the source of the nerves.

114 *a gentleman whom he knew to be an Aristotelian philosopher*: this could be Cesare Cremonini, on whom see note to p. 8. Cremonini defended the Aristotelian position in his *Apologia dictorum Aristotelis de origine et principatu membrorum adversus Galenum* (Venice, 1623), 460–2. Galileo had a copy of this book and annotated it.

perturbed method: a reference to Euclid's *Elements*, book 5, proposition 22, which reads: 'If there be any number of magnitudes whatever, and others equal to them in magnitude, which taken two and two together are in the same ratio, they will also be in the same ratio *ex aequali*'. This proposition is irrelevant to the discussion and serves only to show how pedantic Simplicio can be.

115 *derived from Aristotle*: 'if someone looks through a tube he will see further; in fact, persons in pits or wells sometimes see the stars' (*Generation of Animals*, book 5, section 780b, 20–1).

prophecies of Joachim: Joachim of Fiore (1135–1202) was the most important apocalyptic thinker of the whole medieval period. He is mentioned by Dante in *Paradiso* 12.140–1. Galileo is alluding to an apocryphal work, *Prophecies concerning the Popes*, attributed to Joachim and widely read in the sixteenth and the beginning of the seventeenth century.

116 *guided by their melancholy humour*: since the time of the Greeks, the basic material out of which everything is made was considered to exist in one of four kinds, which were referred to as the elements earth, water, air, and fire. Each humour was associated with one of these elements in the following manner: earth with black bile, water with phlegm, air with blood, and fire with yellow bile. The prevalence of one of these elements was considered the source of the four basic temperaments, which described human behaviour. Too much earth made one melancholic; too much water, phlegmatic; too much air, sanguine; and too much fire, choleric. This is why the melancholic type is prone to sadness and depression or, worse still, to guile and cowardice; the phlegmatic is unemotional and calm; the sanguine is optimistic, courageous, amorous, and outgoing; the choleric is angry and irascible. Galileo blames the behaviour of the alchemists on their melancholic humour.

117 *a well-known philosopher*: this could be Giulio Cesare Lagalla, a professor of philosophy at the University in Rome, who sent Galileo his book *De immortalitate animae* in 1621.

Alexander: Alexander of Aphrodisias, who taught in Athens around the year AD 200, is the most famous of Aristotle's commentators.

118 *confront Hercules . . . when his fury is aroused*: the reference is twofold. On the one hand, 'when his fury is aroused' alludes to the madness of Hercules, caused by Juno, which led him to kill his own children. On the other hand, it refers to the myth of Hercules who, as a slave to Omphale, the Queen of Lydia or Maeonia, was forced for three years to spin wool among her maids, dressed in women's clothes (Ovid, *Ars amatoria*, 2. 217 ff.). Galileo has in mind a passage from Tasso's *Gerusalemme liberata* that he had commented upon in his *Considerazioni al Tasso* (*Opere*, 9, p. 139). In Max Wickert's translation (*The Liberation of Jerusalem*, Oxford World's Classics, 2009), 16.3: 'See here among Maeonian maidens lie | Alcides with his distaff, glib and trim. | He conquered Hell once, held up stars and sky, | now twirls a spindle while Love laughs at him.'

120 *an ancient writer*: a reference to Philolaus and other Pythagoreans who believed that the Earth revolved around a central fire. These ideas are discussed by Aristotle, *On the Heavens*, book 2, chapter 13, section 293a, 17–27; section 293b, 15–31.

122 *a good school of thought*: Galileo is again referring to the ancient school of the Pythagoreans.

'everything that moves . . . some unmoving object': see Aristotle, *On the Motions of Animals*, 698, b8.

125 *over thousands of years*: in his *On the Revolutions*, book 3, chapter 6, Copernicus estimates the precession of the equinoxes to have a period of 25,816 years. The ancient estimate had been 36,000 years.

128 *Primum Mobile*: the tenth and outermost concentric sphere of the universe in Ptolomaic astronomy. It was thought to revolve around the Earth from east to west in 24 hours and to cause the other nine spheres to revolve with it.

130 *his arguments being as follows*: Aristotle, *On the Heavens*, book 2, chapter 14, section 296a, 27–296b, 12.

134 *whose name I think was Christian Wurstisen*: Christian Wurstisen (1544–88) was not from Rostock, the first university where the heliocentric theory was taught, but from Basel. He praises Copernicus in a book published in 1568, but there is no record of his having delivered lectures in Venice or Padua. Note that Sagredo, when referring to him says, 'whose name *I think* was . . .'.

143 *paragraph 97*: *On the Heavens*, book 2, chapter 14, section 296a.

146 *middle term*: the term that appears in both premises of a syllogism. For example: *man* in the following syllogism: a man is mortal (major premise); Socrates is a man (minor premise); Socrates is mortal (conclusion).

153 *vires acquirant eundo*: quoted from Virgil's famous passage about rumour, *Aeneid*, 4.175.

154 *explained a little while ago*: in the First Day, pp. 26–7.

164 *playing with hoops*: these *ruzzole*, as they are called in Italian, are not really hoops but wooden discs about 15 cm in diameter and 2 cm thick. There is a groove in the rim around which a cord is wound and then pulled to set the disc in motion.

165 *Socrates' demon*: Socrates called the source of his inspiration his 'demon'. Sagredo is poking fun at Simplicio and offers to become his source of inspiration by using the Socratic method of questioning.

round hoops run more smoothly than square ones: Galileo is referring to the *Mechanical Questions*, chapter 8, sections 851b, 15–852a, 14. This work was considered to be by Aristotle but is now believed to have been written later.

166 *taught to me by Aristotle*: in the *Mechanical Questions*, chapter 8, section 848a, 3–10.

168 *play with counters*: these counters (*chiose*) were rounded lead objects moulded by children for use as play money.

169 *players at bowls*: the Italian national game of *bocce*, which is played on a rough or irregular surface. A large wooden ball is aimed at the goal, which is a smaller ball, the *pallino*.

170 *our narrative poem*: an allusion to the long-standing debate on the relative merits of Tasso's *Gerusalemme liberata* and Ariosto's *Orlando Furioso*, to which Galileo contributed forcefully in favour of Ariosto's freewheeling, episodic poem in preference to Tasso's attempt (not entirely successful) at writing a monothematic epic. Sagredo continues the allusion on p. 172 when he says, 'I am content to excuse you from telling this story now.' For a discussion of the background see the Introduction to Torquato Tasso, *The Liberation of Jerusalem*, translated by Max Wickert with notes and Introduction by Mark Davie (Oxford World's Classics, 2009). See also note to p. 457.

171 *This is a problem*: the unsolved problem is the law governing the distance covered by freely falling bodies. Galileo provides the solution on p. 234.

172 *the Academician . . . a treatise of his on motion*: the academician is Galileo himself (see note to p. 23). The treatise is *On Naturally Accelerated Motion*, which was eventually published in 1638 at the end of the Third Day of Galileo's *Discourses on Two New Sciences*. See the extract in *Selected Writings*, 382–92.

spirals: *On the Spirals* is one of the works of Archimedes (287–212 BC), the ancient Greek mathematician whom Galileo most admired.

already discussed and established at length: in the First Day; see pp. 30–2.

173 *must terminate . . . Earth*: the erroneous belief that falling bodies must reach the centre of the Earth made it impossible to work out the correct path.

174 *this fancy of mine*: the error was detected by the French mathematician Pierre Fermat and transmitted to Galileo by a mutual friend. Galileo later reached the correct solution that the path of projectiles is a parabola, and he tried to pass off as a jest the explanation given here that the mixture of the straight motion of the falling body and the uniform diurnal motion of the Earth would give rise to a semicircle that ended at the centre of the Earth. For Galileo, a motion is 'horizontal' if it does not carry the moving body towards or away from the centre of the Earth. 'And thus for example', he writes elsewhere, 'a ship having received one single time some impetus would move continuously through a quiet sea around the globe without ever stopping' (Galileo Galilei and Christoph Scheiner, *On Sunspots*, translated by Eileen Reeves and Albert Van Helden (Chicago: University of Chicago Press, 2010), 125).

175 *which you allowed for it at the outset*: see the First Day, p. 49.

180 *Alexandretta*: a port on the Mediterranean in south-eastern Turkey (modern Iskenderun) close to Syria, through which Giovan Francesco Sagredo had travelled en route to the city of Aleppo where he was Venetian consul from 1608 to 1611.

182 *handbook of assertions*: the 'encyclopedia' cited for its 'foolish subtleties' in the marginal note to Sagredo's previous comment. The reference is to the *Encyclopedia Explained and Defended with a Hundred Philosophical Assertions* published in Rome in 1624, from which the notion that a walker's head travels further than his feet is quoted almost word for word (p. 57). The author's name is given as Clemente Clementi, but this is a pseudonym for the Jesuit Leone Santi, who taught mathematics at the Roman College. Galileo gave him a copy of his *Dialogue* when it was published in 1632.

191 *in Copernicus's book*: Galileo is referring to *On the Revolutions of the Heavenly Spheres*, book 1, chapter 12, where the tables are printed at the end. Whereas the ancients used chords we now use sines (a chord is equal to double the sine of half the angle).

195 *all humans naturally desire knowledge*: the first sentence of Aristotle's *Metaphysics*.

198 *including Ptolemy*: Galileo's source is Copernicus's *On the Revolutions*, book 1, chapter 7, where Copernicus paraphrases, rather loosely, Ptolemy's *Almagest*, book 1, chapter 7.

201 *our knowledge . . . remembering*: Galileo returns to the Platonic theme of recollection; see p. 14.

208 *between the secant and the point of contact*: a secant line of a circle is a line that intersects two points on the curve. If the secant is defined by two points, P and Q, with P fixed and Q variable, as Q approaches P along the curve, the secant becomes closer and closer to being the tangent at P, namely it 'just touches' the curve at that point.

230 *two authors*: these are Johann Locher, on whose 'booklet of conclusions' written jointly with Christoph Scheiner see note to p. 98, and Scipione

Chiaramonti, professor of philosophy at the University of Pisa, on whom see note to p. 56.

232 *German miles*: a German mile was about 7.1 km (4.4 miles); hence 12,600 German miles per hour would give 89,460 km/h (55,440 mph).

233 *the modern author just cited*: Galileo is referring to the Locher–Scheiner 'booklet'.

234 *writings which are not yet published*: Galileo is referring to the Fourth Day of his *Two New Sciences*. See the translation by Stillman Drake (Madison, WI: University of Wisconsin Press, 1974).

canna: a unit of length whose value varied in different regions of Italy. In Florence it was 2.9 metres.

235 *a hundred braccia in five seconds*: taken literally this would imply that Galileo accepted 467 cm/sec^2 for g (the acceleration due to gravitation), which is less than half the true value of 980/sec^2. But Galileo gave an improved estimate in the passage he added to the First Day; see p. 33 and note.

236 *56 times the radius of the Earth*: Galileo has modified the value that he used in the *Sidereal Message*, when he gave the distance of the Moon from the Earth as 'almost sixty terrestrial diameters' (see *Selected Writings*, p. 7).

3 hours, 22 minutes, and 4 seconds: Galileo's calculation is erroneous because of his assumption that the rate of acceleration would be constant throughout the fall whereas it varies inversely with the square of the distance. The actual time would be about 4 days and 20 hours.

Let's designate these three numbers: in Galileo's table we find, at the top, three numbers with their corresponding letters (A, B, and C) and a fourth number (25), which is the square of the second number (5). Below, to the right of the vertical line we have the result of the first operation that is mentioned, namely:

$$588{,}000{,}000 \times 25 = 14{,}700{,}000{,}000.$$

The numbers to the left of the vertical line (1, 22, 241, 2422, 24244) are the numbers obtained for the square root of 14,700,000,000 by dividing by 100 the number arrived at on the preceding step. Galileo has omitted to show the following intermediate steps in his calculations:

$\sqrt{147000000}$	12124
1	1
047	22 × 2 = 44
44	
0300	241 × 1 = 241
241	
05900	2422 × 2 = 4844
4844	
105600	24244 × 4 = 96976
96976	
8624	

The last group of numbers to the right of the vertical line and at the bottom of Galileo's table are the reduction of the 12124 seconds to minutes, i. e. 12124 ÷ 60 = 202, and to hours = 202 ÷ 60 = 3. In the text just above the calculations, Galileo gives the more exact figure of 3 hours, 22 minutes, and 4 seconds.

243 *the books we sent for just now*: the Locher–Scheiner 'booklet' and Chiaramonti's *On Three New Stars* that appeared in 1628. See pp. 230 and 288.

one half of the Earth would never see the Sun: Simplicio begins quoting in Latin passages from Locher and Scheiner's *Mathematical Disquisitions* (the 'booklet'), 29.

244 *what are these lovely things here?*: Sagredo is poking fun at a ludicrous illustration of flying birds and falling objects on page 30 of Locher and Scheiner's book (reproduced in Besomi-Helbing, 2, p. 534).

circled around the earth six times, etc.: quoting from Locher and Scheiner, 30–1.

245 *Quandoque bonus . . .* : Simplicio quotes the first two words of the Roman poet Horace's famous phrase, 'quandoque bonus dormitat Homerus' ('sometimes good Homer nods') in his *Ars Poetica*, line 359.

247 *intelligence, either assisting or informing*: the 'assisting intelligences' were angels who guided the planets in their course; the 'informing intelligences' were the internal moving principles of animated beings. It is interesting to note how long the discussion about assisting intelligences lasted. As late as 1651 the most important astronomer of the period, Giovanni Battista Riccioli, argued that the motions of the stars and planets require that they be directed, ultimately by God as the first cause, but that this does not entail that the world has a soul or that some angelic influence is needed to ensure that the proper trajectories are followed. The discussion shows how mathematically rigorous astronomy could be combined with speculation that used conceptual tools that were on their way out. See Flavia Marcacci, *Cieli in Contraddizione: Giovanni Battista Riccioli e il terzo sistema del mondo* (Perugia: Aguaplano, 2018), 197–203.

259 *The error . . . is not there*: this sentence is added from Galileo's marginal note in his copy.

262 *The visual ray . . . at rest*: this sentence is added from Galileo's marginal note in his copy.

266 *more than 2529 miles per hour*: the actual speed (67,000 mph or 107,000 kph) is more than 26 times greater. The distances of the celestial bodies were grossly underestimated at the time. Ptolemy had put the mean distance from the Earth to the Sun at 1210 terrestrial radii (*Almagest*, book 5, chapter 15). The actual average distance, now known as the astronomical unit AU, is 149,597,870 km or 92,955,807 miles. The terrestrial radius at the equator is 6378.1 km or 3963.2 miles. So the correct value is not a distance of 1210 terrestrial radii but 23,439, i.e. more than 20 times greater than Ptolemy's estimate. Aristotle had estimated the distance of the Sun as

18 to 20 times the distance of the Moon, which is even less satisfactory. The Moon's average distance from the Earth is 384,400 km or 238,855 miles, so the Earth is actually almost 390 times further from the Sun than it is from the Moon.

274 *Fingamus . . . et vicissim*: Salviati here reads the Latin original of the passage Simplicio has just translated.

288 *book on the new stars . . . booklet of conclusions*: Chiaramonti's *On Three New Stars* and the Locher–Scheiner 'booklet'. See note to p. 243.

THIRD DAY

291 *rest intervenes at the point of reversal*: Simplicio is quoting Aristotle, *Physics*, book 8, chapter 8, sections 262, a12–263, a2.

the booklet of conclusions: the book by Johann Locher and Christoph Scheiner that was introduced by Simplicio in the First Day (see p. 98 and note) and mentioned several times in the Second Day, although it is only in the Third Day that their arguments are examined at length.

the book on the new stars: Scipione Chiaramonti's book, quoted in the Second Day, p. 260.

292 *against a host of astronomers*: Chiaramonti argued that the supernovae of 1572 and 1604 were below the Moon, against the opinion of astronomers like Tycho Brahe, Cornelius Gemma, Michael Maeslin, Thomas Digges, Taddeus Hagek, Antonio Santucci, and others.

the inane writings of someone like Lorenzini: Antonio Lorenzini da Montepulciano was the author of a tract on the supernova of 1604 which contained a number of silly comments such as that the Moon was a lantern that shed its light only when turned towards us. Johann Kepler in his *De Stella Nova in Pede Serpentari* had challenged Italian astronomers not to let such an ignoramus go unrefuted.

295 *observations made by thirteen astronomers*: Galileo's source for these figures is mainly Tycho Brahe's *Exercises in Astronomy Restored* that appeared in 1602. Thirteen astronomers are mentioned but two of them (Peucer and Schuler) used the same data. The original figures given by Galileo contain mistakes that are explained in Besomi-Helbing, 2, pp. 654–6. The astronomers concerned are Francesco Maurolico (1494–1575), a mathematician and astronomer from Messina in Sicily; Paul Hainzel (1527–81), mayor of Augsburg and an amateur astronomer who was a friend of Tycho Brahe, whom he helped design and construct a large quadrant; Wolfgang Schuler, a professor at the University of Wittenberg who shared data with his friend Caspar Peucer (1525–1602); Tycho Brahe (1546–1601), the great Danish astronomer who made the most accurate observations before the invention of the telescope; William IV Landgrave of Hesse-Kassel (1532–92), who was responsible for the Hessian star catalogue of about a thousand stars; Cornelius Gemma (1535–77), professor of medicine at the University of Leuven and son of the eminent astronomer Gemma Frisius;

Elias Camerarius, professor at the University of Frankfurt; Thaddeus Hagek, physician in Prague who wrote a book on the new star of 1572; Adam Ursinus, an astrologer in Nuremberg who believed that the new star was below the Moon, as did Georg Busch, a German painter and astronomer; Jerónimo Muñoz (*c.*1520–91), professor of mathematics and Hebrew at the University of Valencia, who taught that the new star was a comet.

311 *the size of angle BDC*: Galileo is using the data he found in a table in Copernicus's *On the Revolutions of the Heavenly Spheres* where the values are correctly given as 154° 35' instead of 154° 45' and 42,920 for the sine instead of 42,657 that Galileo provides. Hence the irrelevance of the computations that Galileo goes on to make. There are other such mishaps in the Third Day. For the tables of Copernicus see *On the Revolutions*, 590–621.

312 *as may be seen below*: in the calculations on pp. 312–31 Galileo, following the conventions of his time, omits to show the intermediate steps in the process of division. Thus, in the example below, the divisor (58) is shown on the left, the dividend (347313294) on the right, and the quotient (5988160¼) above. The figure shown below the dividend (5717941) is the result of the successive steps of multiplication and subtraction which are carried out simultaneously, not written down as in modern practice. If these steps were shown in full, they would appear as follows:

```
      5988160¼
58 | 347313294
     573
      511
       473
       092
        349
        0140
          24
```

In the calculation as shown by Galileo only the digits underlined in each line are written in the appropriate column below the dividend, giving the figure of 5717941. This and the other calculations on these pages are not essential to following Galileo's argument and may safely be skipped.

52 radii from the centre of the Earth: it is the case, however, that Copernicus uses two epicycles to calculate the distance of the Moon from the Earth, and arrives at the following results: at apogee it is a little more than 65 terrestrial radii away, and at perigee a little more than 55 (*On the Revolutions*, book 4, chapter 17, pp. 707–9, and chapter 22, p. 714).

319 *chord 4034*: this should be 4304.

320 *sine 36643*: this should be 36623.

321 *30 $^{58672}/_{100000}$, or slightly more than 30½ radii*: the denominator should be 300,000 instead of 100,000, and the end of the sentence should read 'slightly less than 30½ radii'.

326 *azimuths*: the azimuth is the angle between north, measured clockwise around the observer's horizon, and a celestial body (Sun, Moon).

327 *timepiece*: the Italian word *oriuolo* (*orologio* in modern Italian) that Galileo uses usually refers to the wheel clock which he describes in the Fourth Day (see pp. 468–9). It had not yet occurred to Galileo to use the pendulum for the escapement of a clock although he had used it for the measurement of time, as described on p. 469. A few months before his death he hit upon the idea of its application to clocks but because of his blindness he could not carry it into execution. He dictated the design to his son Vincenzio, who made correct drawings but did not complete an actual model. A successful construction of the pendulum clock was made by Christiaan Huyghens, who patented his results in 1657. Galileo's first application of the pendulum was made with an instrument called a pulsilogium (pulse counter) that was devised by Santorio Santori (1561–1636), a physician and colleague. It consisted of a board bearing a peg to which was attached a bob swung on a cord. On the board, at appropriate places, were written various diagnostic descriptions of a patient's pulse. The physician had only to stop the cord with his thumb so as to bring the swinging bob into synchronism with the pulse and read off the diagnosis directly ('sluggish', 'feverish', and so on). Galileo also used a water clock. This consisted in filling a large vessel with water which could escape through a very small orifice into an empty vessel which had previously been dried and weighed. For instance, wishing to determine the speed of a ball when it was dropped along an inclined plane, Galileo would remove his thumb from the orifice at the start of the experiment and replace it when the ball had reached any desired point. By weighing the water which had escaped he could then determine the elapsed time by comparing this with the weight of water escaping in known time. See Fabrizio Bigotti and David Taylor, 'The Pulsilogium of Santorio: New Light on Technology and Measurement in Early Modern Medicine', *Soc Politica*, 11.2 (2017), 53–113.

336 *or is infinite and boundless*: Copernicus states that the size of the Earth compared to the size of the heavens is 'like a point compared to a solid body or a finite magnitude compared to an infinite one', but he does not consider whether the universe itself is finite or infinite (*On the Revolutions*, book 1, chapter 6).

finite, bounded, and spherical: see Aristotle, *On the Heavens*, book 1, chapter 4, sections 271, b1–276, a18, and *Physics* Book 3, chapters 3-8, sections 202, b30–208, a24. Aristotle assumed the universe to be finite, and Kepler also thought the universe must be bounded by some 'tunic or skin' to prevent the sun's warmth from dispersing out over the void. But Galileo believed that it was not a matter that could be settled. Late in life, replying to Fortunio Liceti, who had sent him a book on the question, he wrote: 'Many clever arguments are produced for both sides, but none of them, to my mind, leads to a necessary conclusion, and I remain in great doubt about which of the two answers is the right one. However, one particular argument inclines me more to believe that it is infinite rather than finite, and this is that I cannot imagine the universe as finite or infinite. Since infinity cannot be understood by our finite mind while finiteness can, I feel that my lack of understanding is caused by infinity, which is incomprehensible, rather than to finiteness, which can be comprehended. But, as you state, this is one of those questions that cannot be explained by human reason and are probably similar to predestination, free will, and such others that only Holy Scripture and divine revelation can enlighten a pious mind' (letter to Fortunio Liceti, 24 September 1639; *Opere*, 18, p. 106).

337 *for fear of having to accept them*: the field of view of Galileo's telescope was so narrow that only one quarter of the Moon could be seen at a time. People found this frustrating because they expected to see it whole. The philosopher Cesare Cremonini was candid about his difficulty in focusing the instrument: 'Looking through those lenses', he told a friend, 'just makes me dizzy' (as reported by Paolo Gualdo in his letter to Galileo, 20 July 1611; *Opere*, 11, p. 165). Another professor, Giulio Libri, who had been Galileo's colleague in Padua before moving to Pisa, had the same trouble. When he died at the end of 1610 Galileo joked that although he had failed to see the new stars while on Earth, he might see them on his way to heaven (letter to Paolo Gualdo, 17 December 1610; *Opere*, 10, p. 484). Galileo's bugbears were the pedants who swore by their books instead of looking through the telescope. 'This kind of person', he wrote to Kepler, 'thinks philosophy [used here in the sense of natural philosophy or natural science] is a book like the Aeneid or the Odyssey, and that truth is to be discovered, not in the world or in nature, but by comparing texts (I use their own words)' (letter to Kepler, 19 August 1610; *Opere*, 10, p. 422).

344 *on another occasion*: Salviati is referring to a passage in the Second Day, pp. 243–4, where he criticizes Locher and Scheiner's book for failing to grasp that the year would last one natural day if the Earth did not revolve on its axis.

345 *brighter than any ordinary lamp*: a reference to the telescope.

346 *the Latin poet*: Virgil, *Eclogues*, 2.25–6. The passage in square brackets from here to p. 351 was added from pages in Galileo's autograph inserted in his copy of the first edition. Note that this long digression, introduced by Simplicio with his long-winded quotation of Virgil, comes immediately after Sagredo has impatiently said, 'Come now, Salviati, let's get down to brass tacks . . . I feel as if we are wasting every word we expend on anything else.'

353 *our mutual friend's Letters . . . or his Assayer*: the illumination of Venus is discussed by Galileo in his third letter on sunspots (*Opere*, 5, pp. 195–6; Galileo Galilei and Christoph Scheiner, *On Sunspots*, translated by Eileen Reeves and Albert Van Helden (Chicago: Chicago University Press, 2010), 261–2), and in *The Assayer* (*Opere*, 6, pp. 273 ff.; Galileo Galilei, *Controversies on the Comets of 1618*, translated by Stillman Drake (Philadelphia: University of Pennsylvania Press, 1960), 227 ff.).

359 *move irregularly around its own centre*: Salviati is referring to Ptolemy's equant, a purely mathematical point which allowed Ptolemy to keep the theory of uniform circular motion. Copernicus was praised for having done away with it.

362 *Apollonius of Perga*: Greek geometer and astronomer who lived in Alexandria in the third century BC and is famous for his work on conics.

363 *The first to discover and observe the sunspots*: Johann Goldsmid, better known by his Latinized name Johannes Fabricius (1587–1616), observed the sunspots before Galileo and Scheiner. In a 22-page pamphlet, *De Maculis in Sole Observatis*, published in Wittenberg in 1611, he states that he saw the sunspots on 9 March of that year. See the Introduction, p. xvi.

373 *the third annual motion that Copernicus assigns to the Earth*: Copernicus still believed in celestial spheres and he thought that the Earth, now a planet, was carried about the central Sun by a sphere just like the one previously thought to carry the Sun about a central Earth. If the Earth were firmly fixed in a sphere, its axis would not always stay parallel to the same line through the Sun; instead it would be carried about by the sphere's rotation and would occupy different positions. After the Earth had revolved 180° about the Sun, the Earth's axis would still be tilted 23½° away from the perpendicular, but now in a direction opposite to that in which it had begun. To undo this change in the direction of the axis, caused by the rotation of the sphere that carries the Earth, Copernicus required a third circular motion which applied only to the axis of the Earth, and which carried the north end of the axis westward once each year, thus compensating for the effect on the Earth's axis of the orbital motion. In the figure the sphere (left) carries the Earth as it rotates around the Sun. But this motion does not keep the Earth's axis parallel to itself, so that a conical third motion (right) is required to bring the axis back into line. See the excellent account and the figures in Thomas S. Kuhn, *The Copernican Revolution* (New York: Vintage Books, 1959), 164–5.

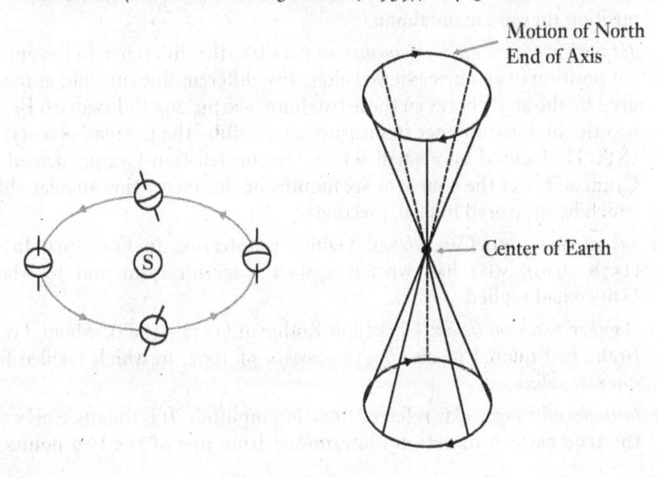

374 *the supreme wisdom*: a periphrasis for God in a well-known line of Dante, *Inferno*, 3.6.

378 *1208 of the Earth's radii*: this number is taken from the Locher–Scheiner booklet (see note to p. 98). Copernicus gives a figure of 1179 terrestrial radii as the maximum distance (apogee). Ptolemy's figure was 1210 terrestrial radii.

379 *Alfraganus, Albategnius, Thabit*: Alfraganus (Al-Fargani) was a famous astronomer at the court in Baghdad in the 9th century; Albategnius (Al-Battani, 858–929) is the best-known astronomer of the medieval Islamic world; Thabit ibn Qurra (838–901) translated from Greek into Arabic works by Apollonius of Perga, Archimedes, Euclid, and Ptolemy.

Clavius: Christopher Clavius (1537–1612) was the most prominent Jesuit astronomer of his day. He helped Galileo get his first appointment at the University of Pisa and then his second at the University of Padua.

380 *the method I adopted*: Galileo described his method in greater detail in the draft of a book that he started writing five years after publishing his *Dialogue* and that bore the title *Le operazioni astronomiche* (see *Opere*, 8, pp. 457–9). Galileo's friend, Paolo Sarpi, used a similar method to determine the size of small bodies and Galileo may have learned it from him, as Libero Sosio suggests; see Paolo Sarpi, *Pensieri naturali, metafisici e matematici*, edited by Luisa Cozzi and Libero Sosio (Milan: Ricciardi, 1996), 746–7.

Vega: the fifth brightest star in the night sky.

383 *from which the visual rays proceed*: in the modern, correct concept of vision we say 'which the visual rays strike'. For the notion of visual rays for Galileo and his contemporaries see note to p. 87.

385 *the new stars in Cassiopeia and Sagittarius*: these are the supernovae of 1572 and 1604 which were discussed at length at the beginning of the Third Day.

391 *ad hominem*: in a way that is directed against a person rather than the position they are maintaining.

did not have accurate instruments: on parallax (the difference in the apparent position of an object viewed along two different lines of sight as measured by the angle between those two lines) see pp. 295 ff. Friedrich Bessel was the first astronomer to measure successfully the parallax of a star in 1838. He focused on a small star in the constellation Cygnus named 61 Cygni and over the course of six months he detected a tiny angular shift, which he measured to be 0.3 seconds.

392 *one of these anti-Copernicans*: Galileo is referring to Francesco Ingoli (1578–1649), who had written against Copernicanism and to whom Galileo had replied.

Another mathematician: Christoph Rothman (1577?–1608), whom Tycho Brahe had quoted in his *Progymnasmata* of 1602, to which Galileo frequently refers.

397 *latitudes of rising*: today referred to as the amplitude. It is the angle between the true eastern direction (determined from one of the two points of

intersection between the celestial equator and the horizon) and the point where a star is seen to rise. Similarly, the angle between the true western direction and the place where a star is seen to set is called the amplitude at setting.

the Gordian knot . . . more than an Alexander: a problem solvable only by bold action. In 333 BC, Alexander the Great on his march through Anatolia reached Gordium, the capital of Phrygia. He was shown the chariot of the ancient founder of the city, Gordius, with its yoke lashed to the pole by means of an intricate knot with its end hidden. According to tradition, this knot was to be untied only by the future conqueror of Asia. Alexander sliced through the knot with his sword.

402 *solstitial colure*: the meridian of the celestial sphere that passes through the two poles and the two solstices.

407 *aligning an alidade*: an alidade is the portion of a graduated instrument, such as a quadrant or astrolabe, carrying the sights or telescope, and showing the degrees cut off on the arc of the instrument.

armillary sphere: an armillary sphere is a model of objects in the sky, consisting of a spherical framework of rings, centred on the Earth or the Sun, that represent lines of celestial longitude and latitude and other astronomically important features, such as the ecliptic.

419 *William Gilbert*: Gilbert (1544–1603) practised as a doctor in London for many years and in 1600 became president of the Royal College of Physicians. He served as physician to Elizabeth I in the last few years of her reign. Galileo was greatly influenced by his book on magnetism, *De Magnete*, which appeared in 1600.

420 *voices that are no longer heard*: a reference to Aristarchus and others in antiquity who considered the possibility of heliocentrism.

a very well-known Peripatetic philosopher: Gilbert's book, *On the Magnet*, was published in Latin in 1600 when Galileo was still teaching in Padua. We do not know who gave him the book but it could have been his colleague, Cesare Cremonini, on whom see note to p. 8.

425 *armature*: following Gilbert, Galileo sought to increase the magnetic power of the lodestone by 'arming' it with caps or pole-pieces of soft iron.

426 *four hammers*: a reference to the legend that Pythagoras discovered the foundations of musical tuning by listening to the difference in pitch of the tone produced by hammers of differing weights when they struck an anvil.

431 *Gilbert has discovered a third motion*: Gilbert describes the effect but he attributes the discovery of the vertical dip of the needle to Robert Norman. See William Gilbert, *On the Loadstone and Magnetic Bodies*, translated by P. Fleury Mottelay (New York: Dover, 1958), book 3, chapter 6, p. 244.

434 *Sacrobosco*: Johannes de Sacrobosco (John Holywood, *c.*1195–*c.*1256) wrote an astronomy textbook, *Tractatus de Sphaera*, that was frequently revised and widely used well into the beginning of the seventeenth century as an introduction to astronomy.

FOURTH DAY

436 *our own Tyrrhenian sea*: Salviati is referring to the coast of his native Tuscany.

437 *three periods . . . tides*: at the beginning of the seventeenth century it was generally recognized that the tides went through four different cycles: the *daily* or *diurnal cycle* with high and low tide recurring at intervals of 12 hours; the *monthly cycle* whereby the tides lag behind 50 minutes each day until they have gone round the clock and are back to their original position; the *half-monthly cycle* with high tides at new and full moon and low tides at quadrature and, finally, the *half-yearly cycle* with greater tides at the equinoxes than at the solstices. Galileo enumerates three periods: the diurnal, whose intervals 'in the Mediterranean . . . are of roughly six hours each', the monthly, which 'appears to derive from the motion of the Moon', and the annual, which 'appears to derive from the Sun'. But he fails to state the differences in the tides when the Moon is new, full, or at quadrature, and although he mentions that the tides at the solstices vary in size from those at the equinoxes, it is only forty pages later (see p. 476), that he states, erroneously, that they are greater at the solstices.

Scylla and Charybdis: in classical mythology, two monsters on either side of the Strait of Messina who embodied the danger to sailors of passing through the strait.

438 *a great Peripatetic philosopher*: Cesare Cremonini (see note to p. 8). In his book *A Defence of what Aristotle said about the Fifth Kind of Essence*, which he published in 1616, he speaks of the tides in a general way (on p. 73), and refers to the last chapter of book 1 of Aristotle's *Meteorology*, where mention is made of the great changes that the earth and the oceans have undergone. There is no specific discussion of the various kinds of tides.

a prelate: Marcantonio de Dominis (1566–1624) taught at the Jesuit School in Padua until 1597. He was made Bishop of Split in Croatia, left the Roman Church to become an Anglican, recanted, and returned to Rome, where he died in prison in 1624, the year of the publication of his book on the tides, *Euripus sive sententia de fluxu et refluxu maris*.

the Moon with its temperate heat . . . rarefied: Girolamo Borro taught medicine and philosophy at Pisa when Galileo was a student there. He invoked the 'temperate heat' of the Moon that acts as an attractive force on the analogy of fire and causes water to rise as it nears the boiling-point. Bernardino Telesio had suggested a more vague relationship between the Sun, the Moon, and the tides. He assumed that the sea rises and tends to boil over when it is heated by the Sun, and that it sets itself in motion to avoid evaporation, thus producing the flow and ebb of the tides. Borro's *Del flusso e reflusso del mare e dell'inondatione del Nilo* appeared in 1561 and was often reprinted, the last time in 1583; Telesio's views are to be found in his *De Rerum Natura*, book 1, chapter 12, a work that was published in 1565 and reprinted several times.

439 *the water only rises . . . here in Venice*: to account for the tides, Marcantonio de Dominis postulated that an attractive force acted from the Moon on the ocean. A common objection to de Dominis' explanation was that high tide does not occur once a day when the Moon is directly above the sea but twice, the second time when the Moon is below the horizon. His theory, like Galileo's own hypothesis, entailed a 24-hour cycle and was rejected for failing to agree with experience. Galileo, of course, could not level this criticism at de Dominis, and he attacked him for failing to realize that water rises and falls only at the extremity and not at the centre of the Mediterranean. De Dominis can hardly be blamed for failing to detect something that was only entailed as a consequence of Galileo's own erroneous theory.

442 *like the breath . . . whale*: this animistic interpretation of the tides on analogy with respiration is set forth in Antonio Ferrari (known as Galateo), *Liber de situ elementorum* (Basel, 1558).

Ancona, Ragusa, or Corfu: Ancona is on the western shore of the Adriatic; Ragusa (modern Dubrovnik) and Corfu are on its eastern shore.

449 *the drawing which we made just now*: of the many ways that water can be made to flow Galileo considered particularly suggestive the to-and-fro motion of water at the bottom of a boat that is alternately speeded up and slowed down. He likens the piling of the water, now at one end and now at the other, to the action of the tide. The analogy is not entirely satisfactory, however, since the acceleration or retardation is shared uniformly by the whole boat whereas, for the flux and reflux of the tides, it is not uniform throughout the sea basins in which they occur. Galileo parries this criticism by asking his readers to imagine that the ecliptic and the equator coincide. A point on the surface of the Earth can be considered to move on an epicycle attached to a deferent representing the Earth's orbit, as in the figure. The epicycle revolves once daily. For half the day the speed of the point is greater than that of the epicycle's centre (the centre of the Earth); for the other half the speed is less. Maximum and minimum velocities occur when a given point is collinear with the centre of both epicycle and deferent. This means that the greatest speed is at midnight and the lesser at noon, and thus entails a twelve-hour period for the tides, not a six-hour period as is actually the case since there are two high and two low tides every day.

450 *a mechanical model*: in his *Dialogue on the Tides* written in 1616 Galileo had written, 'I have a mechanical model that I will disclose at the appropriate time in which the effects of this marvellous composition of movements can be observed in detail' (*Opere*, 5, p. 386). Here the words 'that I will disclose at the appropriate time' have disappeared. This would seem to indicate that he had not been able, in the interval of sixteen years, to translate his idea into practice.

451 *I come, secondly . . . maximum slowness*: see note to p. 449.

452 *Aristotle . . . threw himself into the sea and drowned*: the legend that Aristotle would have drowned himself in despair because he could not understand

the cause of the tides goes back to the early Church Fathers and is found in Justin Martyr, the second-century Christian apologist, and Gregory of Nazianus, a fourth-century Archbishop of Constantinople.

453 *the strait between Scylla and Charybdis*: the Strait of Messina; see note to p. 437.

Atlantic: Galileo wrote 'Ethiopian' for the southern part of the Atlantic Ocean. At the time some maps showed the ocean on both sides of South Africa as the 'Ethiopian Ocean'.

457 *'Even as the Aegean Sea', etc.*: Galileo is quoting Torquato Tasso's *Gerusalemme liberata*, 12.63. In Max Wickert's translation (*The Liberation of Jerusalem*, Oxford World's Classics, 2009): 'Even as the Aegean sea when Aquilo | and Notus cease to blow and churn and pound, | does not fall still, but in the heave and throe | of waves retains the motion and the sound...'. Galileo's flattering reference to Tasso as 'the divine poet' contrasts with his earlier dismissive attitude when he compared him unfavourably to Ariosto. See note to p. 170.

as we have said on another occasion: see pp. 158–9.

459 *Alexandretta*: see note to p. 180.

460 *stronger than those coming from the west*: Galileo's claim to recorded evidence for the Mediterranean runs counter to the fact that the prevailing wind east of Italy is a west wind in all seasons, and not an east wind as Sagredo vouches for.

462 *The Assayer*: see Galileo Galilei, *The Assayer*, translated by Stillman Drake in *The Controversy of the Comets* (Philadephia: University of Pennsylvania Press, 1960), 288–9. Galileo argues that, when someone runs with a lantern, the flame that it encloses is not blown out because the air which surrounds it moves with equal speed.

466 *the wretched Orlando*: in Ariosto's *Orlando Furioso*, Orlando tries to deny the unmistakable evidence that Angelica, the object of his love, has married his rival Medoro. He is driven mad by jealousy when he can no longer deny the truth; hence Sagredo's reply. On Ariosto's poem, see note to p. 170.

told of Aristotle: see p. 452.

annual motion is roughly three times greater: this value is the result of Galileo's mistaken assumption about the distance of the Sun. The actual speed is more than 26 times greater.

470 *a truly remarkable phenomenon*: actually the curve of fastest descent is not the circle but the cycloid, as was shown in 1696 by the Swiss mathematician Johann Bernoulli, who called it the brachistochrone.

474 *so many problems for modern astronomers*: Galileo is aware of the fact that his theory of the tides does not sit easily on the known motion of the Sun, and he is anxious to remind his reader that the motions of the planets, for instance Mars and the Moon, are not perfectly understood. Kepler had

found that the path of Mars is elliptical, but Galileo would have nothing of this.

a marked difference . . . divided by the equinoctial points: the average apparent speed of the Sun around the Earth is about one degree per day but it speeds up and slows down. It takes 187 days to move from the vernal to the autumn equinox (21 March–21 September) and 178 days from the autumn back to the vernal equinox (21 September–21 March).

476 *The additions are equal . . . are the smallest*: Galileo wishes to argue that the inclination of the Earth's axis with respect to its orbit around the Sun entails a modification of his original model. The annual and the diurnal motions are in the same line only at the solstices when their combination produces the greatest acceleration and the greatest retardation. At the equinoxes the two motions are inclined at their maximum angle, and the effect of their combination is consequently least. This is what Galileo's theory entails but the reverse holds true: the equinoctial tides are the most extreme because they receive the maximum effect of the Sun's gravitational pull, something Galileo did not consider.

solstitial colure: the meridian of the celestial sphere that passes through the two poles and the two solstices. See also p. 402.

480 *An ancient mathematician*: Seleucus, a Hellenized Babylonian astronomer who flourished in the second century BC.

481 *similar childish ideas*: in the introduction to his *Astronomia Nova*, published in 1609, the German astronomer Johann Kepler had conjectured that the Moon causes the tides.

The fourth argument . . . says are imperceptible: Galileo hoped that better telescopes would reveal that the motion of the Earth around the Sun had as a consequence a shift in the observed positions of the fixed stars at an interval of six months. This displacement or difference in the apparent position of an object viewed along two different lines of sight is called parallax; see pp. 56, 391 and notes. The stars were actually too far for such a displacement to be noted.

signor Cesare . . . Lincean Academy: in 1631, a year before the publication of the *Dialogue*, Galileo received an essay from Cesare Marsili (1552–1633) in which he declared that he had detected a shift in the meridian line that had been traced on the floor of the church of San Petronio in Bologna, where it can still be seen. Marsili's observations were not conclusive of the motion of the Earth, but Galileo had great hopes that they would be.

483 *a most learned and eminent person*: this person is none less than the Pope, Urban VIII, and it was unfortunate that the argument should have been made by Simplicio, who cuts such a pitiful figure in the *Dialogue*. Worse still, it is immediately ridiculed by Salviati, who had acted as Galileo's spokesman throughout the entire discussion.

we shall never discover all God's handiwork: the reference is to Ecclesiastes 3: 11: 'God has made everything beautiful in its time. He has also set eternity in the hearts of men; yet they cannot fathom what God has done from beginning to end.'

483 *our Academician's new science*: a reference to Galileo's *Two New Sciences* that was smuggled out of Italy and published in Holland in 1638. Galileo was 70 years of age when it appeared but its principal conclusions had been worked out three decades earlier. It is written in a style similar to the *Dialogue*. Salviati, Sagredo, and Simplicio discuss the strength of materials and the science of motion. See the modern translation by Stillman Drake: Galileo, *Two New Sciences* (Madison, WI: University of Wisconsin Press, 1974).

INDEX